THEORETICAL MANTLE DYNAMICS

Geodynamics is the study of the deformation and flow of the solid Earth and other planetary interiors. Focusing on the Earth's mantle, this book provides a comprehensive, mathematically advanced treatment of the continuum mechanics of mantle processes and the craft of formulating geodynamic models to approximate them. Topics covered include slow viscous flow, elasticity and viscoelasticity, boundary-layer theory, long-wave theories including lubrication theory and shell theory, two-phase flow, and hydrodynamic stability and thermal convection. A unifying theme is the utility of powerful general methods (dimensional analysis, scaling analysis, and asymptotic analysis) that can be applied in many specific contexts. Featuring abundant exercises with worked solutions for graduate students and researchers, this book will make a useful resource for Earth scientists and applied mathematicians with an interest in mantle dynamics and geodynamics more broadly.

NEIL M. RIBE is a Senior Researcher with the French National Center for Scientific Research (CNRS), working at the FAST laboratory of the University of Paris-South in Orsay. He has held visiting positions at the Institut de Physique du Globe in Paris, the Earthquake Research Institute of the University of Tokyo, and the University of Cambridge. His research interests include fluid mechanics, geodynamics, and the history of science, and he has contributed chapters in the *Treatise on Geophysics* and *Annual Review of Fluid Mechanics*.

THEORETICAL MANTLE DYNAMICS

NEIL M. RIBE
French National Center for Scientific Research

CAMBRIDGE
UNIVERSITY PRESS

University Printing House, Cambridge CB2 8BS, United Kingdom

One Liberty Plaza, 20th Floor, New York, NY 10006, USA

477 Williamstown Road, Port Melbourne, VIC 3207, Australia

314–321, 3rd Floor, Plot 3, Splendor Forum, Jasola District Centre, New Delhi – 110025, India

79 Anson Road, #06–04/06, Singapore 079906

Cambridge University Press is part of the University of Cambridge.

It furthers the University's mission by disseminating knowledge in the pursuit of education, learning, and research at the highest international levels of excellence.

www.cambridge.org
Information on this title: www.cambridge.org/9781107174467
DOI: 10.1017/9781316795897

© Neil M. Ribe 2018

This publication is in copyright. Subject to statutory exception and to the provisions of relevant collective licensing agreements, no reproduction of any part may take place without the written permission of Cambridge University Press.

First published 2018

Printed and bound in Great Britain by Clays Ltd, Elcograf S.p.A.

A catalogue record for this publication is available from the British Library.

Library of Congress Cataloging-in-Publication Data
Names: Ribe, Neil M., 1955– author.
Title: Theoretical mantle dynamics / Neil M. Ribe, Centre National de la Recherche Scientifique (CNRS), Paris.
Description: Cambridge, United Kingdom ; New York, NY : Cambridge University Press, 2018. | Includes bibliographical references and index.
Identifiers: LCCN 2018030249| ISBN 9781107174467 (hardback ; alk. paper) | ISBN 1107174465 (hardback ; alk. paper)
Subjects: LCSH: Earth (Planet)–Mantle. | Geodynamics.
Classification: LCC QE509.4 .R53 2018 | DDC 551.1/16–dc23 LC record available at https://lccn.loc.gov/2018030249

ISBN 978-1-107-17446-7 Hardback

Cambridge University Press has no responsibility for the persistence or accuracy of URLs for external or third-party internet websites referred to in this publication and does not guarantee that any content on such websites is, or will remain, accurate or appropriate.

To Anne, Isabelle and Aline, *avec toute ma tendresse*

Contents

Preface		*page* xi
Acknowledgements		xv
List of Abbreviations		xvi
1	Formulating Geodynamic Model Problems: Three Case Studies	1
	1.1 Heat Transfer from Mantle Diapirs	1
	1.2 Subduction	4
	1.3 Plume–Lithosphere Interaction	6
2	Dimensional and Scaling Analysis	9
	2.1 Buckingham's Π-Theorem and Dynamical Similarity	9
	2.2 Nondimensionalization	13
	2.3 Scaling Analysis	18
	Exercises	20
3	Self-Similarity and Intermediate Asymptotics	22
	3.1 Conductive Heat Transfer	22
	3.2 Self-Similar Buoyant Thermals	25
	3.3 Classification of Self-Similar Solutions	27
	3.4 Intermediate Asymptotics with Respect to Parameters: The Rayleigh–Taylor Instability	28
	Exercises	31
4	Slow Viscous Flow	33
	4.1 Basic Equations and Theorems	33
	4.2 Potential Representations for Incompressible Flow	39
	4.3 Classical Exact Solutions	43
	4.4 Superposition and Eigenfunction Expansion Methods	51
	4.5 The Complex-Variable Method for 2-D Flows	56

	4.6	Singular Solutions and the Boundary Integral Representation	57
	4.7	The Singularity Method	66
	4.8	Slender-Body Theory	68
	4.9	Flow Driven by Internal Loads	72
		Exercises	84
5	Elasticity and Viscoelasticity	86
	5.1	Correspondence Principles	86
	5.2	Loading of an Elastic Lithosphere	88
	5.3	Surface Loading of a Stratified Viscoelastic Sphere	91
		Exercises	95
6	Boundary-Layer Theory	96
	6.1	The Boundary-Layer Approximation	96
	6.2	Solution of the Boundary-Layer Equations Using Variable Transformations	98
	6.3	The Method of Matched Asymptotic Expansions	101
	6.4	A Plume from a Point Source of Buoyancy	105
	6.5	Boundary Layers with Strongly Variable Viscosity	111
		Exercises	118
7	Long-Wave Theories, 1: Lubrication Theory and Related Techniques	121
	7.1	Lubrication Theory	122
	7.2	Plume–Plate and Plume–Ridge Interaction Models	128
	7.3	Long-Wave Analysis of Thermal Boundary-Layer Instability	133
	7.4	Effective Boundary Conditions from Thin-Layer Flows	136
	7.5	Conduit Solitary Waves	142
		Exercises	145
8	Long-Wave Theories, 2: Shells, Plates and Sheets	147
	8.1	Theory of Thin Viscous Sheets in General Coordinates	148
	8.2	Thin-Shell Theory in Lines-of-Curvature Coordinates	159
	8.3	Geodynamic Applications of Thin-Shell Theory	161
	8.4	Immersed Viscous Sheets	169
		Exercises	173
9	Theory of Two-Phase Flow	175
	9.1	Geometrical Properties of Two-Phase Media	176
	9.2	Conservation Laws	177
	9.3	The Geodynamic Limit $\eta_f \ll \eta_m$	183
	9.4	One-Dimensional Model Problems	185

9.5	Solutions in Two and Three Dimensions	189
9.6	Instabilities of Two-Phase Flow	189
	Exercises	191

10 Hydrodynamic Stability and Thermal Convection 192
    10.1 Rayleigh–Taylor Instability 192
    10.2 Rayleigh–Bénard Convection 197
    10.3 Order-Parameter Equations for Finite-Amplitude
         Rayleigh–Bénard Convection 200
    10.4 Convection at High Rayleigh Number 210
    Exercises 217

11 Convection in More Realistic Systems 219
    11.1 Compressible Convection and the Anelastic Liquid Equations 219
    11.2 Convection with Temperature-Dependent Viscosity 226
    11.3 Convection in a Compositionally Layered Mantle 230
    11.4 Convection with a Phase Transition 235
    Exercises 239

12 Solutions to Exercises 241

References 292
Index 310

Preface

Mantle dynamics, as the term is used in this book, is the study of how the Earth's mantle deforms and flows over long ($\geq 10^2$–10^3 years) time scales. The fundamental concept underlying the theory of mantle dynamics is the notion of a continuous medium. In reality, the rocks of the Earth's mantle are aggregates of discrete mineral grains with characteristic sizes on the order of millimetres. By contrast, the typical length scales of the deformation associated with mantle convection are on the order of 10–10000 km, some 10^7–10^{10} times larger than the grain size. This means that convective motions are not directly influenced by the grain-scale structure of the mantle, but only by the average properties of material elements comprising very large numbers of grains. From a dynamic point of view, therefore, the mantle can be regarded as a continuous medium characterized by physical properties such as density and viscosity that vary smoothly as functions of position. The branch of mechanics that deals with media of this kind is called continuum mechanics.

The basis of continuum mechanics is a set of general conservation laws for mass, momentum and energy, which are usually formulated as partial differential equations. However, because those equations govern all deformations and flows, they describe none in particular and must therefore be supplemented by material constitutive relations and boundary and initial conditions that are appropriate for a particular phenomenon of interest. The result, often called a model problem, is the ultimate object of study in continuum mechanics.

Once posed, a model problem can be solved in one or more of three ways. The first is to construct a physical analog in the laboratory and let nature do the solving. The experimental approach has long played a central role in mantle dynamics, enabling the discovery of many hitherto unexpected phenomena (Davaille and Limare, 2015). The second possibility is to solve the model problem numerically on a computer, which is the dominant approach in present-day mantle dynamics (Ismail-Zadeh and Tackley, 2010). The third approach, the subject of this book, is to solve the problem analytically.

Analytical approaches are, of course, most effective when the model problem at hand is relatively simple and lack some of the flexibility of the best experimental and numerical methods. However, they compensate for this by providing a degree of understanding and insight that no other method can match. Furthermore, an analytical approach is often a necessary preparatory step to subsequent numerical work. Examples treated in this book include Green functions and propagator matrices for Stokes flow driven by internal loads in a spherical annulus (§ 4.9.1, 4.9.2), the governing equations for two-phase flow (§ 9) and the anelastic liquid equations (§ 11.1). Finally, analytical methods play a critical role in the interpretation of experimental and numerical results. For example, dimensional analysis is required to ensure proper scaling of experimental and numerical results to the Earth, and local scaling analysis applied to numerical output can reveal underlying laws that are obscured by numerical tables and graphical images. For all these reasons, the central role that analytical approaches have always played in mantle dynamics is unlikely to diminish.

This book is intended as a combined monograph and textbook for advanced graduate students and researchers working in the field of mantle dynamics. Its seed was a chapter entitled 'Analytical Approaches to Mantle Dynamics' in *Treatise on Geophysics* (Ribe, 2015). The present book more than doubles the length of that chapter by incorporating four types of new material. First, there are expanded derivations and explanations of many topics and results that were merely mentioned in the chapter for lack of space. Examples include the reversibility of Stokes flow, dissipation theorems, non-Newtonian corner flow, viscous eddies, the Stokeslet, slender-body theory, the lubrication theory equations, plume–plate interaction models, effective boundary conditions from thin-layer flows, conduit solitary waves, the thin viscous-sheet equations in two dimensions, matching conditions at a fluid–fluid interface, convection with temperature-dependent viscosity, convection in a compositionally layered mantle and convection with a phase transition. Second, many entirely new topics have been added, including the singularity method for Stokes flow, surface loading of an elastic lithosphere, plumes from a point source of buoyancy in fluids with both constant and temperature-dependent viscosity, the theory of thin viscous sheets in general coordinates, the theory of two-phase flow and compressible convection and the anelastic liquid approximation. Third, nearly 40 new explanatory figures have been included, more than tripling the number that appeared in the *Treatise* chapter. Finally, the book includes some 50 exercises at various levels of difficulty, all with completely worked-out solutions.

Given the advanced nature of the book, a certain amount of preliminary knowledge will be helpful. On the physical side, the main prerequisite is familiarity with basic concepts of general fluid dynamics and low Reynolds number flow at the level of § 2.1–2.3, 3.1–3.4, 3.6–3.7, 4.1–4.2 and 4.7–4.9 of Batchelor (1967), and with the elementary theory of elasticity at the level of chapter 1 of Landau and Lifshitz (1986). Familiarity with geodynamical applications of continuum mechanics (e.g., Turcotte and Schubert, 2014) is welcome but not obligatory. On the mathematical side, the level of the book is admittedly high, but no higher than the best work in the field requires. Frequently used concepts and techniques include Cartesian tensors, linear algebra and the theory of matrices, vector calculus, ordinary and partial differential equations, Fourier series and the Fourier transform and (to a lesser extent) integral equations and the theory of complex variables. More recondite topics such as the differential geometry of surfaces and covariant continuum mechanics also make an appearance, but the necessary elements are reviewed in some detail before they are needed.

However, even if you are not familiar with some of the topics just listed, be assured that I have endeavoured to make the book as widely accessible as possible. I have done this, first, by starting each derivation from first principles, without presupposing that you already know them. Second, I have made an effort to ensure that all derivations are 'followable' from the first step to the last, even if some of the intermediate algebraic steps are only described instead of being written out explicitly. The third means is the many exercises with worked-out solutions, which are specifically designed to help you master the mathematical methods required by the material. Certain of the exercises will be easier if you use a symbolic manipulation package, such as Mathematica© or Maple©, and are indicated by the character string [SM]. You will also be able to construct your own additional exercises by filling in intermediate algebraic steps in the various derivations presented.

Theoretical approaches to mantle dynamics are quite diverse and call for a correspondingly broad and comprehensive treatment. However, it is equally important to highlight the common structures and styles of argumentation that give theoretical mantle dynamics its unity. In this spirit, I begin with a discussion of the craft of formulating geodynamic model problems, focusing on three paradigmatic phenomena (heat transfer from magma diapirs, subduction and plume–lithosphere interaction) that will subsequently reappear in the course of the book treated by different methods. Thus heat transfer from diapirs is treated using dimensional analysis (§ 2.1 and 2.2), scaling analysis (§ 2.3, § 6.5.2) and boundary-layer theory (§ 6.2.2); subduction using dimensional analysis (§ 2.1), scaling analysis based on thin-sheet theory (§ 8.4.1) and the boundary-integral approach (§ 8.4.2); and plume–lithosphere interaction using lubrication theory and scaling analysis (§ 7.1, 7.2). Moreover, the discussions of these and other phenomena are organized as much as possible around three recurrent themes. The first is the importance of scaling arguments (and the scaling laws to which they lead) as tools for understanding physical mechanisms and applying model results to the Earth. Examples of scaling arguments can be found in § 2.3, 3.4, 6.3, 6.4, 6.5.1, 6.5.2, 7.1, 7.2, 8.1.3, 8.4.1,

10.3.3, 10.3.4, 10.4.1, 10.4.4, 10.4.5, 11.1 and 11.2. The second is the ubiquity of self-similar behaviour in geophysical flows, which typically occurs in parts of the spatiotemporal model domain that are sufficiently far from the inhomogeneous initial or boundary conditions that drive the flow to be uninfluenced by the details of those conditions (§ 3.1, 3.2, 3.3, 4.3.3, 4.3.4, 6.2.1, 6.5, 6.5.1, 7.1, 7.2, 10.4.4). The third theme is asymptotic analysis, in which the smallness or largeness of some key parameter in the model problem is exploited to simplify the governing equations, often via a reduction of their dimensionality (3.1, 4.8, 6.1, 6.3, 6.4, 6.5.1, 7.1, 7.3, 7.4, 7.5, 8, 10.3.3, 10.3.4, 10.4.2, 10.4.3, 10.4.4). While these themes by no means encompass everything the book contains, they can serve as threads to guide the reader through what might otherwise appear a trackless labyrinth of miscellaneous methods.

A final aim of the book is to introduce some less-familiar methods that deserve to be better known among geodynamicists. Examples include the use of Papkovich–Fadle eigenfunction expansions (§ 4.4.1) and complex variables (§ 4.5) for 2-D Stokes flows, solutions of the Stokes equations in bispherical coordinates (§ 4.4.3), thin-sheet theory in general nonorthogonal coordinates (§ 8.1) and multiple-scale analysis for modulated convection rolls (§ 10.3).

Throughout this book, unless otherwise stated, Greek indices range over the values 1 and 2; Latin indices range over 1, 2 and 3; and the standard summation convention over repeated subscripts is assumed. Index notation (e.g., u_i, σ_{ij}) and coordinate-free notation ($\mathbf{u}$, $\boldsymbol{\sigma}$) are used interchangeably as convenience dictates. For simplicity, both u_i and $\mathbf{u}$ are called vectors and both σ_{ij} and $\boldsymbol{\sigma}$ tensors, although strictly speaking u_i and σ_{ij} are the components of $\mathbf{u}$ and $\boldsymbol{\sigma}$, respectively. The notations $(x, y, z) = (x_1, x_2, x_3)$ for Cartesian coordinates and $(u, v, w) = (u_1, u_2, u_3)$ for the corresponding velocity components are equivalent. Unit vectors in given coordinate directions are denoted by symbols $\mathbf{e}_x$, $\mathbf{e}_r$, etc. Partial derivatives are denoted either by subscripts or by the symbol ∂, and $\partial_i = \partial/\partial x_i$. Thus, e.g.,

$$T_x = \partial_x T = \partial_1 T = \frac{\partial T}{\partial x} = \frac{\partial T}{\partial x_1}. \tag{0.1}$$

The symbol ∇_h is the 2-D gradient operator with respect to the coordinates (x, y) or (in a few cases) the spherical colatitude and longitude (θ, ϕ). The symbols $\Re[\ldots]$ and $\Im[\ldots]$ denote the real and imaginary parts, respectively, of the bracketed quantities. A 'free-slip' surface is one on which both the normal velocity and the shear traction vanish.

Acknowledgements

It is a pleasure to thank numerous colleagues and friends who were instrumental in helping this book come to fruition. First of all, I am grateful to Ulrich Christensen, Anne Davaille, John Lister and Stephen J. S. Morris for discussions over many years of several of the topics covered here. As noted earlier, this book began its life as a much shorter chapter in *Treatise on Geophysics*; I thank the volume editor, David Bercovici, for encouraging me to expand my chapter into a book, and Elsevier Scientific for permission to reuse material from it. Detailed and constructive reviews of the chapter by Bercovici, Morris and Yanick Ricard led to substantial improvements in the presentation that have been incorporated into the book. The original book proposal was reviewed by Hans-Peter Bunge and Alik Ismail-Zadeh, to whom I am grateful for their positive evaluations. I also owe a debt of gratitude to Bercovici, Davaille, Lister, Morris and Alessandro Forte for reading sections of the book manuscript and suggesting corrections and improvements. Several people including Davaille, Lister, Ricard, Shun-Ichiro Karato and Jerry Mitrovica provided indispensable references, computer programs and data for figures. Zhonghai Li gave generously of his time to create the cover image. I would also like to thank my editors and the production staff at Cambridge University Press, Emma Kiddle, Zoë Pruce, Chloe Quinn, Niranjana Harikrishnan and Karen Slaght for their efficient handling of the manuscript and their counsel at every step of the way. Owing to the contributions of all these people, the present book contains many fewer errors than it otherwise would; any that remain are entirely my own responsibility. Finally, I would like to thank the French 'Programme National de Planétologie' (INSU-CNRS/CNES) for faithful financial support over many years.

Abbreviations

1-D	one-dimensional
2-D	two-dimensional
3-D	three-dimensional
BEM	boundary-element method
BL	boundary layer
BVP	boundary-value problem
CMB	core–mantle boundary
GIA	glacial isostatic adjustment
IMP	intermediate matching principle
LHS	left-hand side
LSA	linear stability analysis
MEE	method of eigenfunction expansions
MMAE	method of matched asymptotic expansions
ODE	ordinary differential equation
PDE	partial differential equation
PMM	propagator matrix method
RHS	right-hand side
RII	reactive infiltration instability
RBC	Rayleigh–Bénard convection
RTI	Rayleigh–Taylor instability
SBT	slender-body theory
TBL	thermal boundary layer

1
Formulating Geodynamic Model Problems: Three Case Studies

An ideal geodynamic model respects two distinct criteria: it is sufficiently simple that the essential physics it embodies can be easily understood, yet sufficiently complex and realistic that it can be used to draw conclusions about the earth. It is seldom easy to satisfy both criteria together, and so most geodynamicists tend to emphasize one or the other, according to temperament and education.

However, there is a way to overcome this dilemma: to investigate not just a single model, but rather a hierarchical series of models of gradually increasing complexity and realism. Such an investigation – whether carried out by one individual or by many – is a cumulative one in which the initial study of highly simplified models provides the physical understanding required to guide the formulation and investigation of more complex models. In many cases, the simpler models in such a series can be solved analytically, whereas the subsequent more realistic models require numerical or experimental approaches. To illustrate how the hierarchical approach works in practice, I have chosen three exemplary geodynamic phenomena as case studies: heat transfer from magma diapirs, subduction and the interaction of mantle plumes with the lithosphere. The following discussions emphasize models at the simpler end of the spectrum that are amenable to analytic methods and omit mathematical detail to keep the focus on the conceptual structure of the hierarchical approach.

1.1 Heat Transfer from Mantle Diapirs

Our first example is the ascent of a hot blob or diapir of magma through the lithosphere, a possible mechanism for the formation of island-arc volcanoes (Marsh and Carmichael, 1974; Marsh, 1978). The goal of this model is to determine how far the diapir can move through the colder surrounding material before losing so much of its excess heat that it solidifies. Figure 1.1 illustrates a series of model problems that can be used to investigate this question.

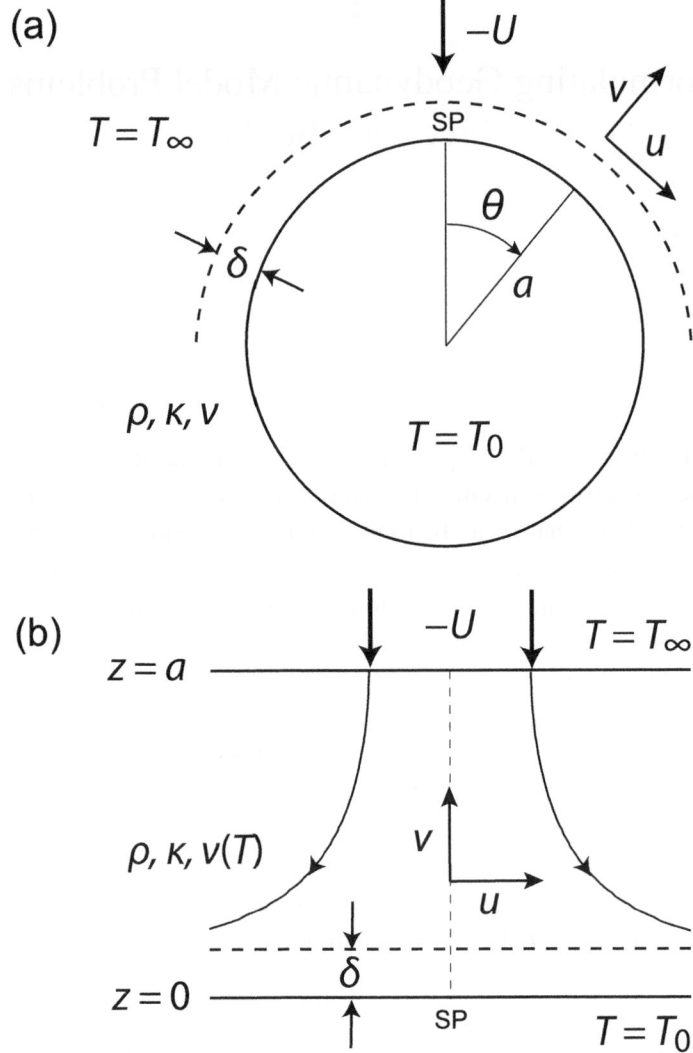

Figure 1.1 Models for the heat transfer from an ascending magma diapir (§ 1.1). (a) Original model in spherical geometry. A spherical diapir of radius a and constant temperature T_0 ascends in an infinite fluid with density ρ, kinematic viscosity ν, thermal diffusivity κ and temperature T_∞ far from the diapir. The viscosity of the fluid inside the diapir is supposed $\ll \nu$. The figure is drawn in the reference frame of the diapir, so that the fluid far from it moves downward with a constant speed $-U$. The colatitude measured from the leading stagnation point (SP) is θ, and the components of the velocity in the colatitudinal and radial directions are u and v, respectively. In the limit $Ua/\kappa \gg 1$, temperature variations in the hemisphere $\theta \leq \pi/2$ are confined to a BL of thickness $\delta \ll a$. The viscosity ν may be constant (Marsh, 1978) or temperature-dependent (Morris, 1982). (b) Stagnation flow model of Morris (1982). The surface of the hot sphere is replaced by the plane $z = 0$, and the far-field streaming velocity $-U$ is imposed as a boundary condition at $z = a$.

1.1 Heat Transfer from Mantle Diapirs

Probably the simplest model that still retains much of the essential physics (Marsh, 1978) can be formulated by assuming that (1) the diapir is spherical and has a constant radius; (2) the diapir's interior temperature is uniform and (3) does not vary with time; (4) the ascent speed and (5) the temperature of the lithosphere far from the diapir are constants; (6) the lithosphere is a uniform viscous fluid with constant physical properties and (7) the viscosity of the diapir is much less than that of the lithosphere. The result is the model shown in Figure 1.1a, in which an effectively inviscid fluid sphere with radius a and temperature T_0 ascends at constant speed U through a fluid with constant density ρ, thermal diffusivity κ, kinematic viscosity ν and constant temperature $T = T_\infty \equiv T_0 - \Delta T$ far from the sphere. An analytical solution for the rate of heat transfer q from the diapir (§ 6.2.2) can now be obtained if one makes the additional (and realistic) assumption (8) that the Péclet number $Pe \equiv Ua/\kappa \gg 1$, in which case the temperature variations around the leading hemisphere of the diapir are confined to a thin boundary layer (BL) of thickness $\delta \ll a$ (Figure 1.1a). One thereby finds (see § 2.3 for the derivation)

$$q \sim k_c a \Delta T P e^{1/2}, \qquad (1.1)$$

where k_c is the thermal conductivity.

While the model just described provides a first estimate of how the heat transfer scales with the ascent speed and the radius and excess temperature of the diapir, it is far too simple for direct application to Earth. A more realistic model can be obtained by relaxing assumptions (3) and (5), allowing the temperatures of the diapir and the ambient lithosphere to vary with time. If these variations are slow enough, the heat transfer at each instant will be described by a law of the form (1.1), but with a time-dependent excess temperature $\Delta T(t)$. A model of this type was proposed by Marsh (1978), who obtained a solution in the form of a convolution integral for the evolving temperature of a diapir ascending through a lithosphere with a prescribed far-field temperature $T_{\text{lith}}(t)$.

A different extension of the simple model of Figure 1.1a, also suggested by Marsh (1978), begins from the observation that the viscosity of mantle materials decreases strongly with increasing temperature. A hot diapir will therefore be surrounded by a thin BL of softened lithosphere, which will act as a lubricant and increase the diapir's ascent speed. The effectiveness of this mechanism depends on whether the BL is thick enough and/or has a viscosity low enough, to carry a substantial fraction of the volume flux $\sim \pi a^2 U$ that the sphere must displace in order to move. Formally, this model is obtained by replacing the constant viscosity ν in Figure 1.1a by one that depends exponentially on temperature as $\nu = \nu_0 \exp(-T/\Delta T_r)$, where ΔT_r is a rheological temperature scale.

While this new variable-viscosity model is more realistic and dynamically richer than the original model, its spherical geometry makes an analytical solution difficult. However, closer examination reveals that the spherical geometry is not in fact essential: all that matters is that the flow outside the softened BL varies over a characteristic length scale a that greatly exceeds the BL thickness. This recognition led Morris (1982) to study a simpler model in which the flow around the sphere is replaced by a stagnation flow between two planar boundaries $z = 0$ and $z = a$ (Figure 1.1b). The model equations now admit 1-D solutions $T = T(z)$ and $v = v(z)$ for the temperature and the vertical velocity, respectively, which can be determined using the method of matched asymptotic expansions (§ 6.3) in the limit of large viscosity contrast $\Delta T/\Delta T_r \gg 1$ (Morris, 1982). Because the general scaling relationships revealed by the 1-D solution apply equally well to the original spherical geometry, they can be exploited to simplify the governing equations in spherical coordinates, which can then be solved analytically for certain limiting cases (Morris, 1982; Ansari and Morris, 1985). Further discussion of these problems will be found in § 6.5.2.

1.2 Subduction

Our second example is the subduction of oceanic lithosphere. Faced with the task of devising the simplest possible model for subduction, it makes sense to begin with a purely kinematic approach in which flow is driven by imposed boundary velocities. A minimal list of parameters for such a model comprises a parameter to specify the overall geometry and a velocity to characterize the motion of the slab and the oceanic plate. Figure 1.2a shows an influential model of this type proposed by McKenzie (1969). This 'corner flow' model comprises two wedge-shaped regions containing fluid with a constant viscosity, bounded by rigid surfaces that meet at a corner. The dip of the surface representing the subducting slab is α. The slab and the oceanic plate move away from and towards the corner, respectively, with speed U_0. The overriding plate is motionless. The lines with arrows are typical streamlines for the flow in the two wedges.

An unrealistic aspect of the Newtonian corner flow model is that the stress in the fluid has a nonintegrable singularity $\propto r^{-1}$ at the corner, implying that an infinite force is required to drive subduction. This can be remedied by extending the model to non-Newtonian shear-thinning fluids in which the viscosity decreases with increasing stress. Fenner (1975) showed that corner flows with non-Newtonian rheology can be determined analytically in certain cases. Subsequently, Tovish et al. (1978) extended Fenner's results to the subduction geometry of Figure 1.2a. While the stress is still singular at the corner, the singularity is now integrable. Solutions for non-Newtonian corner flows are derived in § 4.3.3.

Despite their simplicity, corner flow models have been widely used in geodynamic studies where an analytical expression for the subduction-induced mantle

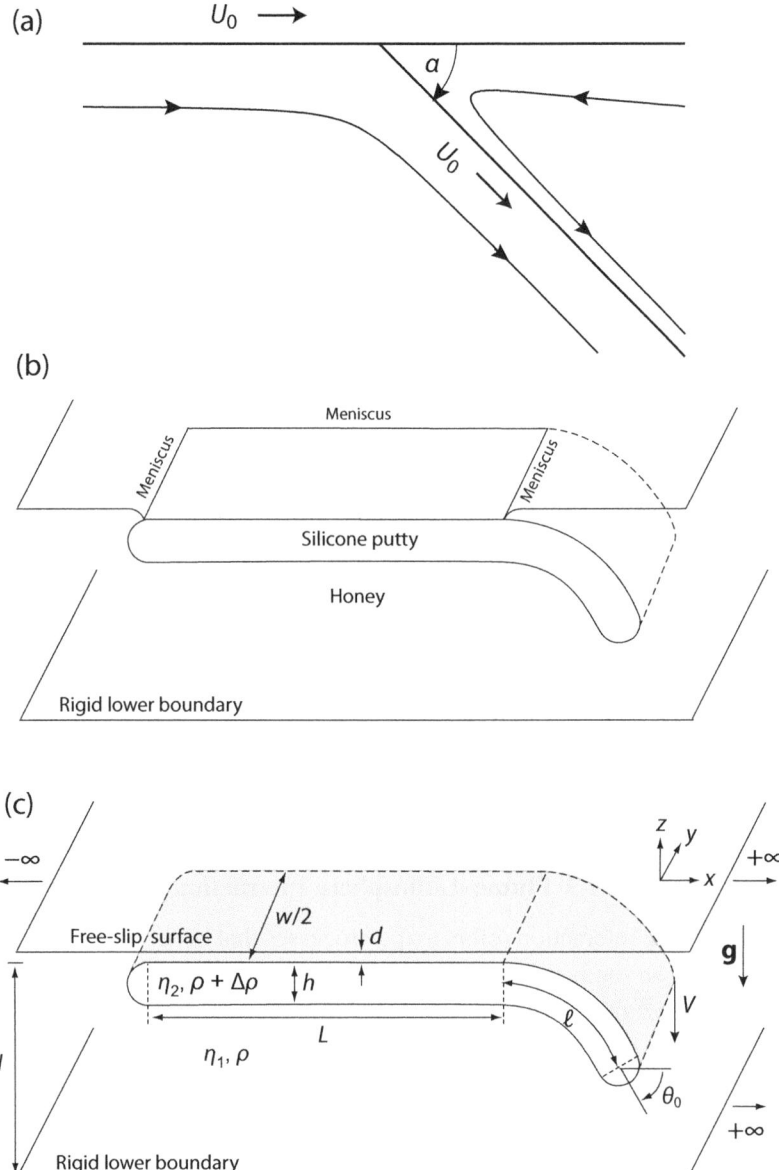

Figure 1.2 Models for the subduction of oceanic lithosphere (§ 1.2). (a) Kinematic corner flow model of McKenzie (1969). (b) Vertical cutaway view of the laboratory configuration studied by Bellahsen et al. (2005). Only the portion of the sheet behind the vertical symmetry plane is shown. The flat portion of the sheet is prevented from sinking by surface tension between the honey and the air acting across a meniscus. (c) Vertical cutaway view of the three-dimensional model of Li and Ribe (2012). Sinking of the flat portion of the sheet is prevented by the presence of a lubrication layer of thickness d between the sheet and the upper free-slip surface.

flow is needed. Nevertheless, they have a number of obvious shortcomings, including their two-dimensionality, the oversimplified geometry of the slab, the stress singularity at the corner and the neglect of the driving buoyancy force and of the viscous force that resists the bending of the plate. The decisive step forward that overcame all these shortcomings was taken in the context of laboratory experiments, using the setup shown schematically in Figure 1.2b (Bellahsen et al., 2005). Earth's upper mantle is represented by a layer of honey $\approx$ 11 cm thick in a transparent tank, and the plate is represented by a thin ($\approx$ 1 cm) sheet of denser silicone putty. The sheet is initially placed flat on top of the honey, where it is prevented from sinking by the surface tension between the honey and the air acting across a meniscus. The experiment is launched by pushing one edge of the sheet down into the honey and letting it subduct freely.

Inspired by these experiments, Li and Ribe (2012) proposed a model (Figure 1.2c) in which surface tension is replaced by a thin lubrication layer of mantle fluid above the sheet. Lubrication theory (§ 7.1) states that the normal stress in the thin layer greatly exceeds the tangential stress. Accordingly, the lubrication layer serves the same purpose as surface tension, which is to prevent the flat part of the sheet from sinking while allowing it to move sideways freely in response to the pull of the slab. However, the advantage of the lubrication layer from a theoretical point of view is that it removes the three-phase (air + honey + putty) contact line. The model then becomes amenable to solution by the semi-analytical boundary-element method (§ 4.6.4). The model of Figure 1.2c will be discussed in more detail in § 8.2.

1.3 Plume–Lithosphere Interaction

Plume–lithosphere interaction refers to the processes that occur after a rising mantle plume impinges on the base of the lithosphere. Because the plume fluid is buoyant relative to its surroundings, it will spread beneath the lithosphere, eventually forming a shallow pool whose lateral dimensions greatly exceed its thickness.

Figure 1.3 shows a series of fluid dynamical models that have been used to study plume–lithosphere interaction, beginning with the kinematic model of Sleep (1987) (Figure 1.3a). Sleep's insight was that the flow associated with a plume rising beneath a moving plate can be regarded as the sum of two parts: a (horizontal) radial flow representing buoyant plume fluid emanating from a steady localized source at the top of the plume conduit and an ambient mantle wind in the direction of the plate motion. Fluid from the source can travel only a finite distance upstream against the wind before being blown back downstream again, leading to the formation of a stagnation point (labelled SP in Figure 1.3a) at which the wind speed just equals the speed of radial outflow from the source. The stagnation streamline that passes through this point (heavy line in Figure 1.3a) divides the (x, y) plane into an inner

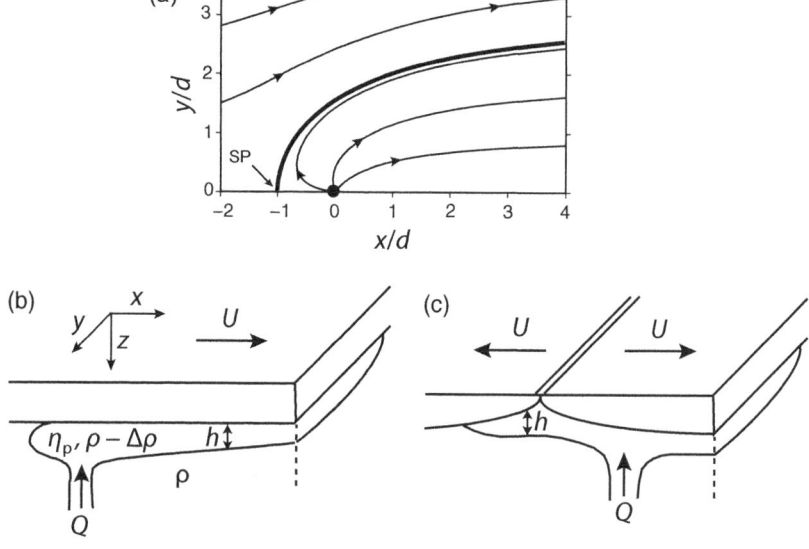

Figure 1.3 Models for plume–lithosphere interaction (§ 1.3). (a) Steady streamlines for the 2-D kinematic model of Sleep (1987). The source is indicated by the black circle, the heavy solid line is the stagnation streamline and d is the distance between the source and the stagnation point SP. (b) Spreading of a pool of buoyant plume fluid supplied at a volumetric rate Q beneath a rigid lithosphere moving at speed U relative to the plume stem (Olson, 1990). The plume fluid has viscosity η_p and density $\rho - \Delta\rho$, where ρ is the density of the ambient mantle. (c) Same as (b), but beneath two plates separated by a spreading ridge with half-spreading rate U (Ribe et al., 1995).

region containing fluid from the source and an outer region containing fluid brought in from upstream by the wind. The stagnation streamline resembles the shape of the topographic swell around the Hawaiian Island chain (Richards et al., 1988).

While the model of Sleep (1987) nicely illustrates the kinematics of plume–plate interaction, it neglects the (driving) buoyancy force and (resisting) viscous force that control the spreading of the plume pool. We now seek the simplest possible model that embodies these dynamics. We first replace the continuous variation of fluid properties by a two-fluid structure, comprising a plume-fed pool with thickness $h(x,y)$, viscosity η_p and density $\rho - \Delta\rho$ spreading in an ambient fluid with viscosity η_m and density ρ. We suppose that the plume stem supplies fluid at a constant volumetric flux Q, at a point (the hotspot) that is fixed relative to a plate moving at a constant speed U. Finally, we assume that η_m/η_p, while large, is nevertheless small enough that the resistance of the ambient mantle to the spreading of the pool can be neglected. The result is the 'refracted plume' model of Olson (1990) (Figure 1.3b). Olson's model is in essence a dynamically self-consistent extension of the kinematic model of Sleep (1987). The refracted plume model can be generalized still further while retaining its analytical character by allowing the plume material to have a more realistic non-Newtonian (shear-thinning) rheology (Asaadi et al., 2011).

As a final illustration, Figure 1.3c shows a further extension of the refracted plume model in which the uniform plate is replaced by two plates separated by a spreading ridge. Despite the increased complexity of this plume-ridge interaction model, some analytical results can still be obtained by scaling analysis (Ribe et al., 1995). Plume–plate and plume–ridge interaction models are discussed in more detail in § 7.2.

2
Dimensional and Scaling Analysis

The goal of studying model problems in continuum mechanics is typically to determine functional relations, called scaling laws, that link certain 'target' parameters of interest to the other parameters on which they depend. Two powerful methods that are appropriate for this purpose are dimensional analysis and scaling analysis.

2.1 Buckingham's Π-Theorem and Dynamical Similarity

Dimensional analysis starts from the principle that the validity of physical laws cannot depend on the system of units (SI, cgs, etc.) in which they are expressed. An important consequence of this principle is the Π-theorem of Buckingham (1914).

Suppose that there exists a (generally unknown) functional relationship among N parameters $P_1\ P_2, \ldots, P_N$ that characterize a physical system of interest, viz.

$$\text{fct}_1(P_1, P_2, \ldots, P_N) = 0. \tag{2.1}$$

In continuum mechanics, the parameters P_n typically include material properties (density, viscosity, Young's modulus, etc.), the dimensions of the system (depth, width, etc.), the parameters that drive the flow or deformation (the gravitational acceleration, a velocity or stress imposed at a boundary, etc.), spatial coordinates (if one is concerned with the properties of the flow at a particular point) and the time (if the problem is unsteady). These parameters usually have dimensions, but can sometimes be dimensionless, e.g., the angle between the walls of a fluid-filled wedge. Let $M < N$ be the number of the parameters P_n that have independent physical dimensions. The dimensions in a set are said to be independent if no one of them can be expressed in terms of (i.e., as products of powers of) the others. Thus the dimensions of length, time and mass are independent, but the dimensions of length, time and velocity are not. Note that physical systems can never have $M = N$ because a dimensionally consistent functional relationship among the dimensionally independent parameters P_n would then be impossible. In most cases,

M is just the number of independent primary units that enter into the problem, e.g., $M = 3$ for mechanical problems involving units of m, kg and s and $M = 4$ for thermomechanical problems involving temperature (units K) in addition. However, there are exceptions. For example, suppose we wish to determine how the shear wave speed c of an isotropic elastic medium depends on its density ρ and its shear modulus μ. Naively counting the primary units (m, kg, s) involved in these quantities, we would conclude that $M = 3$, leading to the impossible result $N = M$. However, counting the number of parameters with independent dimensions gives the correct result $M = 2$. There is therefore a single dimensionless group $c^2 \rho / \mu$, which must be equal to a constant, implying $c \propto (\mu/\rho)^{1/2}$.

Given the numbers N and M as defined earlier, Buckingham's Π-theorem states that there exist $N - M$ independent dimensionless combinations (called dimensionless groups) of the original dimensional physical parameters P_n. The theorem gets its name from the tradition of denoting these groups by the Greek letter Π with a subscript $i = 1, 2, \ldots, N - M$. The theorem implies that the original dimensional functional relation (2.1) is equivalent to a dimensionless relation among the groups Π_i, viz.,

$$\text{fct}_2(\Pi_1, \Pi_2, \ldots, \Pi_{N-M}) = 0. \tag{2.2}$$

The requirement that the functional relation (2.2) should not depend on the systems of units chosen (e.g., SI vs. cgs) requires that each dimensionless group Π_i be a product of powers of the dimensional parameters P_n; no other functional form preserves the value of the dimensionless group when the system of units is changed. The function fct_2 in (2.2), by contrast, can have any form and need not be simple (e.g., a power law). While the total number of independent groups Π_i is fixed ($\equiv N - M$), the definitions of the individual groups are arbitrary and can be chosen as convenient. A more detailed discussion and proof of the Π-theorem can be found in Barenblatt (1996), chapter 1.

The Π-theorem is the basis for the concept of dynamical similarity, according to which two physical systems behave similarly (i.e., proportionally) if they have the same values of the dimensionless groups Π_i that define them. The crucial point is that two systems may have identical values of Π_i even though they are of very different size, i.e., even if the values of the dimensional parameters P_n are very different. Dynamical similarity is thus a natural generalization of the concept of geometrical similarity, whereby (e.g.) two triangles of different sizes are similar if they have the same values of the dimensionless parameters (angles and ratios of sides) that define them. Geometrical similarity is a necessary, but not a sufficient, condition for dynamical similarity. The recognition of this principle is due to Galileo, who discussed the strength of materials and structures in his *Dialogues Concerning Two New Sciences* (1638). In particular, Salviati (the character representing Galileo) noted that a small ship will remain intact when it is about to be launched, whereas a geometrically similar but larger ship will collapse under its own weight unless special precautions are taken.

The importance of dynamical similarity in geodynamics is that it allows results obtained in the laboratory or on a computer to be applied to another system (Earth) with very different scales of length, time, etc. Its power derives from the fact that the function fct_2 in (2.2) has only $N - M$ arguments, M fewer than the original functional relation fct_1 among the N dimensional parameters P_i. Thus an experimentalist or numerical analyst who seeks to determine how a target dimensional parameter P_1 depends on the other $N - 1$ parameters need not vary all those parameters individually; it suffices to vary only $N - M - 1$ dimensionless parameters. The number of parameters to be varied is $N - M - 1$ and not $N - M$, because one of the groups Π_i contains the target output parameter. Consequently, if the variation of a given dimensional parameter requires ≈ 10 samplings, then use of the Π-theorem reduces the effort involved in searching the parameter space by a factor $\approx 10^M$ (Barenblatt, 1996). By the same token, the Π-theorem makes possible a far more economical representation of experimental or numerical data. As an example, suppose that we have $N = 5$ dimensional parameters $P_1, P_2, \ldots, P_5$ from which $N - M = 2$ independent dimensionless groups Π_1 and Π_2 can be formed. To represent our data without the help of the Π-theorem, we would need many shelves (one for each value of P_5) each containing many books (one for each value of P_4) each containing many pages (one for each value of P_3) each containing a plot of P_2 vs. P_1. By using the Π-theorem, however, we can collapse the whole library onto a single plot of Π_2 vs. Π_1.

As a simple illustration of the Π-theorem, consider again the model for heat transfer from a hot sphere (Figure 1.1a). Suppose that we wish to determine the radial temperature gradient β (proportional to the local conductive heat flux) as a function of position on the sphere's surface. A list of all the relevant parameters includes the following eight (units in brackets): β [K m^{-1}], a [m], U [m s^{-1}], ρ [kg m^{-3}], ν [m^2 s^{-1}], κ [m^2 s^{-1}], ΔT [K] and θ [dimensionless]. Note that T_∞ does not appear in the preceding list. This is because none of the fluid properties depends on temperature, so that the absolute temperature plays no role in the problem. Thus T_∞ can be set to zero with no loss of generality. Moreover, ρ can be eliminated immediately because it is the only parameter that involves units of mass: no dimensionless group containing it can be defined. The remaining parameters are $N = 7$ in number, $M = 3$ of which (e.g., a, ΔT and U) have independent units, so $N - M = 4$ independent dimensionless groups can be formed. It is usually good practice to start by defining a single group containing the target parameter (β here). While inspection usually suffices, one can also proceed more formally by writing the group (Π_1, say) as a product of the target parameter and unknown powers of any set of M parameters with independent dimensions, e.g., $\Pi_1 = \beta a^{n_1} \Delta T^{n_2} U^{n_3}$. The requirement that Π_1 be dimensionless then implies $n_1 = 1$, $n_2 = -1$ and $n_3 = 0$. Additional groups are then obtained by applying the same procedure to the remaining dimensional parameters in the list (κ and ν in this case). Finally, any remaining parameters in the list that are already dimensionless (θ in this case) are used as groups by themselves. For the hot sphere, the result is

$$\frac{\beta a}{\Delta T} = \text{fct}\left(\frac{Ua}{\kappa}, \frac{Ua}{\nu}, \theta\right), \qquad (2.3)$$

where fct is an unknown function. The groups $Ua/\kappa \equiv Pe$, $Ua/\kappa \equiv Re$ and $\beta a/\Delta T \equiv \mathcal{N}$ are traditionally called the Péclet number, the Reynolds number and the (local) Nusselt number, respectively (the last to be distinguished from the global Nusselt number $Nu \equiv \int_S \mathcal{N} \, dS$ that measures the total heat flux across the sphere's surface S). As noted earlier, the definitions of the dimensionless groups in a relation like (2.3) are not unique. Thus one can replace any group by the product of itself and arbitrary powers of the other groups, e.g., Pe in (2.3) by the Prandtl number $\nu/\kappa \equiv Pe/Re$. Furthermore, it often happens that the target parameter ceases to depend on a dimensionless group whose value is very large or very small. For example, in a very viscous fluid such as the mantle, $Re \ll 1$ because inertia is negligible, and so Re no longer appears as an argument in (2.3). However, it is not universally true that a function of dimensionless groups tends to a finite limit when one of the groups tends to zero or infinity, and the possible nonexistence of the limit needs to be kept in mind.

A second example of dimensional analysis is furnished by the model of free subduction shown in Figure 1.2c. The first step is to choose a target parameter. We choose the time-dependent sinking speed $V(t)$ of the slab, defined as the vertical component of the velocity of the slab's leading end in the vertical symmetry plane. The sinking speed depends on the following parameters (units in square brackets): the layer depth H [m], the initial slab thickness h [m], the initial plate length L [m], the initial slab length ℓ [m], the slab width w [m], the thickness of the lubrication layer d [m], the initial dip of the slab's leading end θ_0 [dimensionless], the viscosities η_2 and η_1 of the sheet and the ambient fluid [kg m^{-1}s^{-1}], the buoyancy $g\Delta\rho$ [kg m^{-2}s^{-2}] and the time t [s]. The densities ρ_1 and ρ_2 appear only in the combination $g\Delta\rho$ because fluids in Stokes flow have gravitational mass but effectively zero inertial mass (§ 4). The unknown functional relationship among these $N = 12$ parameters has the general form

$$V = \text{fct}_1\left(H, h, L, \ell, w, d, \theta_0, \eta_1, \eta_2, g\Delta\rho, t\right). \qquad (2.4)$$

Now the number of parameters with independent dimensions (e.g., V, h and η_1) is $M = 3$. We therefore have $N - M = 9$ dimensionless groups, and the unknown functional relationship among them can be written as

$$\frac{V}{V_{\text{Stokes}}} = \text{fct}_2\left(\frac{H}{h}, \frac{L}{h}, \frac{\ell}{h}, \frac{w}{h}, \frac{d}{h}, \theta_0, \frac{\eta_2}{\eta_1}, \frac{\eta_1 t}{hg\Delta\rho}\right). \tag{2.5}$$

where $V_{\text{Stokes}} = h\ell g\Delta\rho/\eta_1$ is a characteristic Stokes sinking speed for a body with lateral dimension ℓ and thickness h. Other scales for the velocity could have been chosen, such as $h^2 g\Delta\rho/\eta_1$, but it is generally good practice to use scales that have a physical significance for the problem at hand. We shall return to the example of free subduction in Chapter 7.

2.2 Nondimensionalization

When the equations governing the dynamics of the problem at hand are known, another method of dimensional analysis becomes available: nondimensionalization. I illustrate this using the previous example of a hot sphere.

The first step is to write down the governing equations, together with all the relevant initial and boundary conditions. Because the problem is both steady and axisymmetric, the dependent variables are the velocity $\mathbf{u}(r,\theta)$, the pressure $p(r,\theta)$ and the temperature $T(r,\theta)$, where r and θ are the usual spherical coordinates. If viscous dissipation of energy is negligible, the governing equations and boundary conditions are

$$\nabla \cdot \mathbf{u} = 0, \quad \mathbf{u} \cdot \nabla T = \kappa \nabla^2 T, \quad \mathbf{u} \cdot \nabla \mathbf{u} = -\rho^{-1}\nabla p + \nu \nabla^2 \mathbf{u}, \tag{2.6a}$$

$$T(a,\theta) - T_0 = T(\infty,\theta) - T_\infty = \mathbf{u}(a,\theta) = \mathbf{u}(\infty,\theta) + U\mathbf{e}_z = 0, \tag{2.6b}$$

where $\mathbf{e}_z$ is a vertical unit vector. The next step is to define dimensionless variables using appropriate problem-specific scales. The best scales to use are not always immediately obvious, and in such cases a reliable technique is to nondimensionless all the variables using unknown scales whose definitions will be specified later. Thus in the case of the hot sphere, we define dimensionless (primed) variables as

$$\mathbf{x} = L^*\mathbf{x}', \quad \mathbf{u} = U^*\mathbf{u}', \quad p = p_\infty + P^*p', \quad T = T_\infty + T^*T', \tag{2.7}$$

where the scales L^*, U^*, P^* and T^* are all unknown. The far-field pressure p_∞ is included for completeness but has no effect on the dynamics. Substituting (2.7) into (2.6) and immediately dropping the primes to avoid notational overload, we obtain

$$\frac{U^*}{L^*}\nabla \cdot \mathbf{u} = 0, \quad \frac{U^*T^*}{L^*}\mathbf{u} \cdot \nabla T = \frac{\kappa T^*}{L^{*2}}\nabla^2 T, \tag{2.8a}$$

$$\frac{U^{*2}}{L^*}\mathbf{u} \cdot \nabla \mathbf{u} = -\frac{P^*}{\rho L^*}\nabla p + \frac{\nu U^*}{L^{*2}}\nabla^2 \mathbf{u}, \tag{2.8b}$$

$$T^*T\left(\frac{a}{L^*},\theta\right) - \Delta T = T^*T(\infty,\theta) = 0, \tag{2.8c}$$

$$U^*\mathbf{u}\left(\frac{a}{L^*},\theta\right) = U^*\mathbf{u}(\infty,\theta) + U\mathbf{e}_z = 0. \tag{2.8d}$$

The final step is to choose the unknown scales to simplify the equations and boundary conditions as much as possible. Inspection of the boundary conditions (2.8c) and (2.8d) immediately suggests the choices $L^* = a$, $U^* = U$ and $T^* = \Delta T$. The remaining unknown scale P^* can now be chosen by setting the coefficient of $-\nabla p$ in (2.8b) ($= P^*/\rho L^*$) equal to either the coefficient of the inertia term $\mathbf{u}\cdot\nabla\mathbf{u}$ ($= U^{*2}/L^*$) or the coefficient of the viscous term $\nabla^2\mathbf{u}$ ($= \nu U^*/L^{*2}$). The first choice yields $P^* = \rho U^2$ and makes the dimensionless pressure of order unity in the limit of fast (inertia-dominated) flow. The second choice yields $P^* = \rho\nu U/a$ and makes the dimensionless pressure of order unity in the limit of slow flow with negligible inertia. With this second choice, we obtain the dimensionless boundary-value problem (BVP)

$$\nabla\cdot\mathbf{u} = 0, \quad Pe\,\mathbf{u}\cdot\nabla T = \nabla^2 T, \quad Re\,\mathbf{u}\cdot\nabla\mathbf{u} = -\nabla p + \nabla^2\mathbf{u}, \tag{2.9a}$$

$$T(1,\theta) - 1 = T(\infty,\theta) = \mathbf{u}(1,\theta) = \mathbf{u}(\infty,\theta) + \mathbf{e}_z = 0, \tag{2.9b}$$

where Pe and Re are the Péclet and Reynolds numbers defined in § 2.1. If we had made the first choice $P^* = \rho U^2$, the momentum equation in (2.9a) would have been

$$\mathbf{u}\cdot\nabla\mathbf{u} = -\nabla p + Re^{-1}\nabla^2\mathbf{u}. \tag{2.10}$$

The dimensionless BVP obtained using $P^* = \rho U^2$ thus contains exactly the same dimensionless groups (Pe and Re) as the one obtained using $P^* = \rho\nu U/a$. This is a special case of the more general principle that all possible nondimensionalizations of the equations governing a given physical system yield the same number of independent dimensionless groups, as required by the Π-theorem.

Now because Pe and Re are the only dimensionless parameters appearing in (2.9a) and (2.9b), the dimensionless temperature must have the form $T = \text{fct}(r,\theta,Pe,Re)$. Differentiating this with respect to r and evaluating the result on the surface $r = 1$ to obtain the quantity $\beta a/\Delta T$, we find the same result (2.3) as we did using the Π-theorem.

Whether one chooses to do dimensional analysis using the Π-theorem or nondimensionalization depends on the problem at hand. The Π-theorem is of course the only choice if the governing equations are not known, but its effective use then requires a good intuition of what the relevant physical parameters are. When the governing equations are known, nondimensionalization is usually the best choice, as the relevant physical parameters appear explicitly in the equations and boundary/initial conditions. Further detailed examples of nondimensionalization can be found in Ribe and Davaille (2013).

2.2 Nondimensionalization

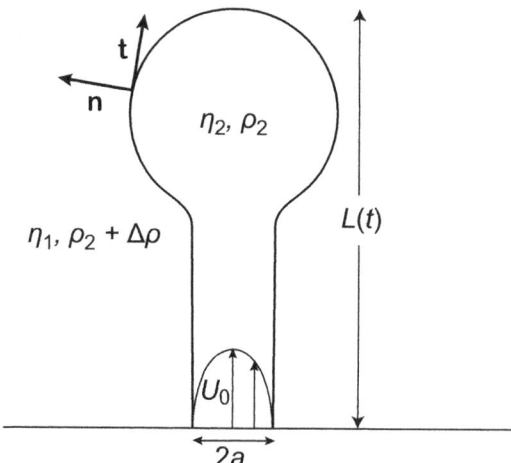

Figure 2.1 A creeping plume with viscosity η_2 and density ρ_2 ascending in an ambient fluid with viscosity η_1 and density $\rho_2 + \Delta\rho$. The plume fluid is injected through a hole of radius a in the bottom of a tank and has a parabolic (Poiseuille) velocity profile with maximum velocity U_0. The time-dependent height of the plume is $L(t)$. This model is discussed in § 2.2.

However, there are cases where it makes sense to use the Π-theorem even when the governing equations are known. This is best explained by way of an example: the slow (inertia-free) rise of a buoyant 'creeping plume' with viscosity η_2 in an ambient fluid with viscosity η_1 (Olson and Singer, 1985). In the laboratory, such plumes can be generated by injecting the plume fluid through a hole of radius a in the bottom of a tank containing the ambient fluid (Figure 2.1). The fluid issues from the hole with a parabolic (Poiseuille) velocity profile with maximum value U_0 and generates a plume of height $L(t)$, where t is the time since the beginning of the injection. We assume that the plume is axisymmetric at all times.

To understand why the Π-theorem is the right choice for this problem, we have first to sketch the method of nondimensionalization to see why it is misleading despite its formal correctness. The velocities $\mathbf{u}^{(1)}$ and $\mathbf{u}^{(2)}$ of the two fluids satisfy the incompressibility condition and the Stokes equations, viz.,

$$\nabla \cdot \mathbf{u}^{(i)} = 0, \qquad \eta_i \nabla^2 \mathbf{u}^{(i)} = \nabla p_i, \qquad (2.11)$$

where p_i is the modified pressure, related to the total pressure by $p_i = p_i^{\text{total}} + \rho_i g x_3$. The matching conditions satisfied by the velocity, the tangential stress and the normal stress on the interface S between the two fluids are

$$\mathbf{u}^{(1)} = \mathbf{u}^{(2)} \quad (\mathbf{x} \in S), \qquad (2.12a)$$

$$\eta_1 \mathbf{n} \cdot \mathbf{E}^{(1)} \cdot \mathbf{t} = \eta_2 \mathbf{n} \cdot \mathbf{E}^{(2)} \cdot \mathbf{t} \quad (\mathbf{x} \in S), \qquad (2.12b)$$

$$-p_1 + 2\eta_1 \mathbf{n} \cdot \mathbf{E}^{(1)} \cdot \mathbf{n} = -p_2 + 2\eta_2 \mathbf{n} \cdot \mathbf{E}^{(2)} \cdot \mathbf{n} - g\Delta\rho x_3 \quad (\mathbf{x} \in S). \qquad (2.12c)$$

In the preceding conditions, **n** is a unit vector normal to the interface pointing from the plume to the ambient fluid, and **t** is the unit vector tangent to the interface in any azimuthal plane (because the plume is axisymmetric, the other tangent vector $\mathbf{n} \times \mathbf{t}$, which points in the azimuthal direction, is irrelevant). Also, $\mathbf{E}^{(i)}$ is the strain rate tensor in fluid i, and $\Delta\rho = \rho_1 - \rho_2$. The boundary condition on the bottom of the tank is

$$\mathbf{u}^{(1)} = \mathbf{0} \quad (r > a), \tag{2.13a}$$

$$\mathbf{u}^{(2)} = U_0 \left[1 - \left(\frac{r}{a}\right)^2 \right] \mathbf{e}_z \quad (r \leq a), \tag{2.13b}$$

where r is the radial distance from the center of the hole. The kinematic condition expressing the fact that S is a material surface is

$$\frac{D\mathbf{x}}{Dt} = \mathbf{u} \quad (\mathbf{x} \in S), \tag{2.14}$$

where D/Dt is the convective derivative and $\mathbf{u} = \mathbf{u}^{(1)} = \mathbf{u}^{(2)}$ is the (unique) velocity on the interface. Finally, there is an initial condition stating that the interface between the two fluids is flat across the hole at $t = 0$. For simplicity, we have assumed that the vertical walls of the tank and the upper surface of the ambient fluid are at sufficiently large distances from the plume that they have no influence on it.

The preceding equations can now be nondimensionalized using a, U_0, $\eta_1 U_0/a$ and a/U_0 as scales for length, velocity, pressure and time, respectively. We need write down only the equations whose form changes as a result of the nondimensionalization. These are the Stokes equations,

$$\nabla^2 \mathbf{u}^{(1)} = \nabla p_1, \qquad \frac{\eta_2}{\eta_1} \nabla^2 \mathbf{u}^{(2)} = \nabla p_2, \tag{2.15}$$

the matching condition on the tangential stress,

$$\mathbf{n} \cdot \mathbf{E}^{(1)} \cdot \mathbf{t} = \frac{\eta_2}{\eta_1} \mathbf{n} \cdot \mathbf{E}^{(2)} \cdot \mathbf{t} \quad (\mathbf{x} \in S), \tag{2.16}$$

the matching condition on the normal stress,

$$-p_1 + 2\mathbf{n} \cdot \mathbf{E}^{(1)} \cdot \mathbf{n} = -p_2 + 2\frac{\eta_2}{\eta_1} \mathbf{n} \cdot \mathbf{E}^{(2)} \cdot \mathbf{n} - \frac{a^2 g \Delta\rho}{\eta_1 U_0} x_3 \quad (\mathbf{x} \in S), \tag{2.17}$$

and the boundary condition across the injection nozzle,

$$\mathbf{u}^{(2)} = \left(1 - r^2\right) \mathbf{e}_z \quad (r \leq 1). \tag{2.18}$$

In equations (2.15)–(2.18), all variables are dimensionless, even though that is not indicated by the notation. These equations contain two dimensionless groups, η_2/η_1

2.2 Nondimensionalization

and $a^2 g \Delta\rho / \eta_1 U_0$. In addition, our dimensionless target parameter is L/a, which depends on the dimensionless time $U_0 t/a$. The dimensionless scaling law for the plume height therefore has the form

$$\frac{L}{a} = \text{fct}\left(\frac{\eta_2}{\eta_1}, \frac{a^2 g \Delta\rho}{\eta_1 U_0}, \frac{U_0 t}{a}\right). \tag{2.19}$$

Although the law (2.19) is perfectly correct, it is potentially misleading because it was derived algorithmically without applying any physical insight. Laboratory experiments show that the dynamics and morphology of starting plumes only become interesting after the plume has ascended a distance that greatly exceeds the hole radius a. At such distances, the top of the plume is no longer influenced by the finite radius of the hole, which ceases to be a relevant length scale for the dynamics. The hole then acts as a point source characterized by a single parameter, the volumetric flux $Q \equiv \pi a^2 U_0 / 2$ that it supplies. The height L of the plume now depends only on the parameters Q, t, η_1, η_2 and $g\Delta\rho$, where the densities again appear only in the combination $g\Delta\rho$ because the fluids have negligible inertial mass. Applying the Π-theorem to the aforementioned list of six parameters (including L), three of which have independent dimensions, we find

$$\Pi_1 = \text{fct}(\Pi_2, \Pi_3), \tag{2.20a}$$

$$\Pi_1 = L\left(\frac{g\Delta\rho}{\eta_1 Q}\right)^{1/4}, \quad \Pi_2 = Q^{1/4}\left(\frac{g\Delta\rho}{\eta_1}\right)^{3/4} t, \quad \Pi_3 = \frac{\eta_2}{\eta_1}. \tag{2.20b}$$

Finally, in the limit $\eta_2/\eta_1 \ll 1$ corresponding to a low-viscosity plume, we anticipate that the plume viscosity η_2 is no longer relevant, in which case (2.20) reduces to $\Pi_1 = \text{fct}(\Pi_2)$. This last prediction is verified by laboratory measurements in the limit $\eta_2/\eta_1 \ll 1$, which define a universal curve on a plot of Π_1 vs. Π_2 (Figure 2.2; Olson and Singer, 1985). This simple result would have been much harder to find starting from (2.19), which can lead the experimenter to suppose incorrectly that a is always a relevant length scale. Olson and Singer (1985) in fact do their dimensional analysis in a somewhat different way and use the parameters

$$\Pi_1^{\text{OS}} = \left(\frac{\eta_1}{Q g \Delta\rho}\right)^{1/2} \frac{L}{t}, \quad \Pi_2^{\text{OS}} = \left(\frac{g\Delta\rho}{\eta_1}\right)^{1/2} Q^{1/6} t^{2/3}. \tag{2.21}$$

which are related to the parameters in (2.20b) by $\Pi_1^{\text{OS}} = \Pi_1/\Pi_2$ and $\Pi_2^{\text{OS}} = \Pi_2^{2/3}$. The two approaches are therefore entirely equivalent.

The preceding example illustrates nicely both the power and the limitations of dimensional analysis. Its power resides in its capacity to reduce substantially the number of parameters that an experimentalist or numerical analyst must vary in order to understand a physical system. But when performed in a blindly algorithmic

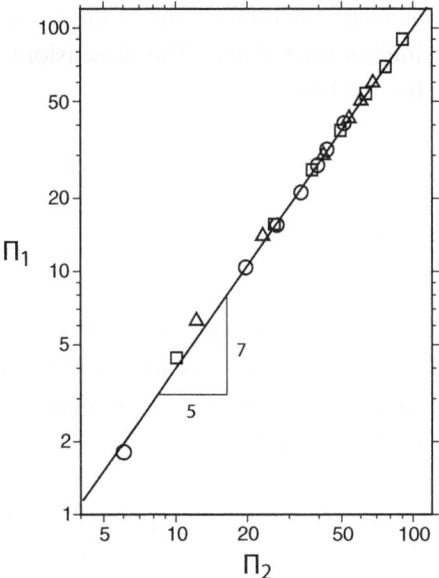

Figure 2.2 Experimentally determined scaling law for the height $L(t)$ of a creeping plume in the limit of negligible plume viscosity $\eta_2/\eta_1 \to 0$ (Olson and Singer, 1985). The dimensionless groups Π_1 and Π_2 are defined by (2.20b).

way, it can lead the unwary practitioner to retain dynamically irrelevant parameters or (in cases without known governing equations) to overlook relevant parameters. The best approach is a flexible and opportunistic one that combines formal dimensional analysis with physical insight in continual dialogue with one another.

2.3 Scaling Analysis

Except when $N - M = 1$, dimensional analysis yields a relation involving an unknown function of one or more dimensionless arguments. To determine the functional dependence itself, methods that go beyond dimensional analysis are required. The most detailed information is provided by a full analytical or numerical solution of the problem, but finding such solutions is rarely easy. Scaling analysis is a powerful intermediate method that provides more information than dimensional analysis while avoiding the labor of a complete solution. It proceeds by estimating the orders of magnitude of the different terms in a set of governing equations, using both known and unknown quantities and then exploiting the fact that the terms must balance (the definition of an equation!) to determine how the unknown quantities depend on the known.

To illustrate this, we consider once again the problem of determining the local Nusselt number $\mathcal{N}$ for the hot sphere, but now in the specific limit $Re \ll 1$ and

$Pe \gg 1$. Recall that Re measures the ratio of advection to diffusion of gradients in velocity ($\equiv$ vorticity), while Pe does the same for gradients in temperature. In the limit $Re \ll 1$, advection of velocity gradients is negligible relative to diffusion everywhere in the flow field, and **u** is given by the classic Stokes-Hadamard-Rybczynski solution for slow viscous flow around a sphere of another fluid (§ 4.3.2). When $Pe \gg 1$, temperature gradients are transported by advection with negligible diffusion everywhere except in a thin thermal boundary layer (TBL) of thickness $\delta(\theta) \ll a$ around the leading hemisphere, where advection and diffusion are of the same order. Because radial temperature gradients greatly exceed surface-tangential gradients within this layer, the temperature distribution there is described by the simplified boundary-layer forms of the continuity and energy equations (see § 6.1)

$$a \sin \theta v_r + (u \sin \theta)_\theta = 0, \tag{2.22a}$$

$$a^{-1} u T_\theta + v T_r = \kappa T_{rr}, \tag{2.22b}$$

where $u(r, \theta)$ and $v(r, \theta)$ are the tangential (θ-) and radial (r-) components of the velocity, respectively, and subscripts indicate partial derivatives. Equation (2.22) are obtained from (6.3) by setting $x = a\theta$ and $r = a \sin \theta$.

We begin by determining the relative magnitudes of the velocity components u and v in the BL. While these can be found directly from the Stokes-Hadamard-Rybczynski solution, it is more instructive to do a scaling analysis of the continuity equation (2.22a). Now $v_r \sim \Delta v / \delta$, where Δv is the change in v across the BL, but because $v(a, \theta) = 0$, $\Delta v = v$. Similarly, $u_\theta \sim \Delta u / \Delta \theta$, where Δu is the change in u over an angle $\Delta \theta$ of order unity from the forward stagnation point $\theta = 0$ towards the equator $\theta = \pi/2$. But because $u(r, 0) = 0$, $\Delta u = u$. The continuity equation therefore implies

$$v \sim (\delta/a) u. \tag{2.23}$$

We turn now to the left-hand side (LHS) of the energy equation (2.22b), whose two terms represent advection of temperature gradients in the tangential and radial directions, respectively. Now the radial temperature gradient $T_r \sim \Delta T / \delta$ greatly exceeds the tangential gradient $a^{-1} T_\theta \sim \Delta T / a$, but this difference is compensated by the smallness of the radial velocity $v \sim (\delta/a) u$, and so the terms representing tangential and radial advection are of the same order. The balance of advection and diffusion in the BL is therefore $a^{-1} u T_\theta \sim \kappa T_{rr}$, which together with $T_{rr} \sim \Delta T / \delta^2$ implies

$$\delta^2 \sim \kappa a / u. \tag{2.24}$$

It remains only to determine an expression for u, which depends on the ratio γ of the viscosity of the sphere to that of the surrounding fluid. We consider the limiting cases $\gamma \ll 1$ (a sphere with a free-slip surface) and $\gamma \gg 1$ (an effectively rigid sphere). Because the fluid outside the sphere has constant viscosity, **u** varies smoothly over a length scale $\sim a$. Within the TBL, therefore, **u** can be approximated by the first term of its Taylor series expansion in the radial distance $r - a \equiv \zeta$ away from the sphere's surface. If the sphere is free-slip, $u_\zeta|_{\zeta=0} = 0$, implying that $u \sim U$ is constant across the TBL to lowest order. If, however, the sphere is rigid, $u|_{\zeta=0} = 0$, and $u \sim (\zeta/a)U$. The tangential velocity u at the outer edge $\zeta \sim \delta$ of the TBL is therefore

$$u \sim (\delta/a)^n U, \tag{2.25}$$

where $n = 0$ for a free-slip sphere and $n = 1$ for a rigid sphere. Now substitute (2.25) into (2.24) and note that $\mathcal{N}$ is just the ratio of the distances over which heat diffuses in the absence ($= a$) and the presence ($= \delta$) of motion, i.e., $\mathcal{N} \sim a/\delta$. We thus obtain

$$\mathcal{N} \sim Pe^{\frac{1}{n+2}} f_n(\theta), \tag{2.26}$$

where $f_n(\theta)$ ($n = 0$ or 1) are unknown functions. Thus when $Pe \gg 1$, $\mathcal{N} \sim Pe^{1/2}$ if the sphere is free-slip and $\mathcal{N} \sim Pe^{1/3}$ if it is rigid. $\mathcal{N}$ is greater in the former case because the tangential velocity u, which carries the heat away from the sphere, is $\sim U$ across the whole TBL.

Exercises

2.1 In Figure 1.3b, the buoyant plume fluid forms a pool at the base of the lithosphere, the width of which is $W(x)$ where x is the distance downstream from the hotspot. The width W depends on x, Q, U, η_p, g and $\Delta\rho$ where g is the gravitational acceleration. Apply Buckingham's Π-theorem to the foregoing list of parameters.

2.2 Convection in a fluid layer of depth d is driven by uniform internal heating at a rate H per unit mass. The viscosity of the fluid depends exponentially on temperature according to

$$\eta = \eta_0 \exp[-(T - T_0)/\Delta T_r],$$

where T_0 is the constant temperature on the top boundary $z = d$ of the layer and ΔT_r is a the temperature difference required to change the viscosity by a factor e. Both boundaries are free-slip, and there is no heat flux ($\partial_z T = 0$) through the bottom boundary $z = 0$. The fluid's reference density ρ_0, thermal expansivity α, thermal diffusivity κ and heat capacity c_p are all constants. The equations governing the flow are

$$\nabla \cdot \mathbf{u} = 0, \tag{2.27a}$$

$$0 = -\nabla p - \rho_0 g \alpha (T - T_0) \mathbf{e}_z + \nabla \cdot \left\{ \eta \left[\nabla \mathbf{u} + (\nabla \mathbf{u})^T \right] \right\}, \tag{2.27b}$$

$$\frac{DT}{Dt} = \kappa \nabla^2 T + \frac{H}{c_p}, \tag{2.27c}$$

where the superscript ^{T} indicates the tensor transpose. Nondimensionalize the preceding equations and boundary conditions in some appropriate way, and interpret the dimensionless groups that appear.

2.3 Consider anew the heat transfer from a hot sphere of radius a moving at speed U in a viscous fluid with thermal diffusivity κ. Suppose, however, that the sphere has a porous surface and contains a sink that sucks fluid through that surface at a volumetric rate Q (m^3 s^{-1}). Assume that the suction is radially symmetric, and that the porous surface of the sphere acts as a no-slip boundary for the (low Reynolds number) flow outside. Perform a scaling analysis of this situation to determine the global Nusselt number Nu, and identify the dimensionless parameter that governs the transition between the low-suction and high-suction limits.

3
Self-Similarity and Intermediate Asymptotics

In geodynamics and in continuum mechanics more generally, one often encounters functions that exhibit the property of scale-invariance or self-similarity. As an illustration, consider a function $f(y, t)$ of two arbitrary variables y and t. The function f is self-similar if it can be written in the form

$$f(y, t) = G(t) F\left(\frac{y}{\delta(t)}\right), \tag{3.1}$$

where F, G and δ are functions of a single argument. The combination $\eta \equiv y/\delta(t)$ is called the similarity variable. Self-similarity simply means that curves of f vs. y for different values of t can be obtained from a single universal curve $F(\eta)$ by stretching its abscissa and ordinate by factors $\delta(t)$ and $G(t)$, respectively.

Self-similarity is closely connected with the concept of intermediate asymptotics (Barenblatt, 1996). In many physical situations, one is interested in the behaviour of a system at intermediate times, long after it has become insensitive to the details of the initial conditions but long before it reaches a final equilibrium state. The behaviour of the system at these intermediate times is often self-similar, as I now illustrate using a simple example of conductive heat transfer (Barenblatt, 1996, § 2.1).

3.1 Conductive Heat Transfer

Consider the 1-D conductive heat transfer in a rod $y \in [0, L]$ in which the initial temperature is zero everywhere except in a heated segment of length h centered at $y = y_0$ (Figure 3.1). The width of the heated segment is much smaller than the distance to either end of the rod, and the segment is much closer to the left end than to the right end, i.e., $h \ll y_0$ and $y_0 \ll L - y_0$. The ends of the rod are held at zero temperature, and its sides are insulated.

3.1 Conductive Heat Transfer

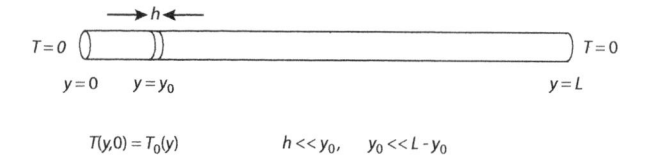

Figure 3.1 Model for 1-D conductive heat transfer in a rod $y \in [0, L]$ in which the initial temperature is zero everywhere except in a heated segment of length h centered at $y = y_0$. The width of the heated segment is much smaller than the distance to either end of the rod, and the segment is much closer to the left end than to the right end. Both ends of the rod are held at zero temperature, and its sides are insulated. This model is discussed in § 3.1.

The equation and initial and boundary conditions governing the temperature $T(y, t)$ in the rod are

$$T_t = \kappa T_{yy}, \tag{3.2a}$$

$$T(y, 0) = T_0(y), \quad T(0, t) = T(L, t) = 0, \tag{3.2b}$$

where $T_0(y)$ is the concentrated initial temperature distribution. While it is relatively easy to solve (3.2a) and (3.2b) numerically for an arbitrary initial temperature $T_0(y)$, such an approach would not reveal the essential fact that the solution has two distinct intermediate asymptotic, self-similar stages. The first obtains long after the temperature distribution has forgotten the details of the initial distribution $T_0(y)$, but long before it feels the influence of the left boundary condition $T(0, t) = 0$, i.e., for (roughly) $0.1 h^2 / \kappa \ll t \ll 0.1 y_0^2 / \kappa$. In this time range, the rod appears effectively infinite, and the integrated temperature anomaly

$$Q = \int_{-\infty}^{\infty} T(y, t) dy \tag{3.3}$$

is constant. The temperature T can depend only on Q, κ, t and $y - y_0$, and only three of these five parameters have independent dimensions. Applying the Π-theorem with $N = 5$ and $M = 3$, we find

$$T = \frac{Q}{(\kappa t)^{1/2}} F_1(\eta), \quad \eta = \frac{y - y_0}{(\kappa t)^{1/2}}, \tag{3.4}$$

which is of the general self-similar form (3.1). Substituting (3.4) into (3.2a) and (3.3), we find that F_1 satisfies

$$2F_1'' + \eta F_1' + F_1 = 0, \quad \int_{-\infty}^{\infty} F_1 d\eta = 1. \tag{3.5}$$

Upon solving (3.5) subject to $F_1(\pm\infty) = 0$, (3.4) becomes

$$T = \frac{Q}{2\sqrt{\pi\kappa t}} \exp\left[-\frac{(y-y_0)^2}{4\kappa t}\right]. \tag{3.6}$$

The second intermediate asymptotic stage occurs long after the temperature distribution has begun to be influenced by the left boundary condition $T(0, t) = 0$, but long before the influence of the right boundary condition $T(L, t) = 0$ is felt, or $y_0^2/\kappa \ll t \ll 0.1(L - y_0)^2/\kappa$. During this time interval, the rod is effectively semi-infinite, and the temperature satisfies the boundary conditions

$$T(0, t) = T(\infty, t) = 0. \tag{3.7}$$

The essential step in determining the similarity solution is to identify a conserved quantity. Multiplying (3.2a) by y, integrating from $y = 0$ to $y = \infty$, and then taking the time derivative outside the integral sign, we obtain

$$\frac{d}{dt}\int_0^\infty yT dy = \kappa \int_0^\infty yT_{yy} dy. \tag{3.8}$$

However, the right-hand side (RHS) of (3.8) is zero, as can be shown by integrating by parts, applying the conditions (3.7) and noting that $yT_y|_{y=\infty} = 0$ because $T_y \to 0$ more rapidly (typically exponentially) than $y \to \infty$. The temperature moment

$$M = \int_0^\infty yT dy \tag{3.9}$$

is therefore constant, and because the initial temperature distribution is effectively a delta-function concentrated at $y = y_0$, $M = Qy_0$. Now in the time interval in question, the influence of the temperature distribution that existed at the time $\approx 0.1 y_0^2/\kappa$ when the heated region first reached the near end $y = 0$ of the rod will no longer be felt. The temperature will therefore no longer depend on y_0, but only on M, κ, y and $t - t_0$, where t_0 is the effective starting time for the second stage, to be determined later. Applying the Π-theorem as before, we find

$$T = \frac{M}{\kappa(t-t_0)} F_2(\eta), \quad \eta = \frac{y}{\sqrt{\kappa(t-t_0)}}. \tag{3.10}$$

Now substitute (3.10) into (3.2a) and (3.9), and solve the resulting equations subject to $F_2(0) = F_2(\infty) = 0$, whereupon (3.10) becomes

$$T = \frac{My}{2\sqrt{\pi}[\kappa(t-t_0)]^{3/2}} \exp\left[-\frac{y^2}{4\kappa(t-t_0)}\right]. \tag{3.11}$$

The final step is to determine the starting time $t_0 = -y_0^2/6\kappa$ (Barenblatt, 1996, p. 74). Because $t_0 < 0$, i.e., before the rod was originally heated, it represents a virtual starting time with respect to which the behaviour of the second stage is self-similar.

3.2 Self-Similar Buoyant Thermals

Self-similar solutions can also exist for problems involving more than one spatial variable in addition to the time. A geodynamically relevant example is the solution of Whittaker and Lister (2008a) for a buoyant thermal, a localized but diffusing distribution of heat and buoyancy in an unbounded uniform viscous fluid. If inertia and viscous dissipation of energy are negligible, the governing equations are

$$\nabla \cdot \mathbf{u} = 0, \tag{3.12a}$$

$$\nu \nabla^2 \mathbf{u} = \frac{1}{\rho_0} \nabla p + g[1 - \alpha(T - T_0)]\mathbf{e}_z, \tag{3.12b}$$

$$\frac{\partial T}{\partial t} + \mathbf{u} \cdot \nabla T = \kappa \nabla^2 T, \tag{3.12c}$$

where α is the thermal expansion coefficient and T_0 and ρ_0 are the temperature and density of the fluid far from the thermal. The total (conserved) buoyancy of the thermal is

$$B = g\alpha \int (T - T_0) \mathrm{d}^3 x, \tag{3.13}$$

where the volume integral is taken over all space. The preceding equations admit a similarity solution in which lengths scale as $(\kappa t)^{1/2}$, velocities as $(\kappa/t)^{1/2}$ and the temperature anomaly as $B/\alpha g(\kappa t)^{3/2}$. The corresponding similarity transformation is (Whittaker and Lister, 2008a)

$$T(\mathbf{x},t) - T_0 = \frac{B}{\alpha g(4\kappa t)^{3/2}} \Theta(\mathbf{X}), \tag{3.14a}$$

$$\mathbf{u}(\mathbf{x},t) = \left(\frac{\kappa}{4t}\right)^{1/2} \mathbf{U}(\mathbf{X}), \tag{3.14b}$$

$$p(\mathbf{x},t) - p_0 + \rho_0 g z = \frac{\rho_0 \nu}{4t} P(\mathbf{X}), \tag{3.14c}$$

where $\mathbf{X} = \mathbf{x}/(4\kappa t)^{1/2}$ is a vector similarity variable for position. Substitution of (3.14) into (3.12) and (3.13) yields

$$\nabla \cdot \mathbf{U} = 0, \tag{3.15a}$$

$$\nabla^2 \mathbf{U} = \nabla P - Ra \, \Theta \mathbf{e}_z, \tag{3.15b}$$

$$\nabla \cdot [(\mathbf{U} - 2\mathbf{X})\Theta] = \nabla^2 \Theta, \tag{3.15c}$$

$$\int \Theta \mathrm{d}^3 X = 1, \tag{3.15d}$$

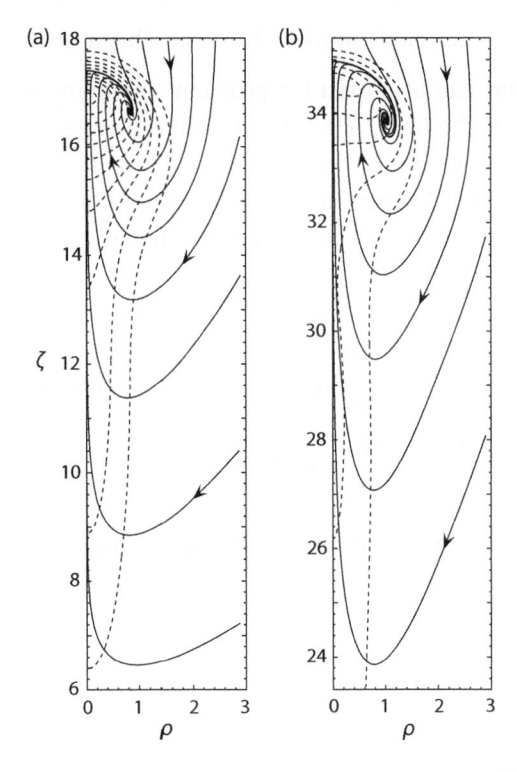

Figure 3.2 Pathlines (solid lines) and temperature contours (dashed) for self-similar thermals with (a) $Ra = 1000$ and (b) 3000, from Whittaker and Lister (2008a). The pathlines are for the effective advection velocity $\mathbf{U} - 2\mathbf{X}$. The contour interval for the temperature is 0.008. See § 3.2 for discussion. Figure reproduced from Figure 7 of Whittaker and Lister (2008a) by permission of Cambridge University Press.

where $Ra = B/(\kappa\nu)$ is a Rayleigh number, and the gradients are taken with respect to $\mathbf{X}$. The appropriate boundary conditions are the vanishing of $\mathbf{U}$, P and Θ at infinite distance from the thermal. The solution can be obtained analytically if $Ra \ll 1$, but must be determined numerically otherwise. For $Ra \ll 1$, the temperature distribution is a spherically symmetric Gaussian function, while for $Ra \gg 1$ it comprises a quasispherical head trailed by a wake of length $(Ra \ln Ra)^{1/2} (\kappa t)^{1/2}$ that contains most of the buoyancy (Figure 3.2). Because $Ra \gg 1$ in Earth's mantle, the head plus wake structure of the thermal in this limit could in principle be geodynamically relevant. However, Peng and Lister (2014) examined the transient behaviour of an initially spherical thermal with $Ra \gg 1$ and concluded that it can rise $O(100)$ times its initial diameter before attaining the self-similar state. Self-similar buoyant thermals are therefore unlikely to exist in Earth's mantle.

3.3 Classification of Self-Similar Solutions

The solutions (3.6) and (3.11) and the numerical solution of (3.15) are examples of what Barenblatt (1996) calls self-similar solutions of the first kind, for which dimensional analysis (in some cases supplemented by scaling analysis of the governing equations) suffices to find the similarity variable. They are distinguished from self-similar solutions of the second kind, for which the similarity variable can only be found by solving an eigenvalue problem. We will encounter solutions of the latter type in the sections on viscous corner eddies (§ 4.3.4) and viscous gravity currents (§ 7.1).

An example of a self-similar solution that does not fit naturally into either class is the impulsive cooling of a half-space deforming in pure shear. Suppose that the half-space $y \geq 0$ has temperature $T = 0$ initially, and that at time $t = 0$ the temperature at its surface $y = 0$ is suddenly decreased by an amount ΔT. The 2-D velocity field in the half-space is $\mathbf{u} = \dot{\epsilon}(x\mathbf{e}_x - y\mathbf{e}_y)$, where $\dot{\epsilon}$ is the constant rate of extension of the surface $y = 0$ and $\mathbf{e}_x$ and $\mathbf{e}_y$ are unit vectors parallel to and normal to the surface, respectively. Given this velocity field, a temperature field $T = T(y, t)$ that is independent of the lateral coordinate x is an allowable solution of the governing advection-diffusion equation $T_t + \mathbf{u} \cdot \nabla T = \kappa \nabla^2 T$, which takes the form

$$T_t - \dot{\epsilon} y T_y = \kappa T_{yy} \tag{3.16}$$

subject to the conditions $T(y, 0) = T(\infty, t) = T(0, t) + \Delta T = 0$. The limit $\dot{\epsilon} = 0$ corresponds to the classic problem of the impulsive cooling of a static half-space.

Neither dimensional analysis nor scaling analysis is sufficient to determine the similarity variable, which does not have the standard power-law monomial form. However, the solution can be found via a generalized form of the familiar separation-of-variables procedure often used to solve partial differential equations (PDEs) such as Laplace's equation. Note first that the amplitude ΔT of the temperature in the half-space is a constant. This implies that the function $G(t)$ in the similarity transformation (3.1) must be independent of time, whence $T(y, t) = \Delta T F(y/\delta(t))$. Substituting this expression into (3.16) and bringing all terms involving $\delta(t)$ to the left-hand side (LHS), we obtain

$$\frac{\delta(\dot{\delta} + \dot{\epsilon}\delta)}{\kappa} = -\frac{F''}{\eta F'}, \tag{3.17}$$

where dots and primes denote differentiation with respect to t and $\eta \equiv y/\delta(t)$, respectively. Now the LHS of (3.17) is a function of t only, whereas the RHS depends on y through the similarity variable η. Equation (3.17) is therefore consistent only if both sides are equal to a constant λ^2, which is positive because $\dot{\delta} > 0$. The solutions for δ and F subject to the conditions $\delta(0) = F(\infty) = F(0) - 1 = 0$ are

$$F = \text{erfc}\frac{\lambda y}{\sqrt{2}\delta}, \quad \delta = \lambda\left\{\frac{\kappa}{\dot{\epsilon}}[1 - \exp(-2\dot{\epsilon}t)]\right\}^{1/2}. \tag{3.18}$$

Evidently λ cancels out when the solution for δ is substituted into the solution for F because different values of λ merely correspond to different (arbitrary) definitions of the thermal layer thickness δ. With $\dot{\epsilon} = 0$ and the conventional choice $\lambda = \sqrt{2}$, we recover the well-known solution $\delta = 2\sqrt{\kappa t}$ for a static half-space. However, when $\dot{\epsilon} > 0$, the BL thickness approaches a steady-state value $\delta = (2\kappa/\dot{\epsilon})^{1/2}$ for which the downward diffusion of temperature gradients is balanced by upward advection, and the similarity variable involves an exponential function of time. The only reliable way to find such nonstandard self-similar solutions is the separation-of-variables procedure outlined earlier. But the same procedure works just as well for problems with similarity variables of standard form and will therefore be used throughout this book.

In conclusion, I note that similarity transformations can also be powerful tools for reducing and interpreting the output of numerical models. As a simple example, suppose that some such model yields values of a dimensionless parameter W as a function of two dimensionless groups Π_1 and Π_2. Depending on the physics of the problem, it may be possible to express the results in the self-similar form

$$W(\Pi_1, \Pi_2) = F_1(\Pi_1) F_2\left(\frac{\Pi_1}{F_3(\Pi_2)}\right), \tag{3.19}$$

where F_1–F_3 are functions to be determined numerically. A representation of the form (3.19) is not guaranteed to exist, but when it does, it provides a compact way of representing multidimensional numerical data by functions of a single variable that can be fit by simple analytical expressions. An example of the use of this technique for a problem involving five dimensionless groups is the lubrication theory model for plume–ridge interaction of Ribe and Delattre (1998).

3.4 Intermediate Asymptotics with Respect to Parameters: The Rayleigh–Taylor Instability

The concept of intermediate asymptotics is not limited to self-similar behaviour of systems that evolve in time, but also applies in a more general way to functions of one or more parameters that exhibit simple (typically power-law) behaviour in some asymptotically defined subregion of the parameter space. Because power-law scaling usually results from a simple dynamical balance between two competing effects, the identification of intermediate asymptotic limits that have this form is crucial for a physical understanding of the system in question.

The dynamical significance of intermediate asymptotic limits and the role that scaling arguments play in identifying them are nicely illustrated by the

3.4 Intermediate Asymptotics

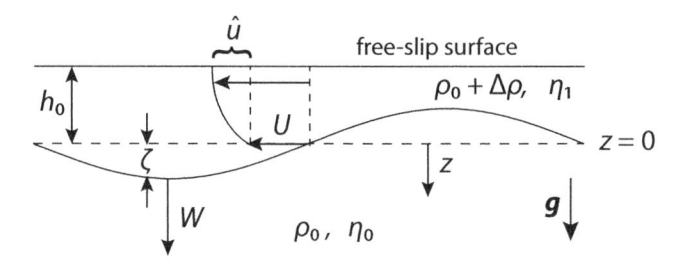

Figure 3.3 Rayleigh–Taylor instability of a layer of fluid with density $\rho_0 + \Delta\rho$ and viscosity η_1 above a half-space of fluid with density ρ_0 and viscosity η_0. The upper surface of the layer is free-slip. The initial thickness of the dense layer is h_0, and the deformation of the interface is ζ. The maximum values of the horizontal and vertical velocities at the interface $z = \zeta$ are U and W, respectively, and $\hat{u}$ is the magnitude of the change in horizontal velocity across the layer. This model is discussed in § 3.4.

Rayleigh–Taylor instability (RTI) of a fluid layer with density $\rho_0 + \Delta\rho$, viscosity η_0 and thickness h_0 above a fluid half-space with density ρ_0 and viscosity η_0 (Figure 3.3). The following discussion is adapted from Canright and Morris (1993).

Linear stability analysis of this problem (see § 10.1) shows that an infinitesimal sinusoidal perturbation $\zeta = \zeta_0 \sin kx$ of the initially flat interface $z = 0$ in Figure 3.3 grows exponentially at a rate (Whitehead and Luther, 1975)

$$s = s_1 \frac{\gamma}{2\epsilon} \left[\frac{\gamma(C-1) + S - 2\epsilon}{\gamma^2(S+2\epsilon) + 2\gamma C + S - 2\epsilon} \right], \tag{3.20}$$

where $s_1 = gh_0\Delta\rho/\eta_1$, $\gamma = \eta_1/\eta_0$, $\epsilon = h_0 k$, $C = \cosh 2\epsilon$ and $S = \sinh 2\epsilon$. Here we shall consider only the long-wavelength limit $\epsilon \ll 1$. To find the limiting form of (3.20), it is necessary to expand the numerator and the denominator separately in powers of ϵ while recalling that γ can be arbitrarily large or small. The result is

$$\frac{s}{s_1} \sim \frac{\epsilon\gamma(2\epsilon + 3\gamma)}{2(2\epsilon^3 + 3\gamma + 6\epsilon\gamma^2)}. \tag{3.21}$$

The next step is to determine the intermediate asymptotic limits of (3.21). Each of these limits corresponds to a pair of dominant terms, one in the numerator and one in the denominator. By testing all six possible pairs, one finds (Exercise 3.4) that (3.21) has four intermediate asymptotic limits: $\gamma \ll \epsilon^3$, $\epsilon^3 \ll \gamma \ll \epsilon$, $\epsilon \ll \gamma \ll \epsilon^{-1}$ and $\gamma \gg \epsilon^{-1}$. The essential dynamics associated with each limit are summarized in columns 3–5 of Table 3.1. Column 3 shows the ratio of the amplitudes of the vertical (W) and horizontal (U) components of the velocity at the interface (Figure 3.3). As γ increases, the motion of the interface changes from dominantly vertical in limit 1 to dominantly horizontal in limits 3 and 4. The ratio of shear deformation to plug flow in the layer is measured by the ratio $\hat{u}/U$ (column 4),

Table 3.1. *Rayleigh–Taylor instability: intermediate asymptotic limits*

Limit	γ	W/U	$\hat{u}/U$	Balancing pressure	s
1	$\ll \epsilon^3$	ϵ^2/γ	ϵ/γ	$p_0 \sim \eta_0 kW$	$\Delta\rho g/2k\eta_0$
2	$\epsilon^3 \ll \gamma \ll \epsilon$	ϵ^2/γ	ϵ/γ	$p_1 \sim \eta_1 \hat{u}/h_0^2 k \sim \eta_1 W/h_0^3 k^2$	$\Delta\rho g h_0^3 k^2/3\eta_1$
3	$\epsilon \ll \gamma \ll \epsilon^{-1}$	ϵ	ϵ/γ	$p_1 \sim \eta_1 \hat{u}/h_0^2 k \sim \eta_0 W/h_0^2 k$	$\Delta\rho g h_0^2 k/2\eta_0$
4	$\gg \epsilon^{-1}$	ϵ	ϵ^2	$p_1 \sim \eta_1 kU \sim \eta_1 W/h_0$	$\Delta\rho g h_0/4\eta_1$

where $\hat{u}$ is the change in horizontal velocity across the layer (Figure 3.3). The layer deforms mainly by shear in limits 1 and 2 and by plug flow in limits 3 and 4.

The growth rate is determined by whether the interfacial buoyancy $\sim g\Delta\rho\zeta_0$ is supported by the normal stress σ_{zz} in the layer or by that in the half-space. However, in Stokes flow the deviatoric part of the normal stress never exceeds the order of magnitude of the pressure, and so our problem reduces to determining which pressure (p_1 in the layer or p_0 in the half-space) balances the buoyancy. While the pressures can be calculated directly from the analytical solution of the problem (§ 10.1), it is more revealing to obtain them via a scaling analysis of the horizontal component $\partial_x p = \eta \nabla^2 u$ of the momentum equation. In the half-space, the only length scale is k^{-1}, so that $\partial_x \sim \partial_z \sim k$. The continuity equation then implies $u \sim w$. The magnitude of $u \sim w$ is set by the larger of the two components of the velocity at the interface, viz., $u \sim [U, W]$, where $[\ldots]$ denotes the maximum of the enclosed quantities. Turning now to the layer, we note that the horizontal and vertical length scales are different, so that $\partial_x \sim k$ and $\partial_z \sim h_0^{-1}$. Moreover, $\nabla^2 u \sim [\hat{u}/h_0^2, k^2 U]$ is the sum of terms arising from the shear and plug flow components of u. We thereby find

$$p_1 \sim \frac{\eta_1}{h_0^2 k}[\hat{u}, \epsilon^2 U], \quad p_0 \sim \eta_0 k[U, W]. \tag{3.22}$$

Column 5 of Table 3.1 gives the expression for the pressure that balances the buoyancy in each limit, and column 6 shows the corresponding growth rate $s = W/\zeta_0$.

In view of the pressure scales (3.22) and those for W/U and $\hat{u}/U$ from Table 3.1, the essential dynamics of each of the four asymptotic limits can be summarized as follows. In limit 1, the half-space feels the thin layer as an effectively free (zero shear traction) boundary, the buoyancy is balanced by the pressure $p_0 \sim \eta_0 kW$ in the half-space and s is controlled by the half-space viscosity η_0. Because all the action takes place in the half-space, the layer is purely passive in this limit. In limit 2, the half-space still sees the layer as shear traction-free, but the pressure $p_1 \sim \eta_1 \hat{u}/h_0^2 k$ induced by shear flow in the layer is nevertheless sufficient to balance the buoyancy. Because the layer deforms mostly in shear, $\hat{u}$ is related directly to W via the continuity equation ($W \sim \epsilon \hat{u}$), so s is controlled by the viscosity η_1 of the layer. In limit 3, each fluid feels the shear stress applied by the other. While the buoyancy is still balanced by the shear-induced pressure in the layer, the dominance of plug flow means that $\hat{u}$ and W are no longer related via the continuity equation, but rather by the matching condition on the shear stress. The growth rate is therefore controlled by the half-space viscosity η_0. Finally, in limit 4 the layer feels the half-space as a free boundary, the buoyancy is balanced by the pressure $p_1 \sim \eta_1 kU$ induced by plug flow in the layer, and s is controlled by the layer viscosity η_1. In this limit all the action occurs in the layer, and so the half-space is passive.

You may have noticed that the ratios W/U and $\hat{u}/U$ that appear in Table 3.1 were determined directly from the analytical solution of the problem and were not themselves found by scaling analysis. The reason is that scaling analysis fails to capture a subtle feature of the shear stress at the interface, which together with the continuity equation determines W/U and $\hat{u}/U$. Consider the interfacial shear stress in fluid 0, which is $\eta_0(\partial_z u_0 + \partial_x w_0)|_{z=0}$. The analytical solution shows that the two terms in this expression scale as

$$\partial_z u_0|_{z=0} \sim \frac{6\epsilon + 2\epsilon^3 \gamma}{\chi}, \quad \partial_x w_0|_{z=0} \sim -\frac{3\epsilon^2 + 2\epsilon^3 \gamma}{\chi}, \tag{3.23}$$

where $\chi = 2(6\epsilon + 3\gamma + 2\epsilon^3 \gamma^2)$. Summing the two terms in (3.23) and noting that $\epsilon^2 \ll \epsilon$, we obtain

$$(\partial_z u_0 + \partial_x w_0)|_{z=0} \sim \frac{6\epsilon}{\chi}. \tag{3.24}$$

The sum (3.24) is simpler than either of the separate contributions $\partial_z u_0|_{z=0}$ and $\partial_x w_0|_{z=0}$ because the two terms $\sim \pm 2\epsilon^3 \gamma/\chi$ in (3.23) cancel exactly. The scaling of the total tangential stress is therefore significantly different than that of either of its two components $\partial_z u_0|_{z=0}$ and $\partial_x w_0|_{z=0}$. Scaling analysis cannot detect behaviour of this kind because it provides no information on the numerical prefactors of the different terms in a scaling relation. The present example is thus a salutary reminder that scaling analysis, while powerful, needs to be used circumspectly and with an awareness of its limitations.

Exercises

3.1 [SM] Buckingham's Π-theorem shows that the temperature in the second intermediate asymptotic stage in a locally heated rod has the form (3.10). Substitute this form into (3.2a) and (3.9) to obtain the ordinary differential equation satisfied by $F_2(\eta)$. Solve this equation subject to the appropriate boundary and normalization conditions to obtain (3.11).

3.2 Show that the similarity transformation (3.14) applied to the energy equation (3.12c) leads to (3.15c).

3.3 [SM] For small Rayleigh numbers $Ra \ll 1$, the temperature and velocity fields for a self-similar thermal have the forms

$$\Theta = \Theta^{(0)}(r) + Ra\Theta^{(1)}(r,\theta) + Ra^2\Theta^{(2)}(r,\theta) + \cdots, \quad (3.25a)$$

$$\mathbf{U} = Ra\mathbf{U}^{(1)}(r,\theta) + Ra^2\mathbf{U}^{(2)}(r,\theta) + \cdots, \quad (3.25b)$$

where (r,θ) are spherical polar coordinates. Determine $\Theta^{(0)}(r)$.

3.4 The expression (3.21) has the four intermediate asymptotic limits summarized in column 2 of Table 3.1. Derive the expressions for these limits directly from (3.21).

3.5 In the RTI problem discussed in section § 3.4, the scales for the pressures in the two layers are given by (3.22). Using the scales for the ratios W/U and $\hat{u}/U$ from column 4 of Table 3.1, verify the results of column 5, which shows which pressure (p_0 or p_1) balances the buoyancy in each of the four asymptotic limits.

4
Slow Viscous Flow

Flows with negligible inertia are fundamental in Earth's mantle, where the Reynolds number $Re \approx 10^{-20}$. Fluids flowing without inertia can be thought of as having effectively zero inertial mass. However, because these fluids can flow in response to the buoyancy force associated with lateral density variations, they have nonzero gravitational mass.

A particularly important subclass of inertialess flow, which I shall call slow viscous flow, comprises flows in which the fluid is incompressible, is isothermal and has a rheology with no memory (elasticity). These conditions, while obviously restrictive, are nevertheless sufficiently realistic to have served as the basis for many important geodynamic models. Indeed, it is no exaggeration to say that the theory of slow viscous flow is the very heart of geodynamics.

4.1 Basic Equations and Theorems

The most general equations required to describe the slow viscous flows discussed in this chapter are

$$\partial_j u_j = 0, \tag{4.1a}$$

$$\partial_j \sigma_{ij} + f_i = 0, \tag{4.1b}$$

$$f_i = -\rho \partial_i \chi, \tag{4.1c}$$

$$\partial_{jj}^2 \chi = 4\pi G \rho, \tag{4.1d}$$

$$\sigma_{ij} = -p\delta_{ij} + 2\eta e_{ij}, \quad e_{ij} = \frac{1}{2}(\partial_i u_j + \partial_j u_i), \tag{4.1e}$$

$$\eta = BI^{(1/n)-1}, \quad I = (e_{ij} e_{ij})^{1/2}, \tag{4.1f}$$

where u_i is the velocity vector, σ_{ij} is the stress tensor, f_i is the gravitational body force per unit volume, χ is the gravitational potential, ρ is the density, p is the pressure, η is the viscosity and e_{ij} is the strain rate tensor. The quantity I is an invariant of the strain rate tensor that has units of strain rate (s^{-1}) and is directly related to the so-called second or quadratic invariant I_2 ($\equiv -I^2/2$ for a traceless tensor). Equation (4.1a) is the incompressibility condition. Equation (4.1b) expresses conservation of momentum in the absence of inertia and states that the net force (viscous plus gravitational) acting on each fluid element is zero. Equation (4.1d) is Poisson's equation for the gravitational potential. Equation (4.1e) is the standard constitutive relation for a viscous fluid. Finally, (4.1f) is the strain-rate dependent viscosity for a power-law fluid (sometimes called a generalized Newtonian fluid), where B is the rheological 'stiffness' (units kg m^{-1} s$^{(1/n)-2}$) and n is the power-law exponent. A Newtonian fluid has $n = 1$, while the rheology of dry olivine deforming by dislocation creep is well described by (4.1f) with $n \approx 3.5$ (Bai et al., 1991). A discussion of more complicated non-Newtonian fluids is beyond the scope of this book; the interested reader is referred to Larson (1999).

Viscous flow described by (4.1) can be driven either externally, by velocities and/or stresses imposed at the boundaries of the flow domain, or internally, by buoyancy forces (internal loads) arising from lateral variations of the density ρ. On the scale of the whole mantle, the influence of long-wavelength lateral variations of ρ on the gravitational potential χ (self-gravitation) is significant and cannot be neglected. In modeling flow on smaller scales, however, one generally ignores Poisson's equation (4.1d) and replaces $\nabla \chi$ in (4.1c) by a constant gravitational acceleration $-\mathbf{g}$.

Slow viscous flow exhibits the property of instantaneity: u_i and σ_{ij} at each instant are determined throughout the fluid solely by the distribution of forcing (internal loads and/or boundary conditions) acting at that instant. A corollary is that slow viscous flow is quasi-static, any time-dependence being due entirely to the time-dependence of the forcing or of the system's geometry.

The theory of slow viscous flow is most highly developed for the special case of Newtonian fluids, for which the relation between stress and strain rate is linear. The slow flow of Newtonian fluids is commonly called Stokes flow or low Reynolds number flow and is the subject of several excellent monographs (Ladyzhenskaya, 1963; Happel and Brenner, 1991; Kim and Karrila, 1991; Pozrikidis, 1992; Langlois and Deville, 2014). Relative to general slow viscous flow, Stokes flow exhibits the important additional property of linearity. Linearity implies that for a given geometry, a sum of different solutions (e.g., for different forcing distributions) is also a solution. It also implies that u_i and σ_{ij} are directly proportional to the forcing that generates them, and hence, e.g., that the force acting on a body in Stokes flow is proportional to its speed.

A final important property of general slow viscous flow is its reversibility, which refers to the fact that changing the sign of the forcing terms reverses the signs of u_i

4.1 Basic Equations and Theorems

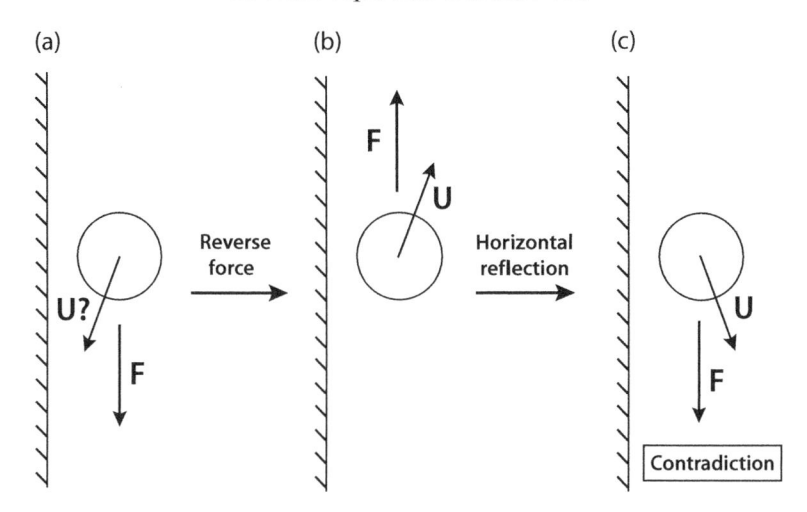

Figure 4.1 Example of a reversibility argument: a sphere moving in the vicinity of a plane wall under a wall-parallel body force has no tendency to migrate toward or away from the wall. See § 4.1 for discussion.

and σ_{ij} for all material particles. The reversibility principle is especially powerful when used in conjunction with symmetry arguments to eliminate or validate some hypothetical possibility. As a first example, consider a sphere falling under gravity in the vicinity of a vertical plane wall (Figure 4.1). We ask whether the sphere will tend to migrate toward the wall (Figure 4.1a). We first reverse the body force **F**, which reverses also the sign of the velocity **U**, leading to the configuration of Figure 4.1b. Next, we reflect this new configuration across a horizontal plane to obtain the configuration of Figure 4.1c. Because reflection involves no change in the physics, the configurations in Figure 4.1a and c must be compatible. However, these two configurations are contradictory: in one the sphere moves toward the wall, in the other, away. The only way to make the two configurations compatible is to suppose that the sphere has no tendency to move either toward or away from the wall, i.e., that **U** is always wall-parallel.

Two somewhat more complicated examples are shown in Figure 4.2. In the first (Figure 4.2a), a body with fore–aft symmetry (e.g., a circular rod with a uniform cross-section) whose axis is inclined to the vertical falls under gravity. We ask whether the body experiences a torque that makes it rotate clockwise. By reflecting the initial configuration across a vertical plane and then a horizontal plane and finally reversing the direction of **F**, we obtain a final configuration that is identical to the initial configuration except for an opposite sense of rotation. The only way to resolve the contradiction is to assume that the body experiences no torque. The situation is different if the body does not have fore–aft symmetry, as in Figure 4.2b.

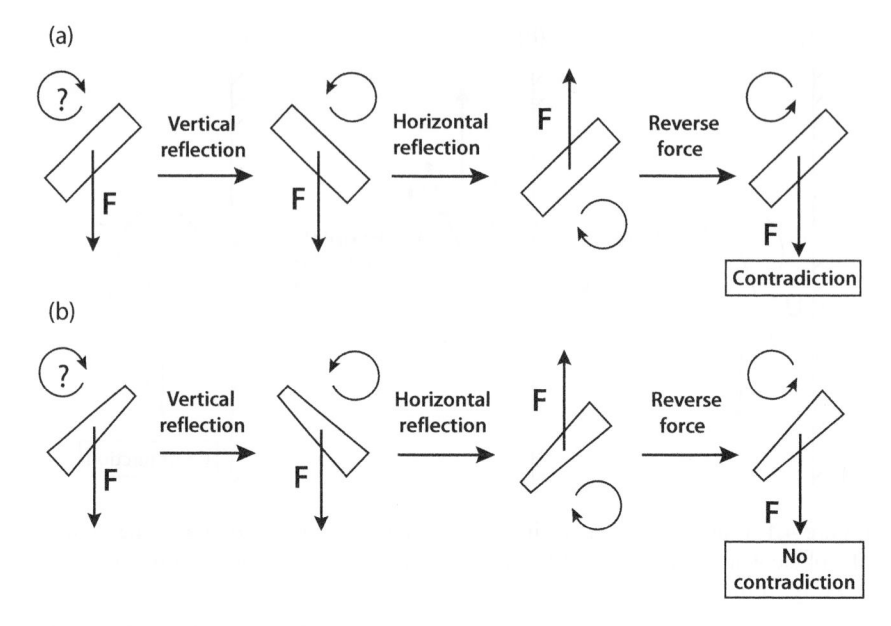

Figure 4.2 Further examples of reversibility arguments. (a) An obliquely oriented body with fore–aft symmetry acted on by a body force experiences no torque. (b) An obliquely oriented body without fore–aft symmetry experiences a nonzero torque. See § 4.1 for discussion.

The sequence of two reflections followed by a reversal of the force produces a final configuration with an opposite sense of rotation, but the orientation of the body is also different. There is therefore no contradiction, from which we conclude that a falling body without fore–aft symmetry can experience a torque.

As a final remark in a more geodynamical context, the reversibility principle implies that the lateral separation of two buoyant spherical diapirs with different radii is the same before and after their interaction (Manga, 1997).

We next turn to several results arising from consideration of the rate of viscous dissipation of energy in a volume. This quantity is

$$D = \int_V \sigma_{ij} e_{ij} dV \equiv \int_V (-p\delta_{ij} + 2\eta e_{ij}) e_{ij} dV = 2\eta \int_V e_{ij} e_{ij} dV, \qquad (4.2)$$

where the continuity equation $e_{jj} = 0$ has been used in the last step. Define also a 'cross-dissipation'

$$D_{\text{cross}} = 2\eta \int_V e^*_{ij} e_{ij} dV \qquad (4.3)$$

where e_{ij} corresponds to a Stokes flow and e^*_{ij} to a general incompressible flow that need not satisfy the Stokes equations. An important lemma is obtained by manipulating the expression for the cross-dissipation as follows:

4.1 Basic Equations and Theorems

$$2\eta \int_V e_{ij}^* e_{ij} dV = \int_V e_{ij}^* \sigma_{ij} dV$$

$$= \int_V \frac{1}{2}(\partial_j u_i^* + \partial_i u_j^*)\sigma_{ij} dV = \int_V \partial_i u_j^* \sigma_{ij} dV$$

$$= \int_V [\partial_i(u_j^* \sigma_{ij}) - u_j^* \partial_i \sigma_{ij}] dV$$

$$= \int_S u_j^* \sigma_{ij} n_i dS + \int_V u_j^* f_j dV. \tag{4.4}$$

The preceding manipulation invokes the symmetry of σ_{ij} (second line), performs an integration by parts (second to third lines) and uses the divergence theorem and the Stokes momentum balance $\partial_i \sigma_{ij} = -f_j$ (third to fourth lines). In coordinate-free notation the lemma can now be written as

$$2\eta \int_V \mathbf{e}^* : \mathbf{e} \, dV = \int_S \mathbf{u}^* \cdot \boldsymbol{\sigma} \cdot \mathbf{n} \, dS + \int_V \mathbf{u}^* \cdot \mathbf{f} \, dV. \tag{4.5}$$

We now prove the minimum dissipation theorem for Stokes flow (Helmholtz, 1868), which states that a Stokes flow $(\mathbf{u}, \mathbf{e})$ with $\mathbf{f} = 0$ on a given domain dissipates less energy than any other incompressible flow $(\mathbf{u}^*, \mathbf{e}^*)$ satisfying the same velocity boundary conditions. Here 'incompressible flow' means any solenoidal vector field, including ones that do not satisfy the Navier–Stokes equations. The starting point is the expression for the (nonnegative) dissipation associated with the difference flow, viz.,

$$2\eta \int_V (\mathbf{e}^* - \mathbf{e}) : (\mathbf{e}^* - \mathbf{e}) \geq 0. \tag{4.6}$$

Now expand the integrand to obtain

$$0 \leq 2\eta \int_V (\mathbf{e}^* : \mathbf{e}^* - \mathbf{e} : \mathbf{e}) dV + 4\eta \int_V \mathbf{e} : (\mathbf{e} - \mathbf{e}^*) dV$$

or

$$0 \leq D^* - D + 4\eta \int_V \mathbf{e} : (\mathbf{e} - \mathbf{e}^*) dV. \tag{4.7}$$

Now the last term vanishes, as we can see by rewriting the lemma (4.5) with $\mathbf{u}^* \to \mathbf{u} - \mathbf{u}^*$, $\mathbf{e}^* \to \mathbf{e} - \mathbf{e}^*$ and $\mathbf{f} = 0$ to obtain

$$2\eta \int_V \mathbf{e} : (\mathbf{e} - \mathbf{e}^*) dV = \int_S (\mathbf{u} - \mathbf{u}^*) \cdot \boldsymbol{\sigma} \cdot \mathbf{n} \, dS. \tag{4.8}$$

But the quantity (4.8) vanishes because $\mathbf{u} = \mathbf{u}^*$ on S by hypothesis. Equation (4.7) thus reduces to $D^* \geq D$, proving the theorem. Note that this theorem merely compares a Stokes flow with other flows that do not satisfy the Stokes equations.

It says nothing about the relative rates of dissipation of Stokes flows with different geometries and/or boundary conditions, and therefore its use as a principle of selection among such flows is not justified.

The lemma (4.5) can also be used to prove the uniqueness of Stokes flow (Helmholtz, 1868). Let $\mathbf{u}$ and $\mathbf{u}^*$ be two Stokes flows with the same body force and satisfying the same boundary conditions:

$$\mathbf{f} = \mathbf{f}^*; \quad \mathbf{u} = \mathbf{u}^* \quad \text{on } S. \tag{4.9}$$

Then the difference flow $\mathbf{u}' = \mathbf{u} - \mathbf{u}^*$ will satisfy the Stokes equations with body force and boundary conditions

$$\mathbf{f}' = 0; \quad \mathbf{u}' = 0 \quad \text{on } S. \tag{4.10}$$

The lemma (4.5) then gives

$$2\eta \int_V \mathbf{e}' : \mathbf{e}' \, dV = \int_S \mathbf{u}' \cdot \boldsymbol{\sigma}' \cdot \mathbf{n} \, dS + \int_V \mathbf{u}' \cdot \mathbf{f}' \, dV, \tag{4.11}$$

which vanishes by the conditions (4.10). The difference flow thus has zero dissipation, which means that it can be at most a rigid-body rotation plus a translation. However, such a difference flow is excluded by the boundary condition $\mathbf{u}' = 0$. The two flows must therefore be identical throughout the domain, implying that Stokes flow is unique.

A further important theorem that can be demonstrated using the lemma (4.5) is the Lorentz reciprocal theorem (Lorentz, 1907), which relates two different Stokes flows $\mathbf{u}^* \equiv \mathbf{u}_1$ and $\mathbf{u} \equiv \mathbf{u}_2$. This theorem is the starting point for the boundary integral representation of Stokes flow derived in § 4.6.4. Now the left-hand side (LHS) of (4.5) is invariant to interchange of the subscripts 1 and 2, and so the right-hand side (RHS) must be also. Equating the two different forms of the RHS yields the integral form of the reciprocal theorem,

$$\int_S \mathbf{u}_1 \cdot \boldsymbol{\sigma}_2 \cdot \mathbf{n} \, dS + \int_V \mathbf{u}_1 \cdot \mathbf{f}_2 \, dV = \int_S \mathbf{u}_2 \cdot \boldsymbol{\sigma}_1 \cdot \mathbf{n} \, dS + \int_V \mathbf{u}_2 \cdot \mathbf{f}_1 \, dV. \tag{4.12}$$

By applying the divergence theorem to the surface integrals in (4.12), one can write the reciprocal theorem in differential form as

$$\nabla \cdot (\mathbf{u}_1 \cdot \boldsymbol{\sigma}_2) + \mathbf{u}_1 \cdot \mathbf{f}_2 = \nabla \cdot (\mathbf{u}_2 \cdot \boldsymbol{\sigma}_1) + \mathbf{u}_2 \cdot \mathbf{f}_1. \tag{4.13}$$

A final important result can be obtained from (4.5) by assuming that the flows $\mathbf{u}^*$ and $\mathbf{u}$ are one and the same, whereupon (4.5) becomes

$$2\eta \int_V \mathbf{e} : \mathbf{e} \, dV = \int_S \mathbf{u} \cdot \boldsymbol{\sigma} \cdot \mathbf{n} \, dS + \int_V \mathbf{u} \cdot \mathbf{f} \, dV. \tag{4.14}$$

4.2 Potential Representations for Incompressible Flow

Equation (4.14) describes the energetics of the flow: it states that the total rate of viscous dissipation is equal to the sum of the rate at which tractions do work on the surface S and the rate at which body forces do work on the volume V.

In an incompressible flow satisfying $\nabla \cdot \mathbf{u} = 0$, only $N - 1$ of the velocity components u_i are independent, where $N = 3$ for general 3-D flows and $N = 2$ for 2-D and axisymmetric flows. This fact allows one to express all the velocity components in terms of derivatives of $N - 1$ independent scalar potentials, thereby reducing the number of independent variables in the governing equations. The most commonly used potentials are the streamfunction ψ (for $N = 2$) and the poloidal scalar $\mathcal{P}$ and the toroidal scalar $\mathcal{T}$ (for $N = 3$). Next we give expressions for the components of $\mathbf{u}$ in terms of these potentials in Cartesian coordinates (x, y, z), cylindrical coordinates (ρ, ϕ, z) and spherical coordinates (r, θ, ϕ), together with the partial differential equations (PDEs) they satisfy for the important special case of constant viscosity.

4.2.1 2-D and Axisymmetric Flows

A 2-D flow is one in which the velocity vector $\mathbf{u}$ is everywhere perpendicular to a fixed direction ($\mathbf{e}_z$, say) in space. The velocity can then be represented in terms of a streamfunction by

$$\mathbf{u} = \mathbf{e}_z \times \nabla \psi = -\mathbf{e}_x \psi_y + \mathbf{e}_y \psi_x = \mathbf{e}_\phi \psi_\rho - \frac{\mathbf{e}_\rho}{\rho} \psi_\phi. \tag{4.15}$$

The PDE satisfied by ψ is obtained by applying the operator $\mathbf{e}_z \times \nabla$ to the momentum equation (4.1b) with the constitutive law (4.1e). If the viscosity is constant and $f_i = 0$, ψ satisfies the biharmonic equation

$$\nabla_h^4 \psi = 0, \quad \nabla_h^2 = \partial_{xx}^2 + \partial_{yy}^2 = \rho^{-1} \partial_\rho (\rho \partial_\rho) + \rho^{-2} \partial_{\phi\phi}^2. \tag{4.16}$$

An axisymmetric flow (without swirl) is one in which $\mathbf{u}$ at any point lies in the plane containing the point and some fixed axis ($\mathbf{e}_z$, say). For this case,

$$\mathbf{u} = \frac{\mathbf{e}_\phi}{\rho} \times \nabla \psi = -\frac{\mathbf{e}_z}{\rho} \psi_\rho + \frac{\mathbf{e}_\rho}{\rho} \psi_z, \tag{4.17a}$$

$$\mathbf{u} = \frac{\mathbf{e}_\phi}{r \sin \theta} \times \nabla \psi = \frac{\mathbf{e}_\theta}{r \sin \theta} \psi_r - \frac{\mathbf{e}_r}{r^2 \sin \theta} \psi_\theta, \tag{4.17b}$$

where ψ is referred to as the Stokes streamfunction. The PDE satisfied by the Stokes streamfunction is obtained by applying the operator $\mathbf{e}_\phi \times \nabla$ to the momentum equation. For a fluid with constant viscosity and no body force, the result is

$$E^4\psi = 0, \quad E^2 = \frac{h_3}{h_1 h_2}\left[\frac{\partial}{\partial q_1}\left(\frac{h_2}{h_1 h_3}\frac{\partial}{\partial q_1}\right) + \frac{\partial}{\partial q_2}\left(\frac{h_1}{h_2 h_3}\frac{\partial}{\partial q_2}\right)\right], \quad (4.18)$$

where (q_1, q_2) are orthogonal coordinates in any half-plane normal to $\mathbf{e}_\phi$, (h_1, h_2) are the corresponding scale factors and h_3 is the scale factor for the azimuthal coordinate ϕ. Thus $(q_1, q_2, h_1, h_2, h_3) = (z, \rho, 1, 1, \rho)$ in cylindrical coordinates and $(r, \theta, 1, r, r\sin\theta)$ in spherical coordinates. Note that the operator E^2 is in general different from the 2-D Laplacian operator with respect to the same coordinates, viz.,

$$\nabla^2 = \frac{1}{h_1 h_2 h_3}\left[\frac{\partial}{\partial q_1}\left(\frac{h_2 h_3}{h_1}\frac{\partial}{\partial q_1}\right) + \frac{\partial}{\partial q_2}\left(\frac{h_1 h_3}{h_2}\frac{\partial}{\partial q_2}\right)\right]. \quad (4.19)$$

The two operators are identical only for 2-D flows ($h_3 = 1$).

4.2.2 3-D Flows

The most commonly used potential representation of 3-D flows in geodynamics is a decomposition of the velocity $\mathbf{u}$ into poloidal and toroidal components (Backus, 1958; Chandrasekhar, 1981). This representation requires the choice of a preferred direction, which is usually taken to be an upward vertical ($\mathbf{e}_z$) or radial ($\mathbf{e}_r$) unit vector. Relative to Cartesian coordinates, the poloidal/toroidal decomposition has the form

$$\begin{aligned}\mathbf{u} &= \nabla \times (\mathbf{e}_z \times \nabla \mathcal{P}) + \mathbf{e}_z \times \nabla \mathcal{T} \\ &= \mathbf{e}_x\left(-\mathcal{P}_{xz} - \mathcal{T}_y\right) + \mathbf{e}_y\left(-\mathcal{P}_{yz} + \mathcal{T}_x\right) + \mathbf{e}_z \nabla_h^2 \mathcal{P},\end{aligned} \quad (4.20)$$

where $\mathcal{P}$ is the poloidal scalar and $\mathcal{T}$ is the toroidal scalar. The associated vorticity $\omega \equiv \nabla \times \mathbf{u}$ is

$$\omega = \mathbf{e}_x\left(\nabla^2 \mathcal{P}_y - \mathcal{T}_{xz}\right) + \mathbf{e}_y\left(-\nabla^2 \mathcal{P}_x - \mathcal{T}_{yz}\right) + \mathbf{e}_z \nabla_h^2 \mathcal{T}. \quad (4.21)$$

Inspection of (4.20) and (4.21) immediately reveals the fundamental distinction between the poloidal and toroidal fields: the former has no vertical vorticity, while the latter has no vertical velocity. Note also that $\mathcal{P}$ and $\mathcal{T}$ are only defined to within arbitrary functions of z, because the velocity associated with such functions is identically zero by (4.20). Similarly, an arbitrary harmonic function $f(x, y)$ (i.e., one satisfying $\nabla_h^2 f = 0$) can be added to $\mathcal{P}$ without changing the velocity.

The PDEs satisfied by $\mathcal{P}$ and $\mathcal{T}$ are obtained by applying the operators $\nabla \times (\mathbf{e}_z \times \nabla)$ and $\mathbf{e}_z \times \nabla$, respectively, to the momentum equation (4.1b), (4.1e). Supposing that the viscosity is constant but retaining a body force $\mathbf{b} \equiv -g\Delta\rho \mathbf{e}_z$ where $\Delta\rho(\mathbf{x})$ is a density anomaly, we find

$$\nabla_h^2 \nabla^4 \mathcal{P} = \frac{g}{\eta}\nabla_h^2(\Delta\rho), \quad \nabla_h^2 \nabla^2 \mathcal{T} = 0. \quad (4.22\text{a,b})$$

The equation for $\mathcal{T}$ is homogeneous, implying that flow driven by internal density anomalies in a fluid with constant viscosity is purely poloidal. The same is true in a fluid whose viscosity varies only as a function of depth, although the equations satisfied by $\mathcal{P}$ and $\mathcal{T}$ are then more complicated than (4.22). When the viscosity varies laterally, however, the equations for $\mathcal{P}$ and $\mathcal{T}$ are coupled, and internal density anomalies drive a toroidal flow that is slaved to the poloidal flow. Toroidal flow will also be driven by any surface boundary conditions having a nonzero vertical vorticity, even if the viscosity does not vary laterally.

A common problem in geodynamics is to determine a poloidal flow in a fluid layer confined between horizontal rigid and/or free-slip boundaries. The conditions satisfied by $\mathcal{P}$ on these boundaries can be determined as follows. The vanishing of the vertical velocity (impermeability) requires $\nabla_h^2 \mathcal{P} = 0$, which means that $\mathcal{P}$ must be a harmonic function of x and y on the boundary. But because $\mathcal{P}$ is only defined to within an arbitrary additive harmonic function, we can replace $\nabla_h^2 \mathcal{P} = 0$ by the simpler condition $\mathcal{P} = 0$. Next, if the surface is rigid then the no-slip condition there requires $\partial_{13}^2 \mathcal{P} = \partial_{23}^2 \mathcal{P} = 0$, which implies that $\partial_3 \mathcal{P}$ is constant on the boundary. But because $\mathcal{P}$ is only defined to within an arbitrary additive function of z, we can set the constant equal to zero with no loss of generality to obtain the simpler condition $\partial_3 \mathcal{P} = 0$. Finally, if the surface is free-slip then the vanishing of the tangential stress there requires $\partial_1(\nabla_h^2 - \partial_{33}^2)\mathcal{P} = \partial_2(\nabla_h^2 - \partial_{33}^2)\mathcal{P} = 0$. However, because $\mathcal{P} = 0$ on the boundary these conditions reduce to $\partial_{133}^3 \mathcal{P} = \partial_{233}^3 \mathcal{P} = 0$, which imply that $\partial_{33}^2 \mathcal{P}$ is constant on the boundary. Setting this constant to zero following the same argument as for a rigid surface, we obtain $\partial_{33}^2 \mathcal{P} = 0$. In summary, the conditions satisfied by $\mathcal{P}$ on horizontal boundaries are

$$\mathcal{P} = \partial_{33}^2 \mathcal{P} = 0 \quad \text{(free-slip)}, \tag{4.23a}$$

$$\mathcal{P} = \partial_3 \mathcal{P} = 0 \quad \text{(rigid)}. \tag{4.23b}$$

You may have noticed that the equation (4.22a) for the poloidal scalar is of fourth order in the vertical coordinate but of sixth order in the horizontal coordinates. To understand why, suppose that we have a flow in a closed box with rigid sidewalls, and consider the boundary conditions on a wall whose normal is in the $\mathbf{e}_1$-direction. The vanishing of the three components of velocity on this wall requires $\partial_{13}^2 \mathcal{P} = \partial_{23}^2 \mathcal{P} = \nabla_h^2 \mathcal{P} = 0$. Because the condition $\partial_{23}^2 \mathcal{P}$ involves only derivatives parallel to the wall, it can be replaced by the simpler condition $\mathcal{P} = 0$. Similarly the condition $\partial_{13}^2 \mathcal{P} = 0$ can be replaced by $\partial_1 \mathcal{P} = 0$. Finally, the condition $\nabla_h^2 \mathcal{P} \equiv (\partial_{11}^2 + \partial_{22}^2) \mathcal{P} = 0$ can be replaced by $\partial_{11}^2 \mathcal{P} = 0$. We thus have three distinct boundary conditions $\mathcal{P} = \partial_1 \mathcal{P} = \partial_{11}^2 \mathcal{P} = 0$, which together with those on the other wall normal to $\mathbf{e}_1$ make six boundary conditions in the $\mathbf{e}_1$-direction. The differential equation for the poloidal scalar must therefore be of sixth order in the horizontal directions. However, for the special case of flow that is horizontally periodic, the operators ∇_h^2 in (4.22a) and (4.22b) can be dropped.

Because of Earth's spherical geometry, the spherical-coordinate form of the poloidal-toroidal representation is particularly important in geodynamics. However, the definitions of $\mathcal{P}$ and $\mathcal{T}$ used by different authors sometimes differ by a sign and/or a factor of r. Following Forte and Peltier (1987),

$$\mathbf{u} = \nabla \times (r\mathbf{e}_r \times \nabla \mathcal{P}) + r\mathbf{e}_r \times \nabla \mathcal{T}$$

$$= \mathbf{e}_\theta \left[-\frac{1}{r}(r\mathcal{P})_{r\theta} - \frac{\mathcal{T}_\phi}{\sin\theta} \right] + \mathbf{e}_\phi \left[-\frac{1}{r\sin\theta}(r\mathcal{P})_{r\phi} + \mathcal{T}_\theta \right] + \frac{\mathbf{e}_r}{r}\mathcal{B}^2\mathcal{P}, \quad (4.24)$$

where

$$\mathcal{B}^2 = \frac{1}{\sin\theta} \frac{\partial}{\partial\theta} \sin\theta \frac{\partial}{\partial\theta} + \frac{1}{\sin^2\theta} \frac{\partial^2}{\partial\phi^2}. \quad (4.25)$$

Another common convention is that of Chandrasekhar (1981), who uses a poloidal scalar $\Phi \equiv -r\mathcal{P}$ and a toroidal scalar $\Psi \equiv -r\mathcal{T}$. The PDEs satisfied by $\mathcal{P}$ and $\mathcal{T}$ in a fluid of constant viscosity in spherical coordinates are obtained by applying the operators $\nabla \times (r\mathbf{e}_r \times \nabla)$ and $r\mathbf{e}_r \times \nabla$, respectively, to the momentum equation (4.1b), (4.1e), yielding

$$\mathcal{B}^2 \nabla^4 \mathcal{P} = \frac{g}{\eta r} \mathcal{B}^2 \Delta\rho, \qquad \mathcal{B}^2 \nabla^2 \mathcal{T} = 0. \quad (4.26)$$

Because the preceding equation for $\mathcal{T}$ is homogeneous, the remarks following (4.22) apply also to spherical geometry. The boundary conditions (4.23) also remain valid in spherical geometry if ∂_3 is replaced by ∂_r.

Two additional quantities of interest are the lateral divergence $\nabla_h \cdot \mathbf{u}$ and the radial component $\mathbf{e}_r \cdot (\nabla \times \mathbf{u})$ of the vorticity, which are

$$\nabla_h \cdot \mathbf{u} = -\mathcal{B}^2 \left[\frac{1}{r^2}(r\mathcal{P})_r \right], \quad \mathbf{e}_r \cdot (\nabla \times \mathbf{u}) = \frac{\mathcal{B}^2 \mathcal{T}}{r}. \quad (4.27)$$

The lateral divergence depends only on the poloidal component of the flow, whereas the radial vorticity depends only on the toroidal component. At Earth's surface, therefore, divergent and convergent plate boundaries (where $\nabla_h \cdot \mathbf{u} \neq 0$) are associated with poloidal flow, while transform faults (where $\mathbf{e}_r \cdot (\nabla \times \mathbf{u}) \neq 0$) reflect toroidal flow. Spherical harmonic decomposition of Earth's surface velocity field reveals that the power of the toroidal component is 50–90% of the power of the poloidal component at all degrees (O'Connell et al., 1991, Figure 4). Toroidal motion this intense can only be generated by strong lateral variations in effective viscosity, raising the possibility that the observed poloidal–toroidal partitioning may place constraints on the rheology of the lithosphere. To quantify this link, Bercovici (1993) proposed a semikinematic source-sink model of surface flow in which toroidal motion is generated by the interaction of lateral viscosity variations with a prespecified poloidal flow. Solution of the model equations using various candidate non-Newtonian rheologies shows that a power-law rheology is not sufficient to produce plate-like toroidal flow with strong shear localization, and that a more exotic self-lubricating rheology is needed (Bercovici, 1998).

4.3 Classical Exact Solutions

The equations of slow viscous flow admit exact analytical solutions in a variety of geodynamically relevant geometries. Some of the most useful of these solutions are the following.

4.3.1 Steady Unidirectional Flow

The simplest conceivable fluid flow is a steady unidirectional flow with a velocity $w(x, y, t)\mathbf{e}_z$, where (x, y) are Cartesian coordinates in the plane normal to $\mathbf{e}_z$. If the viscosity η is constant, w satisfies (Batchelor, 1967, p. 180)

$$\eta \nabla_h^2 w = -G, \tag{4.28}$$

where $-G \equiv p_z$ is a constant pressure gradient. Two special cases are of interest in geodynamics. The first is the steady Poiseuille flow through a cylindrical pipe of radius a, for which the velocity w and the volume flux Q are

$$w = \frac{G}{4\eta}(a^2 - r^2), \quad Q \equiv 2\pi \int_0^a rw\, dr = \frac{\pi G a^4}{8\eta}. \tag{4.29}$$

Poiseuille flow in a vertical pipe driven by an effective pressure gradient $g\Delta\rho$ has been widely used as a model for the ascent of buoyant fluid in the conduit or tail of a mantle plume (e.g., Whitehead and Luther, 1975). The second case is that of a 2-D channel $0 \leq y \leq d$ bounded by rigid walls in which flow is driven by a combination of an applied pressure gradient $-G$ and motion of the boundary $y = d$ with speed U_0 in its own plane. For this case,

$$w = \frac{G}{2\eta}y(d-y) + \frac{U_0 y}{d}. \tag{4.30}$$

Equation (4.30) has been widely applied to model flow in a low-viscosity asthenosphere beneath a moving plate (e.g., Yale and Phipps Morgan, 1998; Höink and Lenardic, 2010; Shiels and Butler, 2015).

4.3.2 Stokes–Hadamard–Rybczynski Solution for a Sphere

Another classic result that is useful in geodynamics is the Stokes–Hadamard–Rybczynski solution for the creeping flow in and around a fluid sphere with radius a and viscosity η_1 at rest in an unbounded fluid with viscosity $\eta_0 = \eta_1/\gamma$ and velocity $-U\mathbf{e}_z$ far from the sphere (Stokes, 1851; Hadamard, 1911; Rybczynski, 1911; Batchelor, 1967). In spherical coordinates (r, θ), the outer ($n = 0$) and inner ($n = 1$) Stokes streamfunctions have the forms $\psi_n = -Ua^2 \sin^2 \theta f_n(r)$, where

$$f_0 = \frac{(2+3\gamma)\hat{r} - \gamma\hat{r}^{-1} - 2(1+\gamma)\hat{r}^2}{4(1+\gamma)}, \quad f_1 = \frac{\hat{r}^2 - \hat{r}^4}{4(1+\gamma)}, \quad (4.31)$$

and $\hat{r} = r/a$. The term $-\hat{r}^2/2$ in the preceding expression for f_0 is the Stokes streamfunction corresponding to the uniform fluid stream with speed $-U$ at $\hat{r} \to \infty$. The drag on the sphere is

$$\mathbf{F} = -4\pi a \eta_2 U \frac{1+3\gamma/2}{1+\gamma} \mathbf{e}_z. \quad (4.32)$$

If the densities of the two fluids differ such that $\rho_1 = \rho_0 + \Delta\rho$, the steady velocity $\mathbf{V}$ of the sphere as it moves freely under gravity is obtained by equating $\mathbf{F}$ to the Archimedean buoyancy force, yielding

$$\mathbf{V} = \frac{a^2 \mathbf{g} \Delta\rho}{3\eta_2} \frac{1+\gamma}{1+3\gamma/2}, \quad (4.33)$$

where $\mathbf{g}$ is the gravitational acceleration. The speed of an effectively inviscid sphere ($\gamma = 0$) is only 50% greater than that of a solid sphere ($\gamma \to \infty$). Moreover, the normal stress jump across the interface is independent of position for any value of γ, implying that the sphere has no tendency to deform even in the complete absence of surface tension (Batchelor, 1967, p. 238).

The expression (4.33) has been widely used to estimate the ascent speed of plume heads (e.g., Whitehead and Luther, 1975) and isolated thermals (e.g., Griffiths, 1986) in the mantle. A related application is to derive a simple criterion for the separation of an incipient mantle plume from its source region (Whitehead and Luther, 1975). In this approach, the plume is idealized as an effectively inviscid sphere whose radius $a(t)$ increases with time as $\dot{a} = Q/(4\pi a^2)$ due to supply of fluid at a constant volumetric rate Q. The model posits that the sphere detaches from its source when its ascent speed (as predicted by Stokes's law) exceeds $\dot{a}$. This prediction agrees well with laboratory experiments in which the plume fluid is supplied at a constant rate through a pipe. However, it is difficult to apply directly to mantle plumes, which are supplied at a nonconstant rate by a low-viscosity BL. This recognition led Bercovici and Kelly (1997) to generalize the model to take into account the progressive draining of the boundary layer, yielding a nonlinear evolution equation for the height of the growing plume disturbance.

4.3.3 Models for Subduction Zones and Ridges

2-D viscous flow in a fluid wedge driven by motion of the boundaries (corner flow) has been widely used to model mantle flow in subduction zones and beneath mid-ocean ridges. Figure 4.3a and b shows the geometry of the simplest corner-flow models of these features, due respectively to McKenzie (1969) and Lachenbruch and Nathenson (1974). The corner-flow model for subduction was already evoked in § 1.2.

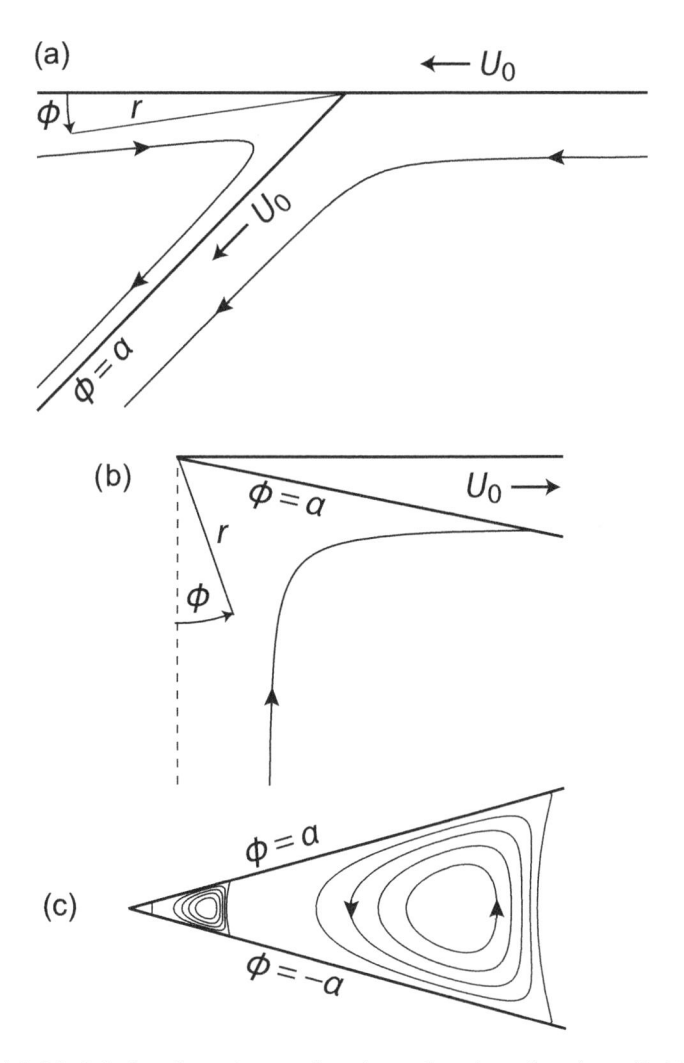

Figure 4.3 Models for slow viscous flow in wedge-shaped regions (§ 4.3.3). (a) subduction zone, (b) mid-ocean ridge and (c) self-similar viscous corner eddies generated by an agency far from the corner. In all models, streamlines (for Newtonian rheology) are shown by solid lines with arrows. These models are discussed in § 4.3.3 and 4.3.4.

The models of Figure 4.3a and b admit analytical solutions for both Newtonian and power-law rheology. The boundary conditions for both models can be satisfied by a self-similar (of the first kind; see § 3.3) streamfunction $\psi = -U_0 r F(\phi)$, where (r, ϕ) are polar coordinates. The only nonzero component of the strain rate tensor $\mathbf{e}$ is $e_{r\phi} = U_0(F'' + F)/2r$, whence the invariant $I \equiv (e_{ij}e_{ij})^{1/2}$ in (4.1f) is simply $\sqrt{2} e_{r\phi}$. The equation satisfied by F is obtained by taking the curl of the Stokes momentum equation $\nabla \cdot \sigma = 0$ with the power-law constitutive relation (4.1e), (4.1f). The result is (Exercise 4.3)

$$\left(\frac{d^2}{d\phi^2} + \frac{2n-1}{n^2}\right)\left[\left(\frac{d^2 F}{d\phi^2} + F\right)^{1/n}\right] = 0, \tag{4.34}$$

where n is the power-law index. The associated pressure, obtained by integrating the $\mathbf{e}_r$-component of the momentum equation with respect to r subject to the condition of vanishing pressure at $r = \infty$, is $p = -\eta U_0 r^{-1}(F''' + F')$, where the viscosity η is given by (4.1f). Because the viscosity itself is proportional to $r^{1-1/n}$, $p \propto r^{-1/n}$. The stress singularity at the corner is therefore integrable when $n > 1$, i.e., when the fluid is shear thinning.

Solving (4.34) for $F'' + F$, we obtain

$$F'' + F = C \cos^n \frac{(2n-1)^{1/2}}{n}(\phi + D) \tag{4.35}$$

where C and D are arbitrary constants. Equation (4.35) can be solved analytically if n is any positive integer. The solution is (Tovish et al., 1978)

$$F = A \sin \phi + B \cos \phi$$
$$+ \frac{C}{2^n(2n-1)^{1/2}}\left[\binom{n}{n/2}\delta + 2\sum_{k=0}^{m}\binom{n}{k}\cos q(\phi + D)/(1 - q^2)\right] \tag{4.36}$$

where A and B are arbitrary constants, $q = (n - 2k)(2n - 1)^{1/2}/n$, $\delta = 0$ and $m = (n-1)/2$ for n odd, $\delta = 1$ and $m = (n-2)/2$ for n even and

$$\binom{n}{n/2}, \quad \binom{n}{k}$$

are binomial coefficients. Equation (4.36) corrects two small misprints in equation (A12) of Tovish et al. (1978), which is not consistent with their equation (A10).

The values of n most relevant to Earth are $n = 1$, which corresponds to deformation by diffusion creep, and $n = 3$, which is close to the value ($n = 3.5$) determined experimentally for dislocation creep in olivine (Bai et al., 1991). In an obvious notation, the general solutions for these two values of n are

4.3 Classical Exact Solutions

$$F_1 = A_1 \sin\phi + B_1 \cos\phi + C_1\phi \sin\phi + D_1\phi \cos\phi, \quad (4.37)$$

$$F_3 = A_3 \sin\phi + B_3 \cos\phi + C_3 H(\phi, D_3), \quad (4.38)$$

respectively, where

$$H(\phi, D_3) = 27 \cos\frac{\sqrt{5}}{3}(\phi + D_3) - \cos\sqrt{5}(\phi + D_3) \quad (4.39)$$

and $A_n - D_n$ are arbitrary constants for $n = 1$ and $n = 3$ that are determined by the boundary conditions. For the ridge model,

$$\{A_1, B_1, C_1, D_1\} = \frac{\{c^2, 0, 0, -1\}}{\alpha - sc}, \quad (4.40)$$

$$\{A_3, B_3, C_3, D_3\} = \left\{-C_3(h_1 s + h'_1 c), 0, \frac{1}{h'_1 s - h_1 c}, \frac{3\pi}{2\sqrt{5}}\right\}, \quad (4.41)$$

where $s = \sin\alpha$, $c = \cos\alpha$, $h_1 = H(\alpha, D_3)$ and $h'_1 = H_\phi(\alpha, D_3)$. For the Newtonian ($n = 1$) subduction model in the wedge $0 \leq \phi \leq \alpha$,

$$\{A_1, B_1, C_1, D_1\} = \frac{\{\alpha s, 0, \alpha c - s, -\alpha s\}}{\alpha^2 - s^2}. \quad (4.42)$$

For the power-law ($n = 3$) subduction model in $0 \leq \phi \leq \alpha$, D_3 satisfies $h_1 - h_0 c - h'_0 s = 0$, and the other constants are

$$C_3 = \left(h'_1 + h_0 s - h'_0 c\right)^{-1}, \quad B_3 = -h_0 C_3, \quad A_3 = -h'_0 C_3, \quad (4.43)$$

where $h_0 = H(0, D_3)$ and $h'_0 = H_\phi(0, D_3)$. If needed, the solutions in the wedge $\alpha \leq \phi \leq \pi$ can be obtained from (4.37) and (4.38) by applying the boundary conditions shown in Figure 4.3a.

An important dynamic parameter that can be calculated from the corner-flow subduction model is the torque about the origin acting on a length L of the subducting plate. The relevant component of the stress tensor is $\sigma_{\phi\phi}$, which is equal to $-p$ because $e_{\phi\phi} \equiv 0$. For the Newtonian ($n = 1$) model, the torque per unit length in the $\mathbf{e}_z$-direction exerted on the subducting plate by the fluid in the wedge $0 \leq \phi \leq \alpha$ is

$$\tau = -\int_0^L r\sigma_{\phi\phi} dr = -\eta U_0 L[F'''(\alpha) + F'(\alpha)]. \quad (4.44)$$

Explicit calculation shows that $\tau < 0$, which means that τ is a lifting torque that tends to rotate the subducting plate clockwise, towards the motionless overriding plate (Figure 4.3a). A similar calculation for the wedge $\alpha \leq \phi \leq \pi$ shows that the torque exerted by the fluid beneath the subducting plate is also a lifting torque. These results are the basis for the hypothesis of Stevenson and Turner (1977) that the angle of subduction is controlled by the balance between the hydrodynamic lifting torque and the opposing gravitational torque acting on the slab. Their results were extended to power-law fluids by Tovish et al. (1978). Hsui and Tang (1988) and Hsui et al. (1990) performed similar analyses using analytical models for the slab thermal structure to calculate the gravitational torque on the slab as a function of dip angle. However, Willemann and Davies (1982) contested the conclusion of Stevenson and Turner (1977) and Tovish et al. (1978) on the basis that normal loads on the slab should be supported locally by slab bending. Using a thin-plate flexural model, Marotta and Mongelli (1998) concluded that slabs subducting eastward should have smaller dips than those subducting westward due to the influence of the global average W-NW trending motion of the lithosphere with respect to the underlying mantle.

Whereas the corner-flow model for a ridge assumes that the ridge is a straight line, real ridges are more complicated, comprising spreading segments connected by transform faults. An analytical approach for such ridges is to use Cartesian coordinates and Fourier transforms in the horizontal directions to calculate the 3-D flow induced in the underlying mantle by a specified surface plate velocity field. Ligi et al. (2008) present solutions of this kind for a spreading ridge with a transform offset above a fluid half-space with constant or layered viscosity, for both infinitely thin plates and plates that thicken with age.

4.3.4 Viscous Eddies

Another important exact solution for Stokes flow describes viscous eddies near a sharp corner (Moffatt, 1964). This solution is an example of a self-similar solution of the second kind (§ 3.3), the determination of which requires the solution of an eigenvalue problem. Here we consider only Newtonian fluids; for power-law fluids, see Fenner (1975).

The flow domain is a 2-D wedge $|\phi| \leq \alpha$ bounded by rigid walls (Figure 4.3c). Flow in the wedge is driven by an agency (e.g., stirring) acting at a distance $\sim r_0$ from the corner. We seek to determine the asymptotic character of the flow near the corner, i.e., in the limit $r/r_0 \to 0$. Because the domain of interest is far from the driving agency, we anticipate that the flow will be self-similar.

The streamfunction satisfies the biharmonic equation (4.16), which admits separable solutions of the form

$$\psi = r^\lambda F(\phi). \tag{4.45}$$

Substituting (4.45) into (4.16), we find that F satisfies

$$F'''' + [\lambda^2 + (\lambda - 2)^2]F'' + \lambda^2(\lambda - 2)^2 = 0, \tag{4.46}$$

where primes denote d/dϕ. For all λ except 0, 1 and 2, the solution of (4.46) is

$$F = A\cos\lambda\phi + B\sin\lambda\phi + C\cos(\lambda - 2)\phi + D\sin(\lambda - 2)\phi \qquad (4.47)$$

where A-D are arbitrary constants. The solutions for $\lambda = 0$, 1 and 2 do not exhibit eddies (the solution with $\lambda = 1$ is the one used in the models of subduction zones and ridges in § 4.3.3).

The most interesting solutions are those for which $\psi(r, \phi)$ is an even function of ϕ, so that $B = D = 0$. Application of the rigid-surface boundary conditions $F(\pm\alpha) = F'(\pm\alpha) = 0$ to (4.47) yields two equations for A and C:

$$A\cos\lambda\alpha + C\cos(\lambda - 2)\alpha = 0, \qquad (4.48a)$$

$$A\lambda\sin\lambda\alpha + C(\lambda - 2)\sin(\lambda - 2)\alpha = 0. \qquad (4.48b)$$

These have a nontrival solution only if the determinant of the coefficient matrix vanishes, i.e., if

$$\sin 2(\lambda - 1)\alpha + (\lambda - 1)\sin 2\alpha = 0. \qquad (4.49)$$

The only physically relevant roots of (4.49) are those with $\Re(\lambda) > 0$, corresponding to solutions that are finite at $r = 0$. When $2\alpha < 146°$, these roots are all complex. Let $\lambda_1 \equiv 1 + p_1 + iq_1$ be the root with the smallest real part, corresponding to the solution that decays least rapidly towards the corner. The dominant term of the streamfunction is then

$$\psi \sim r^{\lambda_1}[A\cos\lambda_1\phi + C\cos(\lambda_1 - 2)\phi]. \qquad (4.50)$$

Using the boundary condition (4.48a) to eliminate the constant C, we obtain

$$\psi \sim A'\left(\frac{r}{r_0}\right)^{\lambda_1}[\cos\lambda_1\phi\cos(\lambda_1 - 2)\alpha - \cos(\lambda_1 - 2)\phi\cos\lambda_1\alpha] \qquad (4.51)$$

where $A' = Ar_0^{\lambda_1}\sec(\lambda_1 - 2)\alpha$.

The structure of the flow can be understood by calculating the azimuthal ($\mathbf{e}_\phi$-direction) component of the velocity on the midplane $\phi = 0$, which is

$$v(r, 0) = -\Re[\psi_r(r, 0)] = -\Re\left\{\lambda_1 A'[\cos(\lambda_1 - 2)\alpha - \cos\lambda_1\alpha]\frac{1}{r}\left(\frac{r}{r_0}\right)^{\lambda_1}\right\}. \qquad (4.52)$$

After some algebraic manipulations (Exercise 4.4), we find that $v(r, 0)$ can be written in the form

$$v(r, 0) = \frac{\gamma}{r}\left(\frac{r}{r_0}\right)^{p_1+1}\sin\left(q_1\ln\frac{r}{r_0} + \epsilon\right) \qquad (4.53)$$

where γ and ϵ are real constants. Equation (4.53) implies that $v(r, 0)$ changes sign an infinite number of times as $r \to 0$. The physical meaning of this striking result is revealed by a contour plot of the streamlines of the flow. Figure 4.3c shows the streamlines for $2\alpha = 30°$, for which $p_1 = 4.22$ and $q_1 = 2.20$. The flow comprises an infinite sequence of self-similar eddies with alternating senses of rotation.

The expression (4.53) can be used to characterize the relative dimensions and intensities of the eddies as a function of the wedge angle 2α. Define the center of each eddy as the point where $v(r, 0) = 0$. This occurs where

$$q_1 \ln \frac{r}{r_0} + \epsilon = -n\pi \quad (n = 0, 1, 2, \ldots) \tag{4.54}$$

or

$$r = r_0 \exp(-\epsilon/q_1) \exp(-n\pi/q_1) \equiv r_n \text{ (say)}. \tag{4.55}$$

Therefore

$$\frac{r_n - r_{n+1}}{r_{n+1} - r_{n+2}} = \exp(\pi/q_1). \tag{4.56}$$

The dimensions of successive eddies diminish geometrically as the corner is approached, with common ratio $\exp(\pi/q_1)$ (= 4.17 for $2\alpha = 30°$). To characterize the relative intensities of the eddies, note that the velocity $v(r, 0)$ has a local maximum at the points

$$r_{n+1/2} = r_0 \exp(-\epsilon/q_1) \exp\left[-\left(n + \frac{1}{2}\right)\pi/q_1\right] \tag{4.57}$$

and that the velocity at these points is

$$v_{n+1/2} = \gamma r_0^{-p_1-1} r_{n+1/2}^{p_1}. \tag{4.58}$$

The intensities of successive eddies thus decrease geometrically towards the corner with common ratio

$$\frac{v_{n+1/2}}{v_{n+3/2}} = \exp(\pi p_1/q_1), \tag{4.59}$$

which ≈ 414 for $2\alpha = 30°$. In the limit $\alpha = 0$ corresponding to flow between parallel planes, $p_1 = 4.21$ and $q_1 = 2.26$, all the eddies have the same size (≈ 2.78 times the channel width) and the intensity ratio ≈ 348.

A geodynamical model that exhibits corner eddies is that of Ribe (1989), in which back-arc spreading at a distance r_0 from the corner of a subduction zone generates an infinite sequence of eddies for $r \ll r_0$. However, these eddies are too weak to have any observable consequences. In geodynamics, corner eddies are significant primarily as a simple model for the tendency of forced viscous flows in domains with large aspect ratio to break up into separate cells. An example (§ 10.4) is steady 2-D cellular convection at high Rayleigh number, in which Stokes flow in the isothermal core of a cell is driven by the shear stresses applied to it by the thermal plumes at its ends. When the aspect ratio (width/depth) $\beta = 1$, the core flow comprises a single cell; but when $\beta = 2.5$, the flow separates into two distinct eddies (Jimenez and Zufiria, 1987, Figure 2).

4.4 Superposition and Eigenfunction Expansion Methods

The linearity of the equations governing Stokes flow is the basis of two powerful methods for solving Stokes flow problems in regular domains: superposition and eigenfunction expansion. In both methods, a complicated flow is represented by infinite sums of elementary separable solutions of the Stokes equations for the coordinate system in question, and the unknown coefficients in the expansion are determined to satisfy the boundary conditions. In the superposition method, the individual separable solutions do not themselves satisfy all the boundary conditions in any of the coordinate directions. In the method of eigenfunction expansions (MEE), by contrast, the separable solutions are true eigenfunctions that satisfy all the (homogeneous) boundary conditions at both ends of an interval in one of the coordinate directions, so that the unknown constants are determined entirely by the boundary conditions in the other direction(s). Let us turn now to some concrete illustrations of these methods in three coordinate systems of geodynamical interest: 2-D Cartesian, spherical and bispherical.

4.4.1 2-D Flow in Cartesian Coordinates

The streamfunction ψ for 2-D Stokes flow satisfies the biharmonic equation (4.16), which has the general solution

$$\psi = f_1(x+iy) + f_2(x-iy) + (y+ix)f_3(x+iy) + (y-ix)f_4(x-iy), \quad (4.60)$$

where f_1–f_4 are arbitrary functions of their (complex) arguments. However, the most useful solutions for applications are the separable solutions that vary periodically in one direction. Setting $\psi \propto \exp ikx$ and solving (4.16) by separation of variables, we find

$$\psi = \left[(A_1 + A_2 y)\exp(-ky) + (A_3 + A_4 y)\exp ky\right]\exp ikx, \quad (4.61)$$

where A_1–A_4 are arbitrary functions of k. An analogous solution is obtained by interchanging x and y in (4.61), and both solutions remain valid if k is complex.

The solution (4.61) has been widely used in geodynamics to describe flows generated by loads that vary sinusoidally in the horizontal direction (e.g., Fleitout and Froidevaux, 1983). The representation of more general flows, however, typically

Slow Viscous Flow

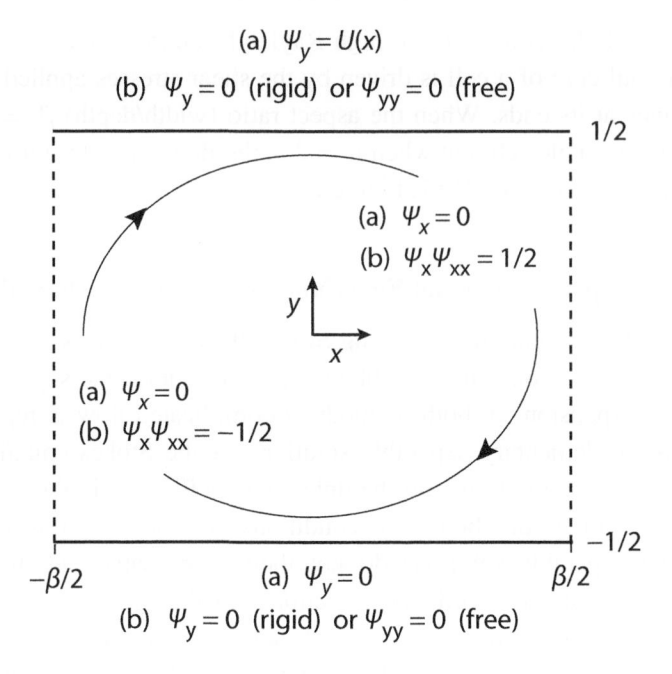

(a) $\Psi_y = U(x)$
(b) $\Psi_y = 0$ (rigid) or $\Psi_{yy} = 0$ (free)

(a) $\Psi_x = 0$
(b) $\Psi_x \Psi_{xx} = 1/2$

(a) $\Psi_x = 0$
(b) $\Psi_x \Psi_{xx} = -1/2$

(a) $\Psi_y = 0$
(b) $\Psi_y = 0$ (rigid) or $\Psi_{yy} = 0$ (free)

Figure 4.4 Geometry and boundary conditions for (a) driven cavity flow and (b) steady cellular convection in a rectangle $x \in [-\beta/2, \beta/2]$, $y \in [-1/2, 1/2]$ with impermeable ($\psi = 0$) boundaries. All lengths are nondimensionalized by the height of the rectangle. See § 4.4.1 for discussion.

requires the use of a superposition or an eigenfunction expansion. To illustrate the use of these two methods, we consider the problem of driven cavity flow in a rectangular domain $x \in [-\beta/2, \beta/2]$, $y \in [-1/2, 1/2]$ with impermeable ($\psi = 0$) boundaries (Figure 4.4). The sidewalls $x = \pm\beta/2$ and the bottom $y = -1/2$ are rigid and motionless, while the upper boundary $y = 1/2$ moves in its own plane with velocity $U(x)$ (Figure 4.4, boundary conditions (a)).

In the superposition method, the solution is represented as a sum of two ordinary Fourier series in the two coordinate directions. The flow is the sum of two parts that are even and odd functions of y, respectively, and the representation for the even part is (Meleshko, 1996)

$$\psi = \sum_{m=1}^{\infty} A_m F(y, p_m, 1) \cos p_m x + \sum_{n=1}^{\infty} B_n F(x, q_n, \beta) \cos q_n y, \tag{4.62a}$$

$$p_m = \frac{2m-1}{\beta}\pi, \quad q_n = (2n-1)\pi, \tag{4.62b}$$

$$F(z, k, r) = \frac{r \tanh(rk/2) \cosh kz - 2z \sinh kz}{2 \cosh(rk/2)}, \tag{4.62c}$$

where A_m and B_n are undetermined coefficients. Due to the choice of the wavenumbers p_m and q_n and the function $F(z, k, r)$ (which vanishes at $z = \pm r/2$), each of the series in (4.62a) satisfies the impermeability condition $\psi = 0$ on all boundaries. Moreover, because the trigonometric systems $\cos p_m x$ and $\cos q_n y$ are complete, the superposition (4.62) (with suitable choices of A_m and B_n) can represent arbitrary distributions of tangential velocity on $y = \pm 1/2$ and $x = \pm \beta/2$.

Unlike the superposition method, the MEE makes use of so-called Papkovich–Fadle eigenfunctions $\phi(x)$ that satisfy all the homogeneous boundary conditions in the direction (x in this case) perpendicular to two motionless walls. For simplicity, suppose that $U(x) = U(-x)$, so that ψ is an even function of x. The even eigenfunctions on the canonical unit interval $x \in [-1/2, 1/2]$ are obtained by substituting $\psi = \phi(x) \exp(\lambda y)$ into the biharmonic equation (4.16) and solving the resulting equation for ϕ subject to the conditions $\phi(\pm 1/2) = \phi_x(\pm 1/2) = 0$, yielding

$$\phi_n(x) = x \sin \lambda_n x - \frac{1}{2} \tan \frac{\lambda_n}{2} \cos \lambda_n x, \qquad (4.63)$$

where λ_n are the first-quadrant complex roots of $\sin \lambda_n + \lambda_n = 0$. The streamfunction for the flow in the cavity with $x \in [-\beta/2, \beta/2]$ (Figure 4.4) can then be written as (Shankar, 1993)

$$\psi = \sum_{n=1}^{\infty} \left\{ A_n \Phi_n \exp\left[-\lambda_n \left(y + \frac{1}{2}\right)\right] + \overline{A}_n \overline{\Phi}_n \exp\left[-\overline{\lambda}_n \left(y + \frac{1}{2}\right)\right] \right.$$
$$\left. + B_n \Phi_n \exp\left[-\lambda_n \left(\frac{1}{2} - y\right)\right] + \overline{B}_n \overline{\Phi}_n \exp\left[-\overline{\lambda}_n \left(\frac{1}{2} - y\right)\right] \right\}, \qquad (4.64)$$

where $\Phi_n = \phi_n(x/\beta)$, overbars denote complex conjugation and the constants A_n and B_n are chosen to satisfy the boundary conditions at $y = \pm 1/2$. The difficulty, however, is that the reduced biharmonic equation $(d^2/dx^2 + \lambda^2)^2 \phi = 0$ satisfied by the eigenfunctions ϕ_n is not self-adjoint, as can be seen by rewriting it in the form

$$\frac{d^2}{dx^2} \begin{pmatrix} \phi \\ -\lambda^{-2} \phi'' \end{pmatrix} = \lambda^2 \begin{pmatrix} 0 & -1 \\ 1 & -2 \end{pmatrix} \begin{pmatrix} \phi \\ -\lambda^{-2} \phi'' \end{pmatrix} \qquad (4.65)$$

and noting that the square matrix on the RHS is not Hermitian. Consequently, the eigenfunctions ϕ_n are not mutually orthogonal, which makes the determination of A_n and B_n a nontrivial matter. One solution is to use a complementary set of adjoint eigenfunctions χ_m which are biorthogonal to the set ϕ_n, although considerable care must then be taken to ensure convergence of the expansion (see Katopodes et al., 2000) for a discussion and references to the relevant literature). A cruder but very effective approach is to determine A_n and B_n via a numerical least-squares procedure that minimizes the misfit of the solution (4.64) to the boundary conditions (Shankar, 1993; Bloor and Wilson, 2006). Shankar (2005) showed how this approach can be extended to an irregular domain by embedding the latter in a larger, regular domain on which a complete set of eigenfunctions exists.

Another geodynamically relevant 2-D Stokes problem that can be solved using superposition and eigenfunction expansion methods is that of the flow within the isothermal core of a vigorous (high-Rayleigh number) convection cell. The boundary conditions for this case (Figure 4.4, conditions (b) comprise either rigid ($\psi_y = 0$) or free-slip ($\psi_{yy} = 0$) conditions at $y = \pm 1/2$ and nonlinear sidewall conditions $\psi_x \psi_{xx} = \pm 1/2$ (derived in § 10.4.2) that represent the shear stresses applied to the core by the buoyant thermal plumes. If the boundaries $y = \pm 1/2$ are free-slip, then the conditions $\psi = \psi_{yy} = 0$ involve only even derivatives of ψ and are satisfied identically if $\psi \propto \cos q_n y$ where q_n is defined by (4.62b). The streamfunction that also satisfies the sidewall impermeability conditions $\psi(\pm \beta/2, y) = 0$ is (Roberts, 1979)

$$\psi = \sum_{n=0}^{\infty} A_n F(x, q_n, \beta) \cos q_n y, \qquad (4.66)$$

where $F(x, q_n, \beta)$ is defined by (4.62c). The constants A_n are then determined iteratively to satisfy the boundary conditions $\psi_x \psi_{xx}(\pm \beta/2, y) = \pm 1/2$. Strictly speaking, (4.66) is an eigenfunction expansion because the functions $\cos q_n y$ satisfy all the homogeneous boundary conditions at $y = \pm 1/2$. In practice, however, the term *eigenfunction expansion* is usually reserved for expressions like (4.64) that involve a sequence of complex wavenumbers λ_n. No such terminological ambiguity applies for a convection cell bounded by rigid surfaces, for which the flow can be represented by the eigenfunction expansion (Roberts, 1979)

$$\psi = \sum_{n=1}^{\infty} \left[A_n \phi_n(y) \cosh \lambda_n x + \overline{A}_n \overline{\phi}_n(y) \cosh \overline{\lambda}_n x \right]. \qquad (4.67)$$

Other examples of the use of superposition and eigenfunction expansion methods for cellular convection problems can be found in Turcotte (1967), Turcotte and Oxburgh (1967), Olson and Corcos (1980), Morris and Canright (1984), Busse et al. (2006) and Morris (2008).

4.4.2 Spherical Coordinates

Lamb (1945) derived a general solution of the equations of Stokes flow in spherical coordinates (r, θ, ϕ). Because the pressure p satisfies Laplace's equation, it may be expressed as a sum of solid spherical harmonics p_l:

$$p = \sum_{l=-\infty}^{\infty} p_l, \quad p_l = r^l \sum_{m=-l}^{l} c_{lm} Y_l^m(\theta, \phi), \qquad (4.68)$$

where Y_l^m are surface spherical harmonics and c_{lm} are complex coefficients. The velocity $\mathbf{u}$ can then be written

$$\mathbf{u} = \sum_{l=-\infty}^{\infty} [\nabla \Phi_l + \nabla \times (\mathbf{x}\chi_l)] + \sum_{\substack{l=-\infty \\ l \neq 1}}^{\infty} \left[\frac{(1/2)(l+3)r^2 \nabla p_l - l\mathbf{x}p_l}{\eta(l+1)(2l+3)} \right], \quad (4.69)$$

where η is the viscosity, $\mathbf{x}$ is the position vector, and Φ_l and χ_l are solid spherical harmonics of the form (4.68) but with different coefficients. The first sum in (4.69) is the solution of the homogeneous Stokes equations $\nabla^2 \mathbf{u} = 0$, $\nabla \cdot \mathbf{u} = 0$, whereas the second sum is the particular solution of $\nabla^2 \mathbf{u} = \nabla p/\eta$. An application of (4.69) is the solution of Gomilko et al. (2003) for steady Stokes flow driven by the motion of one of three mutually perpendicular walls that meet in a corner, a 3-D generalization of the 2-D corner-flow model (Figure 4.3a). The main difficulty in using Lamb's solution is in applying boundary conditions because the elements of (4.69), while complete, do not form an orthogonal basis for vector functions on the surface of a sphere in the way that standard spherical harmonics form an orthonormal basis for scalar functions. Methods for dealing with this problem are discussed in chapter 4 of Kim and Karrila (1991).

4.4.3 Bispherical Coordinates

Flow in a domain bounded by nonconcentric spheres is a useful model for the interaction of one buoyant diapir or plume head with another or with a flat interface (i.e., a sphere of infinite radius). Such flow problems can often be solved analytically using bispherical coordinates (ξ, θ, ϕ), which are related to the Cartesian coordinates (x, y, z) by

$$(x, y, z) = \frac{(a \sin \theta \cos \phi, a \sin \theta \sin \phi, a \sinh \xi)}{\cosh \xi - \cos \theta}, \quad (4.70)$$

where a is a fixed length scale. Surfaces of constant ξ are nonconcentric spheres with their centers on the axis $x = y = 0$, and $\xi = 0$ corresponds to the plane $z = 0$ (Figure 4.5).

In geodynamics, the most useful solutions using bispherical coordinates are axisymmetric ones for which the flow does not depend on the azimuthal angle ϕ. The flow can then be represented using a Stokes streamfunction ψ. The general solutions for ψ and the pressure p are (Stimson and Jeffrey, 1926)

$$\psi = (\cosh \xi - \eta)^{-3/2} \sum_{n=0}^{\infty} C_{n+1}^{-1/2}(\eta) [a_n \cosh \beta_{-1} \xi + b_n \sinh \beta_{-1} \xi$$
$$+ c_n \cosh \beta_3 \xi + d_n \sinh \beta_3 \xi], \quad (4.71a)$$

$$p = (\cosh \xi - \eta)^{1/2} \sum_{n=0}^{\infty} [A_n \exp \beta_1 \xi + B_n \exp(-\beta_1 \xi)] P_n(\eta), \quad (4.71b)$$

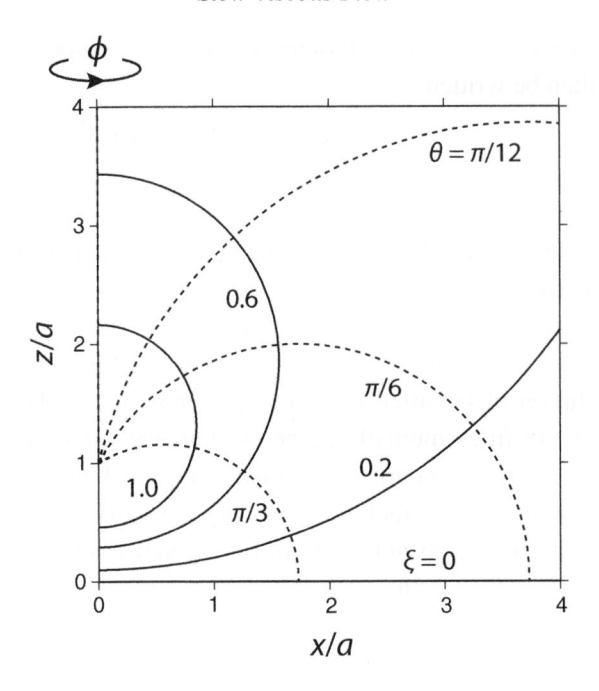

Figure 4.5 Bispherical coordinates (ξ, θ, ϕ) defined by (4.70). Selected lines of constant ξ (solid) and constant θ (dashed) are shown. See § 4.4.3 for discussion.

where $\eta = \cos\theta$, $\beta_m = n + m/2$, $C_{n+1}^{-1/2}$ is a Gegenbauer polynomial of order $n+1$ and degree $-1/2$, P_n is a Legendre polynomial and $a_n - d_n$, A_n and B_n are arbitrary constants. Koch and Ribe (1989) used a solution of the form (4.71) to model the effect of a viscosity contrast on the topography and gravity anomalies produced by the motion of a buoyant and deformable fluid sphere beneath a free surface of a fluid with a different viscosity.

On a higher level of generality, Lee and Leal (1980) determined a complete solution of the Stokes equations in bispherical coordinates and used it to determine the flow due to an arbitrary translational or rotational motion of a solid sphere in the vicinity of a plane (undeformable) fluid–fluid interface. Their solution is appropriate for nonaxisymmetric situations such as two buoyant rising spheres that are displaced laterally with respect to each other.

4.5 The Complex-Variable Method for 2-D Flows

A powerful method for 2-D Stokes flows and analogous problems in elasticity (Muskhelishvili, 1953) is based on the so-called Goursat representation of a biharmonic function ψ in terms of two analytic functions ϕ and χ of the complex variable $z \equiv x + iy$,

4.6 Singular Solutions

$$\psi = \Re[\bar{z}\phi(z) + \chi(z)], \tag{4.72}$$

where an overbar denotes complex conjugation. If ψ is the streamfunction of a 2-D flow, then the velocity components $u = -\psi_y$ and $v = \psi_x$ are

$$v - iu = \phi(z) + z\overline{\phi'(z)} + \overline{\chi'(z)}, \tag{4.73}$$

the vorticity $\omega \equiv \nabla^2\psi$ and the pressure p are

$$\omega + ip/\eta = 4\phi'(z) \tag{4.74}$$

where η is the viscosity, and the components of the stress tensor are

$$\sigma_{xx} = -2\eta\Im[2\phi'(z) - \bar{z}\phi''(z) - \chi''(z)], \tag{4.75a}$$

$$\sigma_{yy} = -2\eta\Im[2\phi'(z) + \bar{z}\phi''(z) + \chi''(z)], \tag{4.75b}$$

$$\sigma_{xy} = 2\eta\Re[\bar{z}\phi''(z) + \chi''(z)]. \tag{4.75c}$$

The Goursat representation reduces the task of solving the biharmonic equation to one of finding two analytic functions that satisfy the relevant boundary conditions. The method is most powerful when used in conjunction with conformal mapping, which allows a flow domain with a complex shape to be mapped onto a simpler one (such as the interior of the unit disk). A remarkable example is the analytical solution of Jeong and Moffatt (1992) for the formation of a cusp above a vortex dipole located at depth d beneath the free surface of a viscous fluid. An educated guess led these authors to try the conformal mapping

$$z = w(\zeta) = a(\zeta + i) + (a + 1)i\frac{\zeta - i}{\zeta + i}, \tag{4.76}$$

where a is a real constant to be determined, and the variables z and ζ have been nondimensionalized by the length d. The function (4.76) maps the fluid half-space with its deformed upper surface onto the unit disk, with the imaged vortex dipole at the center ($\zeta = 0$). In the geodynamically relevant limit of zero surface tension, it turns out that $a = -1/3$. The (dimensional) surface displacement $y(x)$ in this limit satisfies $x^2 y = -(y + 2d/3)^3$, which has an infinitely sharp cusp at $(x, y) = (0, -2d/3)$. Although the model is too idealized for direct geodynamical application, it is relevant to the formation of cusp-like features by entrainment in thermochemical convection (Davaille, 1999).

4.6 Singular Solutions and the Boundary Integral Representation

The Stokes equations admit a variety of singular solutions in which the velocity becomes infinite at one or more points in space. Such solutions are the basis of the boundary integral representation, whereby a Stokes flow in a given domain is expressed in terms of surface integrals of velocities and stresses over the domain boundaries. The dimensionality of the problem is thereby reduced by one (from 3-D to 2-D or from 2-D to 1-D), making possible a powerful numerical technique – the boundary element method – that does not require discretization of the whole flow domain (Pozrikidis, 1992).

The most useful singular solutions fall into two classes: those involving point forces, and those associated with volume sources and sinks. The following discussion is limited to fluids with constant viscosity, but some of the results can be generalized to fluids whose viscosity varies exponentially in one direction (e.g., Zhong, 1996).

4.6.1 Flow Due to Point Forces

The most important singular solution of the Stokes equations is that due to a point force F_i, or Stokeslet, applied at a position $\mathbf{x}'$ in the fluid. The velocity u_i and stress tensor σ_{ij} induced at any point $\mathbf{x}$ satisfy

$$\partial_j u_j = 0, \qquad (4.77a)$$

$$-\partial_i p + \eta \nabla^2 u_i = -F_i \delta(\mathbf{x} - \mathbf{x}'), \qquad (4.77b)$$

where $\delta(\mathbf{x} - \mathbf{x}') = \delta(x_1 - x_1')\delta(x_2 - x_2')\delta(x_3 - x_3')$ and δ is the Dirac delta-function. Here and throughout § 4.6, vector arguments of functions are denoted using boldface vector notation, while other quantities are written using Cartesian tensor (subscript) notation. In an infinite fluid, the required boundary conditions are that $u_i \to 0$ and $\sigma_{ij} \to -p_0 \delta_{ij}$ as $|\mathbf{x} - \mathbf{x}'| \to \infty$, where p_0 is the (dynamically irrelevant) far-field pressure. Because the response of the fluid is proportional to the applied force, we can write the velocity, pressure and stress as

$$u_i = J_{ij} F_j / \eta, \quad p = \Pi_j F_j, \quad \sigma_{ik} = K_{ijk} F_j, \qquad (4.78)$$

where J_{ij}, Π_j and K_{ijk} are Green functions representing the response to a unit force. Substituting these expressions into (4.77b), eliminating the arbitrary vector F_j and setting $\mathbf{x}' = \mathbf{0}$ with no loss of generality, we obtain

$$\partial_i J_{ij} = 0, \qquad (4.79a)$$

$$-\partial_i \Pi_j + \nabla^2 J_{ij} = -\delta_{ij} \delta(\mathbf{x}). \qquad (4.79b)$$

The solutions of (4.79) can be found using a Fourier transform or by reducing (4.79) to Poisson's equation (Pozrikidis, 1992, p. 22). Here I describe the Fourier transform method, which has the advantage that it can be generalized to more

4.6 Singular Solutions

complicated geometries (see Exercise 4.9). Define the three-dimensional Fourier transform and its inverse as

$$\bar{\mathcal{F}}(\mathbf{k}) = \frac{1}{(2\pi)^{3/2}} \int \mathcal{F}(\mathbf{x}) \exp(-i\mathbf{k} \cdot \mathbf{x}) d\mathbf{x}, \quad (4.80a)$$

$$\mathcal{F}(\mathbf{x}) = \frac{1}{(2\pi)^{3/2}} \int \bar{\mathcal{F}}(\mathbf{k}) \exp(i\mathbf{x} \cdot \mathbf{k}) d\mathbf{k}, \quad (4.80b)$$

where an overbar denotes the Fourier transform, $\mathbf{k}$ is the wavevector and the integrals are over all space. The Fourier transforms of (4.79a) and (4.79b) are

$$i k_i \bar{J}_{ij} = 0, \quad (4.81a)$$

$$-i k_i \bar{\Pi}_j - |\mathbf{k}|^2 \bar{J}_{ij} = -\frac{1}{(2\pi)^{3/2}} \delta_{ij}. \quad (4.81b)$$

Now multiply (4.81b) by k_i, use (4.81a) and solve for $\bar{\Pi}_j$ to obtain

$$\bar{\Pi}_j = -\frac{i}{(2\pi)^{3/2}} \frac{k_j}{|\mathbf{k}|^2}. \quad (4.82)$$

Substitute (4.82) back into (4.81b) and solve for $\bar{J}_{ij}$ to obtain

$$\bar{J}_{ij} = \frac{1}{(2\pi)^{3/2}} \frac{1}{|\mathbf{k}|^2} \left(\delta_{ij} - \frac{k_i k_j}{|\mathbf{k}|^2} \right). \quad (4.83)$$

The inverse transform of $\bar{J}_{ij}$ is

$$J_{ij} = \frac{1}{8\pi^3} \int \frac{1}{|\mathbf{k}|^2} \left(\delta_{ij} - \frac{k_i k_j}{|\mathbf{k}|^2} \right) \exp(i\mathbf{x} \cdot \mathbf{k}) d\mathbf{k}$$

$$= \frac{\delta_{ij}}{8\pi^3} \int \frac{\exp(i\mathbf{x} \cdot \mathbf{k})}{|\mathbf{k}|^2} d\mathbf{k} + \frac{1}{8\pi^3} \frac{\partial^2}{\partial x_i \partial x_j} \int \frac{\exp(i\mathbf{x} \cdot \mathbf{k})}{|\mathbf{k}|^4} d\mathbf{k}. \quad (4.84)$$

Now use the inverse transforms

$$\int \frac{\exp(i\mathbf{x} \cdot \mathbf{k})}{|\mathbf{k}|^2} d\mathbf{k} = \frac{2\pi^2}{|\mathbf{x}|}, \quad \int \frac{\exp(i\mathbf{x} \cdot \mathbf{k})}{|\mathbf{k}|^4} d\mathbf{k} = -\pi^2 |\mathbf{x}| \quad (4.85)$$

whereupon J_{ij} takes the form

$$J_{ij} = \frac{1}{8\pi} \left(\frac{2\delta_{ij}}{|\mathbf{x}|} - \frac{\partial^2}{\partial x_i \partial x_j} |\mathbf{x}| \right). \quad (4.86)$$

Finally, note that

$$\frac{\partial^2 |\mathbf{x}|}{\partial x_i \partial x_j} = \frac{\delta_{ij}}{|\mathbf{x}|} - \frac{x_i x_j}{|\mathbf{x}|^3}. \quad (4.87)$$

Using (4.87) in (4.86) and moving the point force to $\mathbf{x}'$ via the transformation $\mathbf{x} \to \mathbf{x} - \mathbf{x}'$, we obtain

$$J_{ij}(\mathbf{r}) = \frac{1}{8\pi}\left(\frac{\delta_{ij}}{r} + \frac{r_i r_j}{r^3}\right), \tag{4.88a}$$

where $\mathbf{r} = \mathbf{x} - \mathbf{x}'$ and $r = |\mathbf{r}|$. The tensor J_{ij} is often called the Oseen tensor. The corresponding Green functions for the pressure and stress are

$$\Pi_j(\mathbf{r}) = \frac{1}{4\pi}\frac{r_j}{r^3}, \qquad K_{ijk}(\mathbf{r}) = -\frac{3}{4\pi}\frac{r_i r_j r_k}{r^5}. \tag{4.88b}$$

The Green functions (4.88) represent the 3-D flow generated by a force $\mathbf{F}$ applied at a single point in space. The analogous situation in 2-D is the flow generated by a line force perpendicular to the flow plane with strength $\mathbf{F}$ per unit length. The 2-D analogs of (4.88) are

$$J_{ij}(\mathbf{r}) = \frac{1}{4\pi}\left(-\delta_{ij}\ln r + \frac{r_i r_j}{r^2}\right), \tag{4.89a}$$

$$\Pi_j(\mathbf{r}) = \frac{1}{2\pi}\frac{r_j}{r^2}, \qquad K_{ijk}(\mathbf{r}) = -\frac{1}{\pi}\frac{r_i r_j r_k}{r^4}, \tag{4.89b}$$

where the indices i, j and k now range over the values 1 and 2. The expression (4.89a) for J_{ij} diverges logarithmically as $r \to \infty$, which is related to the fact that a 2-D Stokes flow around an infinitely long cylinder does not exist (Stokes's paradox). Presently we will see how this paradox is resolved by the presence of an impermeable boundary.

Starting from the 3-D Stokeslet solution, one can use the principle of superposition to construct an infinite variety of additional singular solutions. An example is the flow due to a force dipole, comprising a point force $\mathbf{F}$ at $\mathbf{x}'$ and an equal and opposite force $-\mathbf{F}$ at $\mathbf{x}' - d\mathbf{n}$, where $\mathbf{n}$ is a unit vector directed from the negative to the positive force. The associated velocity field is $u_i(\mathbf{x}) = [J_{ij}(\mathbf{r}) - J_{ij}(\mathbf{r}+d\mathbf{n})]F_j/\eta$. In the limit $d \to 0$ with dF_j fixed,

$$u_i = dF_j n_k G^{FD}_{ijk}/\eta, \tag{4.90a}$$

$$G^{FD}_{ijk}(\mathbf{r}) = -\partial_k J_{ij}(\mathbf{r}) \equiv \frac{1}{8\pi}\left[\frac{\delta_{ij}r_k - \delta_{ik}r_j - \delta_{jk}r_i}{r^3} + \frac{3r_i r_j r_k}{r^5}\right] \tag{4.90b}$$

where G^{FD}_{ijk} is the force-dipole Green function. The force-dipole moment $dF_j n_k$ is sometimes decomposed into symmetric (stresslet) and antisymmetric (rotlet) parts (Kim and Karrila, 1991, § 2.5). The flow due to a stresslet is the far-field limit of the disturbance flow produced by a sphere embedded in a pure straining flow. Similarly, the flow due to a rotlet is the far-field limit of the flow produced by a rotating sphere. Both of these are special cases of the disturbance due to a sphere embedded in a general linear flow, which is the subject of Exercise 4.10.

4.6 Singular Solutions

4.6.2 Flows Due to Point Sources

The basic singular solution of this second class is that associated with a volume source of strength Q at $\mathbf{x}'$, which generates a spherically symmetric flow

$$u_i = \frac{Qr_i}{4\pi r^3}. \tag{4.91}$$

The flow due to a source doublet comprising a source and sink with equal strengths Q separated by a vector $d\mathbf{n}$ pointing from the sink to the source is

$$u_i = Qdn_j G_{ij}^{SD}, \quad G_{ij}^{SD}(\mathbf{r}) = -\partial_j \left(\frac{r_i}{4\pi r^3}\right) \equiv \frac{1}{4\pi}\left(\frac{3r_i r_j}{r^5} - \frac{\delta_{ij}}{r^3}\right). \tag{4.92}$$

The pressure associated with the preceding solutions is zero.

4.6.3 Singular Solutions in the Presence of a Boundary

The flow produced by a point force is modified by the presence of an impermeable wall. Consider a force F_i at a point $\mathbf{x}'$ located a distance d from the wall, and let $\mathbf{n}$ be a unit vector normal to the wall directed towards the side containing $\mathbf{x}'$ (Figure 4.6a). The modified velocity can be written as $u_i = G_{ij}^B F_j/\eta$, where $G_{ij}^B(\mathbf{x}, \mathbf{x}')$ is a Green function that satisfies all the required boundary conditions at the wall. Its general form is

$$G_{ij}^B(\mathbf{x}, \mathbf{x}') = J_{ij}(\mathbf{x} - \mathbf{x}') + G_{ij}^{IM}(\mathbf{x} - \mathbf{x}^{IM}), \tag{4.93}$$

where G_{ij}^{IM} is a Green function that is singular at the image point $\mathbf{x}^{IM} = \mathbf{x}' - 2d\mathbf{n}$ (Figure 4.6a).

The two limiting cases of greatest interest are free-slip and rigid walls. A free-slip wall is equivalent to a plane of mirror symmetry. The modified flow for this case can therefore be constructed simply by adding a reflected Stokeslet with strength $\mathbf{R} \cdot \mathbf{F} \equiv \mathbf{F}^*$ at the image point $\mathbf{x}^{IM}$, where $R_{ij} \equiv \delta_{ij} - 2n_i n_j$ is a reflection tensor that reverses the sign of the wall-normal component of a vector while leaving its wall-parallel components unchanged (Figure 4.6a). Therefore

$$G_{ij}^{IM}(\mathbf{r}^{IM}) = R_{jk} J_{ik}(\mathbf{r}^{IM}), \tag{4.94}$$

where $\mathbf{r}^{IM} = \mathbf{x} - \mathbf{x}^{IM}$. If the wall is rigid, the no-slip and impermeability conditions can be satisfied by adding three different singular solutions at $\mathbf{x}^{IM}$: a Stokeslet with strength $-\mathbf{F}$, a force dipole with moment $2d\mathbf{n}\mathbf{F}^*$ and a source doublet with strength $-d^2\mathbf{F}^*$ (Figure 4.6b). The result is (Blake, 1971)

$$G_{ij}^{IM}(\mathbf{r}^{IM}) = -J_{ij}(\mathbf{r}^{IM}) + 2dR_{jl}n_k G_{ikl}^{FD}(\mathbf{r}^{IM}) - d^2 R_{jk} G_{ik}^{SD}(\mathbf{r}^{IM}), \tag{4.95}$$

where G_{ij}^{FD} and G_{ij}^{SD} are defined by (4.90b) and (4.92), respectively.

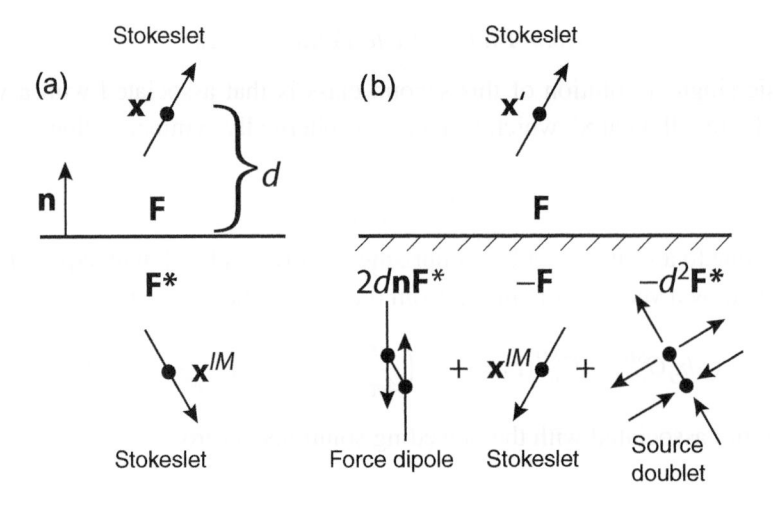

Figure 4.6 Singular solutions required to describe the flow due to a point force **F** located at a point **x**′ a distance d above a plane wall. The boundary conditions on the wall (horizontal lines) are satisfied by adding one or more singular solutions at the image point $\mathbf{x}^{IM} \equiv \mathbf{x}' - 2d\mathbf{n}$, where **n** is the unit vector normal to the wall. The strength or moment of each required singular solution is indicated, and **F*** is the reflection of the vector **F** across the wall. (a) free-slip wall; (b) rigid wall. See § 4.6.3 for discussion. Figure redrawn from figure 3 of Whittaker and Lister (2006a) by permission of Cambridge University Press.

Expressions analogous to those above also apply to 2-D flow in the presence of a boundary. As an example, consider the case of a free-slip wall $x_2 = 0$ bounding the fluid half-space $x_2 < 0$. Using (4.93) and (4.94), we find that the velocity field $u_i = F_j G_{ij}^B / \eta$ is

$$u_i = F_1 \left[J_{i1}(\mathbf{x} - \mathbf{x}') + J_{i1}(\mathbf{x} - \mathbf{x}^{IM}) \right] / \eta$$
$$+ F_2 \left[J_{i2}(\mathbf{x} - \mathbf{x}') - J_{i2}(\mathbf{x} - \mathbf{x}^{IM}) \right] / \eta \qquad (4.96)$$

where $\mathbf{x}^{IM} = \mathbf{x}' + 2d\mathbf{e}_2$ and J_{ij} is given by (4.89a). Rewriting (4.96) in terms of polar coordinates (R, ϕ) and taking the far-field limit $R/d \gg 1$, we obtain

$$\mathbf{u} = \frac{F_1}{2\pi\eta} \left(\cos\phi \, \mathbf{e}_R - \ln\hat{R} \, \mathbf{e}_1 \right) + \frac{F_2}{2\pi\eta} \frac{\cos 2\phi}{\hat{R}} \mathbf{e}_R \qquad (4.97)$$

where $\hat{R} = R/d$ and $\mathbf{e}_R$ is the radial unit vector. Now if the direction of the line force **F** is perpendicular to the wall ($F_1 = 0$), the velocity (4.97) vanishes in the limit $\hat{R} \to \infty$. The problem is therefore well posed, and Stokes's paradox is resolved. If, however, **F** has a nonzero wall-parallel component $F_1 \neq 0$, the velocity diverges logarithmically as $\hat{R} \to \infty$.

4.6 Singular Solutions

The difference between the two cases just discussed illustrates a general principle: solutions of the equations governing 2-D Stokes flow in infinite and semi-infinite domains exist only if the net force on the fluid is zero. This criterion is satisfied for, e.g., a rigid circular cylinder moving perpendicular to a free-slip wall because the integrated normal stress over the wall turns out to be exactly equal and opposite to the drag on the cylinder (Exercise 4.8). There is no corresponding solution for a cylinder moving parallel to a free-slip wall because the shear stresses on the wall that would be required to balance the drag are zero by definition. By contrast, a rigid wall can support both normal and shear stresses. The 2-D Stokes equations therefore have solutions for cylinders moving either perpendicular to or parallel to a rigid wall (Wakiya, 1975).

4.6.4 The Boundary-Integral Representation

The boundary-integral representation for Stokes flow expresses the velocity u_i at any point in a fluid volume V bounded by a surface S in terms of the velocity and traction on S. The starting point is the integral form (4.12) of the Lorentz reciprocal theorem. Let (u_i, σ_{ij}) be the flow of interest in a fluid with no distributed body forces ($f_i = 0$), and let $u_i^* \equiv J_{ij}(\mathbf{x} - \mathbf{x}')F_j/\eta$ and $\sigma_{ik}^* \equiv K_{ijk}(\mathbf{x} - \mathbf{x}')F_j$ be the flow produced by a point force $f_i^* \equiv F_i \delta(\mathbf{x} - \mathbf{x}')$ at the point $\mathbf{x}'$. Substituting these expressions into (4.12) and dropping the arbitrary vector F_i, we obtain

$$\frac{1}{\eta} \int_S J_{ij}(\mathbf{x} - \mathbf{x}')\sigma_{ik}(\mathbf{x})n_k(\mathbf{x}) dS(\mathbf{x}) - \int_V u_j(\mathbf{x})\delta(\mathbf{x} - \mathbf{x}') dV(\mathbf{x})$$
$$= \int_S K_{ijk}(\mathbf{x} - \mathbf{x}')u_i(\mathbf{x})n_k(\mathbf{x}) dS(\mathbf{x}), \tag{4.98}$$

where the normal $\mathbf{n}$ points out of V. Now

$$\int_V u_j(\mathbf{x})\delta(\mathbf{x} - \mathbf{x}') dV = \chi(\mathbf{x}')u_j(\mathbf{x}'), \tag{4.99}$$

where $\chi(\mathbf{x}') = 0$, $1/2$ or 1 depending on whether $\mathbf{x}'$ lies outside V, right on S, or inside V, respectively. To understand the meaning of $\chi(\mathbf{x}')$, it is helpful to regard the Dirac delta function in (4.99) as the limit $\epsilon \to 0$ of a Gaussian distribution with radius ϵ centered on $\mathbf{x}'$. The function $\chi(\mathbf{x}')$ is then the fraction of the (unit) volume of the Gaussian that remains inside V in the limit $\epsilon \to 0$ (Figure 4.7). If for example $\mathbf{x}'$ lies outside V, then the whole Gaussian lies outside V in the limit $\epsilon \to 0$, and so $\chi = 0$. By similar reasoning, $\chi = 1$ if $\mathbf{x}'$ lies inside V. If however $\mathbf{x}'$ lies right on S, then exactly half of the Gaussian remains inside V in the limit $\epsilon \to 0$.

Substituting (4.99) into (4.98) and interchanging the roles of $\mathbf{x}$ and $\mathbf{x}'$, we obtain the boundary-integral representation

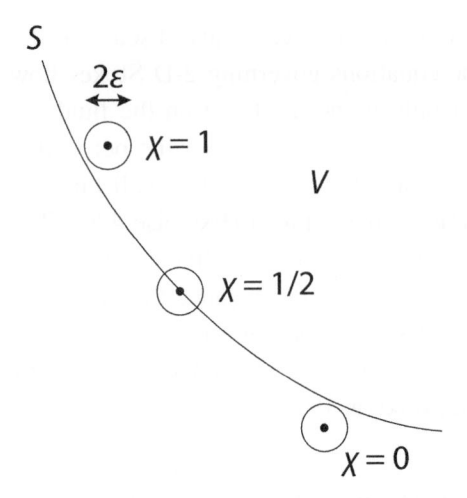

Figure 4.7 Meaning of the function $\chi(\mathbf{x}')$ defined by (4.99). χ is the fraction of the volume of a Gaussian distribution of radius ϵ that remains inside a volume V bounded by a surface S in the limit $\epsilon \to 0$. The values of χ are indicated for $\mathbf{x}'$ inside V (top), right on S (middle) and outside V (bottom).

$$\frac{1}{\eta} \int_{S'} J_{ij}(\mathbf{x}' - \mathbf{x}) \sigma_{ik}(\mathbf{x}') n_k(\mathbf{x}') dS(\mathbf{x}')$$

$$- \int_{S'} K_{ijk}(\mathbf{x}' - \mathbf{x}) u_i(\mathbf{x}') n_k(\mathbf{x}') dS(\mathbf{x}') = \chi(\mathbf{x}) u_j(\mathbf{x}). \quad (4.100)$$

The first integral in (4.100) represents the velocity due to a surface distribution of point forces with density $\sigma_{ik} n_k$. It is called the single-layer potential by analogy to the electrostatic potential generated by a surface distribution of electric charges. The second integral, called the double-layer potential, represents the velocity field generated by the sum of a distribution of sources and sinks and a symmetric distribution of force dipoles (Kim and Karrila, 1991, § 2.4.2).

An important extension of the integral representation (4.100) is to the buoyancy-driven motion of a fluid drop with viscosity $\eta_2 \equiv \gamma \eta_1$ and density $\rho_2 \equiv \rho_1 + \Delta \rho$ in another fluid with viscosity η_1 and density ρ_1 (Pozrikidis, 1990; Manga and Stone, 1993). Let S, V_1 and V_2 be the surface of the drop and the volumes outside and inside it, respectively. We begin by writing separate integral equations of the form (4.100) for each fluid:

$$-\frac{1}{\eta_1} \int_{S'} J_{ij} \sigma_{ik}^{(1)} n_k dS' + \int_{S'} K_{ijk} u_i^{(1)} n_k dS' = \chi_1(\mathbf{x}) u_j^{(1)}(\mathbf{x}), \quad (4.101a)$$

$$\frac{1}{\eta_2} \int_{S'} J_{ij} \sigma_{ik}^{(2)} n_k dS' - \int_{S'} K_{ijk} u_i^{(2)} n_k dS' = \chi_2(\mathbf{x}) u_j^{(2)}(\mathbf{x}), \quad (4.101b)$$

where **n** points out of the drop and the volume (V_1 or V_2) in which a given quantity is defined as indicated by a superscript in parentheses. The arguments of the quantities under the integral signs have been suppressed for brevity, and $dS' = dS(\mathbf{x}')$. Moreover, $\chi_1(\mathbf{x}) = 0$, $1/2$, or 1 if $\mathbf{x}$ is in V_2, right on S, or in V_1, respectively, and $\chi_2(\mathbf{x})$ is defined similarly but with the subscripts 1 and 2 interchanged. Now multiply (4.101b) by γ, add the result to (4.101a) and apply the matching conditions $u_j^{(1)} = u_j^{(2)} = u_j$ and $(\sigma_{ik}^{(1)} - \sigma_{ik}^{(2)})n_k = n_i \Delta\rho g_k x_k'$ on S, where g_k is the gravitational acceleration. The latter condition states that the jump in the nonhydrostatic normal stress across the interface is equal to the difference of the hydrostatic pressures in the two fluids. The result is

$$\chi_1(\mathbf{x})u_j^{(1)}(\mathbf{x}) + \gamma \chi_2(\mathbf{x})u_j^{(2)}(\mathbf{x}) - (1-\gamma)\int_{S'} K_{ijk} u_i n_k dS'$$
$$= -\frac{g_k \Delta\rho}{\eta_1} \int_{S'} J_{ij} n_i x_k' dS'. \qquad (4.102)$$

For points $\mathbf{x}$ on S, (4.102) reduces to

$$\frac{1}{2}u_j(\mathbf{x}) - \frac{1-\gamma}{1+\gamma}\int_{S'} K_{ijk} u_i n_k dS' = -\frac{g_k \Delta\rho}{\eta_1(1+\gamma)}\int_{S'} J_{ij} n_i x_k' dS' \quad (\mathbf{x} \in S), \quad (4.103)$$

which is a Fredholm integral equation of the second kind for the velocity **u** of the interface. Once **u** on S has been determined by solving (4.103), **u** at points in V_1 and V_2 can be determined if desired from (4.102). A generalization of (4.103) to $N \geq 2$ interacting drops was derived by Manga and Stone (1993).

The integral equation (4.103), which must in general be solved numerically, has been used in geodynamics to model systems comprising distinct fluids with different viscosities. Manga and Stone (1993) solved the N-drop generalization of (4.103) using the boundary-element method (BEM; Pozrikidis, 1992) to investigate the buoyancy-driven interaction between two drops in an infinite fluid with a different viscosity. Manga et al. (1993) used a similar method to study the interaction of plume heads with compositional discontinuities in Earth's mantle. The deformation of viscous blobs in a 2-D cellular flow was investigated by Manga (1996), who concluded that geochemical reservoirs can persist undisturbed for relatively long times if they are 10–100 times more viscous than the surrounding mantle. Manga (1997) showed that the deformation-induced mutual interactions of deformable diapirs in a rising cloud causes the diapirs progressively to cluster, but that the rate of this clustering is probably too slow to affect significantly the lateral spacing of rising diapirs in the mantle. Leahy and Bercovici (2010) used the BEM to study the gravitational spreading of a lens of dense low-viscosity melt along the 410 km discontinuity.

A final important application of the BEM is to free (buoyancy-driven) subduction of sheets of viscous fluid immersed in another less dense and less viscous fluid.

The equation to be solved is still (4.103), but the interface S is now the surface (both sides plus the edges) of a sheet whose lateral dimensions greatly exceed its thickness. Morra et al. (2007), Morra et al. (2009) and Morra et al. (2012) developed a multipole-accelerated BEM method for modeling the subduction of several interacting viscous sheets in a spherical annulus beneath a free surface. Morra and Regenauer-Lieb (2006) proposed a hybrid method in which a finite-element solution inside the sheet is coupled to a BEM solution for the outer flow, and Morra et al. (2006) subsequently used it to investigate the causes of the curvature of trenches and island arcs. Butterworth et al. (2012) used the BEM to study the interaction of a subducting plate with an overriding plate and concluded that models without an overriding plate significantly overestimate trench retreat.

Ribe (2010) and Li and Ribe (2012) used the 2-D and 3-D BEM, respectively, to determine scaling laws for the sinking speed and style of free subduction as functions of the sheet/mantle viscosity ratio γ, the initial slab geometry, and (in the 3-D case) the ratio of the mantle depth to the sheet thickness. Figure 4.8a and b shows the temporal evolution of 2-D subducting sheets for two different viscosity contrasts $\gamma = 100$, and 1000, starting from the same initial condition. The subsequent three images are shown at times corresponding to three fixed depths of penetration that are the same for both parts. For $\gamma = 100$, return flow from left to right around the leading edge of the slab prevents it from becoming vertical, whereas the slab with $\gamma = 1000$ bends over backwards to dips exceeding 90°. Figure 4.8c shows three configurations of subducted sheets in 3-D, obtained from BEM solutions in a layer of finite depth bounded below by a rigid surface. For $\gamma = 100$ the slab lies down on the bottom boundary, and the trench moves from right to left ('trench retreating' mode). For $\gamma = 600$, the slab folds after it reaches the bottom boundary. Finally, for $\gamma = 3000$, the slab bends over backwards before reaching the bottom of the layer, and the trench moves from left to right ('trench advancing' mode).

The results shown in Figure 4.8 will be discussed further in § 8 on thin-sheet theory, which provides a powerful interpretive tool for flow in thin viscous sheets.

4.7 The Singularity Method

In the previous section, we saw how the boundary-integral representation of Stokes flow allows the flow in a volume to be expressed in terms of weighted integrals of the velocity and stress over the bounding surface of the volume. Those integrals are of convolution form and involve singular kernels or Green functions for the velocity and stress produced by a point force. Because the singular points of the kernels lie right on the boundary, special care must be taken when evaluating the integrals numerically in the context of the BEM.

4.7 The Singularity Method

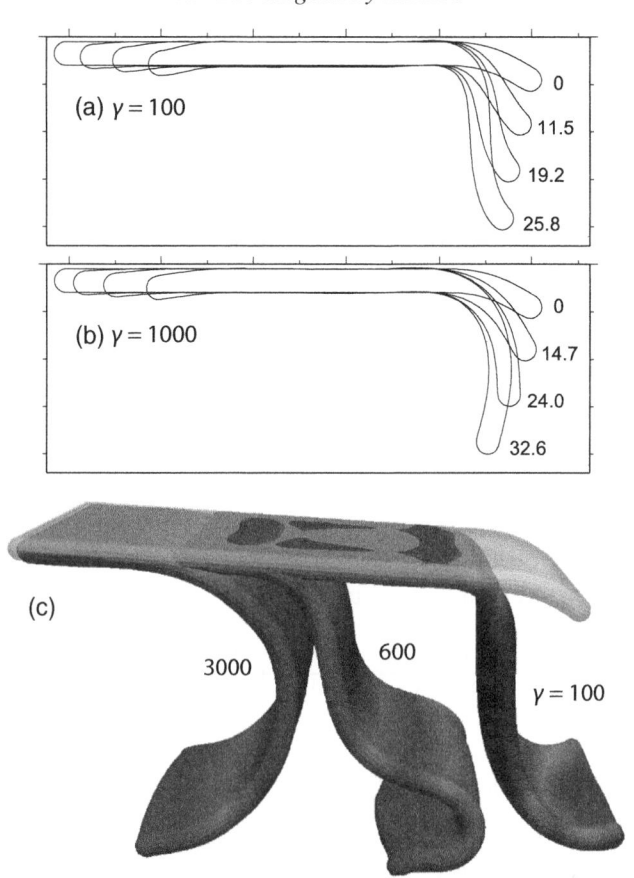

Figure 4.8 Free subduction of viscous sheets. (a), (b): Evolution of 2-D subducting sheets in a semi-infinite ambient fluid as a function of time for two different viscosity ratios γ. Numbers adjoining the different slab images are dimensionless times $\tau = thg\Delta\rho/\eta_1$, where η_1 is the viscosity of the ambient fluid. The weak subduction of the left end of each sheet is a natural consequence of the thin lubrication layer above the sheet (Figure 1.2c). (c) Composite view of three 3-D BEM simulations for subduction in a fluid layer bounded below by a rigid surface. The initial condition for all three simulations is shown in light grey, and the viscosity ratios for each case are indicated. The width of the sheet is $12h$, where h is the sheet's initial thickness, and the layer depth is $13.8h$. See § 4.6.4 for discussion. Parts (a) and (b) reproduced from Figure 2 of Ribe (2010). Part (c) courtesy of Z. Li.

One way to avoid numerically troublesome singularities on the boundaries of the flow domain is the so-called singularity method, wherein a flow of interest is expressed as a sum of singular solutions whose poles lie outside the domain of the flow. As a simple illustration, let us consider anew the problem of Stokes flow around a solid sphere, the solution of which has already been given in § 4.3.2. We work in the reference frame of the sphere, so that the flow at infinite distance from the sphere is a uniform stream U_i. Now the fluid at great distances from the sphere sees the latter as a point force F_i whose magnitude is just minus the drag exerted by the fluid on the sphere. Therefore we expect the flow far from the sphere to be the sum of (at least) a uniform stream U_i and a flow generated by a point force F_i of unknown strength:

$$u_i \approx U_i + \frac{F_j}{8\pi\eta}\left(\frac{\delta_{ij}}{r} + \frac{r_i r_j}{r^3}\right). \tag{4.104}$$

However, the expression (4.104) does not satisfy the no-slip boundary condition $u_i = 0$ on the sphere's surface $r = a$. We therefore seek an additional singular solution that we can add to (4.104) in order to satisfy the boundary condition. Examining the various singular solutions derived in the previous section, we remark that the source doublet (4.92) contains terms $\sim \delta_{ij}$ and $\sim r_i r_j$ that can in principle cancel the similar terms in (4.104) at $r = a$. We therefore add to (4.104) a source doublet of unknown strength G_i to obtain

$$u_i = U_i + \frac{F_j}{8\pi\eta}\left(\frac{\delta_{ij}}{r} + \frac{r_i r_j}{r^3}\right) + \frac{G_j}{4\pi}\left(\frac{3 r_i r_j}{r^5} - \frac{\delta_{ij}}{r^3}\right). \tag{4.105}$$

Cancelling the terms $\sim r_i r_j$ at $r = a$, we find $\mathbf{G} = -a^2 \mathbf{F}/6\eta$. Using this result and evaluating the remaining terms in (4.105) at $r = a$, we obtain

$$0 = u_i(r=a) = \left(U_j + \frac{F_j}{6\pi\eta a}\right)\delta_{ij}, \tag{4.106}$$

which implies Stokes's drag law $\mathbf{F} = -6\pi\eta a \mathbf{U}$.

Figure 4.9 shows the Stokes streamfunctions for the uniform stream (upper left), the Stokeslet (upper right) and the source doublet (lower left), as well as the total flow obtained by summing them (lower right). The total flow field is singular at $r = 0$, as indicated by the contours inside the circle representing the sphere. However, this does not matter because $r = 0$ is outside the domain of the flow.

4.8 Slender-Body Theory

Slender-body theory (SBT) is concerned with Stokes flow around thin rod-like bodies whose length greatly exceeds their other two dimensions. The approach takes its departure from Stokes's paradox: the nonexistence of a solution of the equations for Stokes flow around an infinitely long circular cylinder moving steadily in an unbounded fluid, due to a logarithmic singularity that makes it impossible to satisfy all the boundary conditions (Batchelor, 1967). However, the problem can be regularized in one of three ways: by including inertia, by making the domain bounded

4.8 Slender-Body Theory

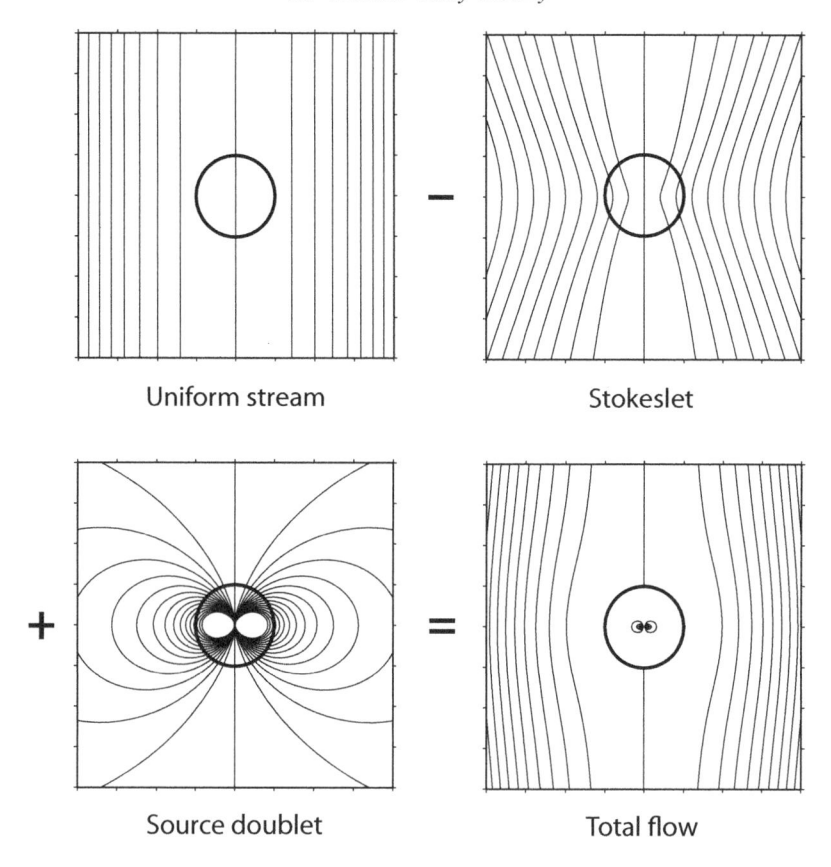

Figure 4.9 The singularity method applied to Stokes flow around a rigid sphere. The complete flow is the sum of a uniform stream, a Stokeslet and a source doublet, for each of which contours of the Stokes streamfunction are shown. The contour intervals are 1.0 (uniform stream), 0.3 (Stokeslet), 0.02 (source doublet) and 0.6 (total flow).

(§ 4.6.3) or by making the length of the cylinder finite. SBT is concerned with the last of these possibilities.

The canonical problem of SBT is to determine the force $\mathbf{F}$ on a rod of length $2L$ and radius $R \ll L$ moving with uniform velocity $\mathbf{U}$ in a viscous fluid (Figure 4.10a). The most rigorous way to solve the problem (Keller and Rubinow, 1976) is to use the method of matched asymptotic expansions (MMAE), which exploits the fact that the flow field comprises two distinct regions characterized by very different length scales. The first or inner region includes points whose radial distance ρ from the rod is small compared to their distance from the rod's nearer end. In this region, the fluid is not affected by the ends of the rod and sees it as an infinite cylinder with radius R. The second, outer region $\rho \gg R$ is at distances from the rod that are

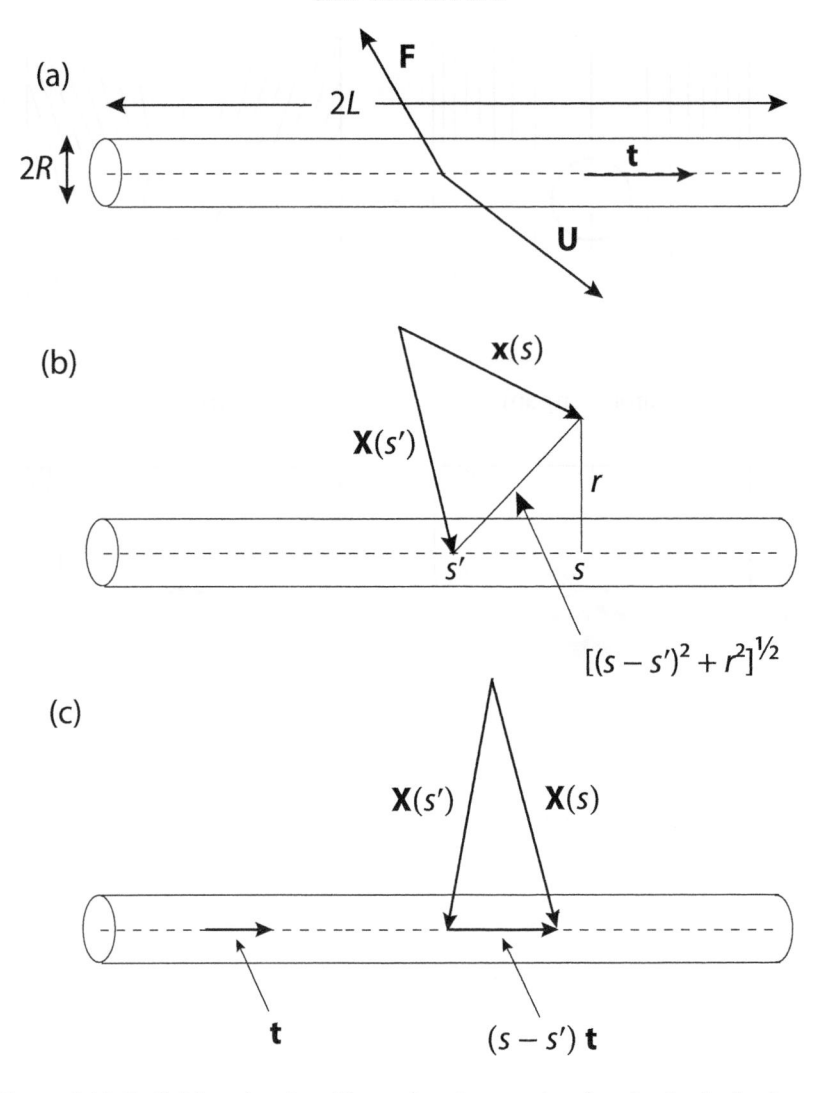

Figure 4.10 Definition sketches illustrating the notation for slender-body theory, discussed in § 4.8.

large compared to its radius. Here, the fluid is unaffected by the rod's finite radius and sees it as a line distribution of point forces with effectively zero thickness. The basic idea of the MMAE is to obtain two different asymptotic expansions for the velocity field that are valid in the inner and outer regions, respectively, and then to match them together in an intermediate or overlap region where both expansions must coincide.

As the details of the matching are rather complicated, my discussion of the MMAE is deferred to § 6.3, where it will be applied in the context of

4.8 Slender-Body Theory

boundary-layer theory. Here I follow instead a simpler approach due to Batchelor (1970), in which the slender rod is modelled as a line distribution of Stokeslets. We shall work in the reference frame of the rod, so that the undisturbed velocity there is $-\mathbf{U}$. The disturbance flow at a point $\mathbf{x}$ due to the rod is

$$\mathbf{v}(\mathbf{x}) = \int_{-L}^{L} \mathbf{f}(s') \cdot \mathbf{J}(\mathbf{x} - \mathbf{X}(s'))ds', \tag{4.107}$$

where s' is the arclength along the rod's axis, $\mathbf{f}(s')$ is the local Stokeslet density (force per unit length) and $\mathbf{X}(s')$ are the Cartesian coordinates of the point s'. Writing out the Green function $\mathbf{J}$ in (4.107) explicitly, we have

$$\mathbf{v}(\mathbf{x}) = \frac{1}{8\pi\eta} \int_{-L}^{L} \mathbf{f}(s') \cdot \left[\frac{\mathbf{I}}{\{(s-s')^2 + r^2\}^{1/2}} + \frac{(\mathbf{x} - \mathbf{X}(s'))(\mathbf{x} - \mathbf{X}(s'))}{\{(s-s')^2 + r^2\}^{3/2}} \right] ds' \tag{4.108}$$

where r is the radial distance from the rod's axis (Figure 4.10b). Now the no-slip boundary condition on the rod's surface requires $\mathbf{v}(\mathbf{x}) = \mathbf{U}$ at $r = R$, or

$$\mathbf{U} = \frac{1}{8\pi\eta} \int_{-L}^{L} \mathbf{f}(s') \cdot \left[\frac{\mathbf{I}}{\{(s-s')^2 + R^2\}^{1/2}} + \frac{(\mathbf{X}(s) - \mathbf{X}(s'))(\mathbf{X}(s) - \mathbf{X}(s'))}{\{(s-s')^2 + R^2\}^{3/2}} \right] ds', \tag{4.109}$$

where the position $\mathbf{x}$ in the second term of the integrand has been replaced by $\mathbf{X}(s)$ with a small error. Now

$$\mathbf{X}(s) - \mathbf{X}(s') = (s - s')\mathbf{t}, \tag{4.110}$$

where $\mathbf{t}$ is the unit tangent vector to the axis (Figure 4.10c). Furthermore, we expect $\mathbf{f}(s')$ to be a slowly varying function of s'. The largest contribution to the integral (4.109) will therefore come from points $s \approx s'$ near the singularity. Accordingly, we can set $\mathbf{f}(s') \approx \mathbf{f}(s)$ and take the latter outside the integral. Equation (4.109) now becomes

$$\mathbf{U} = \frac{\mathbf{f}}{8\pi\eta} \cdot (I_0 \mathbf{I} + I_2 \mathbf{t}\mathbf{t}), \quad I_n = \int_{-L}^{L} \frac{(s-s')^n}{\{(s-s')^2 + R^2\}^{(n+1)/2}} ds'. \tag{4.111a,b}$$

Now in the limit $R/L \ll 1$, $I_0 = 2\ln(2L/R) + O(1)$ and $I_2 \approx I_0 - 2$. Moreover, by Newton's third law the total force exerted on the rod by the fluid is $\mathbf{F} = -2L\mathbf{f}$. Retaining only the dominant term $2\ln(2L/R)$ in the integrals I_0 and I_2, we find that (4.111a) becomes

$$\mathbf{U} = -\frac{1}{8\pi\eta\epsilon L}[\mathbf{F} + (\mathbf{F} \cdot \mathbf{t})\mathbf{t}], \quad \epsilon = \left(\ln\frac{2L}{R}\right)^{-1}. \tag{4.112}$$

Inverting (4.112) for **F** (Exercise 4.11), we find

$$\mathbf{F} = -4\pi\eta\epsilon L[2\mathbf{U} - (\mathbf{U}\cdot\mathbf{t})\mathbf{t}]. \qquad (4.113)$$

The expression (4.113) illustrates the fundamental fact that the force on a body in Stokes flow is controlled by its largest dimension (L in this case). The force **F** comprises both drag (parallel to **U**) and lift (perpendicular to **U**) components in general. This means that a rod falling obliquely under gravity will have a horizontal component to its motion. The lift vanishes only when **U** is perpendicular to **t** (transverse motion) or parallel to **t** (longitudinal motion). The drags for these two cases are

$$\mathbf{F}_{\text{transverse}} = -8\pi\eta\epsilon L\mathbf{U}, \quad \mathbf{F}_{\text{longitudinal}} = -4\pi\eta\epsilon L\mathbf{U}. \qquad (4.114\text{a,b})$$

For a given velocity, the drag on a rod moving transversely is exactly twice the drag on one moving longitudinally. A corollary is that a rod falling under gravity moves twice as fast in longitudinal motion as in transverse motion. Further details and extensions of SBT can be found in Batchelor (1970), Cox (1970), Keller and Rubinow (1976) and Johnson (1980).

Geodynamically relevant applications of SBT include Olson and Singer (1985), who used (4.113) to predict the rise velocity of buoyant quasi-cylindrical ('diapiric') plumes. Koch and Koch (1995) used an expression analogous to (4.113) for an expanding ring to model the buoyant spreading of a viscous drop beneath the free surface of a much more viscous fluid, with application to volcanic features on Venus. Whittaker and Lister (2006a) presented a model for a creeping plume above a planar boundary from a point source of buoyancy, in which they modelled the flow outside the plume as that due to a line distribution of Stokeslets (see § 6.4 for a detailed discussion.) Whittaker and Lister (2008a) used SBT to determine the structure of the self-similar wake trailing a rising buoyant thermal in Stokes flow, a problem that was discussed earlier in § 3.2. Whittaker and Lister (2008b) used SBT to calculate the trajectories of plumes rising and widening diffusively in a background shear flow and found reasonably good agreement with laboratory experiments.

4.9 Flow Driven by Internal Loads

In Earth's mantle, flow is driven by gravity acting on the lateral density variations that arise from anomalies in temperature and/or chemical composition. Because inertia is negligible in the mantle, the flow at each instant is determined entirely by the distribution of density anomalies ('internal loads') at that instant. This principle is the basis of a class of Stokes flow models in which an instantaneous 3-D mantle

4.9 Flow Driven by Internal Loads

flow field is calculated for a distribution of internal loads inferred from seismic tomography and/or the distribution of subducted slabs.

The aim of most internal loading models is to infer the radial profile of viscosity in the mantle, which strongly controls the flow field produced by a given distribution of internal loads. Now an internal loading model delivers not only a flow field, but also a variety of observable quantities such as the dynamically supported surface topography and the geoid anomaly over Earth's surface. Accordingly, one can estimate the viscosity profile by adjusting it until the best fit between model predictions and observations is obtained.

The principle of this approach is explained schematically in Figure 4.11 for an internal loading distribution comprising a single dense spherical 'sinker' located at mid-mantle depth. The stresses associated with the flow induced by the sinker act on Earth's surface and on the CMB, pulling both of these surfaces downward against the upward gravitational restoring force. The total geoid anomaly at the surface is therefore the sum of three contributions: a positive one ($\Delta G_0 > 0$) due to the sinker itself, and two negative ones ($\Delta G_1 < 0$, $\Delta G_2 < 0$) due to the boundary deflections. However, the gravitational effects of these three mass anomalies are determined not only by their relative sizes, but also by their distances from the equipotential surface that defines the geoid. Thus for a given mass anomaly, the geoid contribution of the surface deflection is the largest, that of the CMB deflection is the smallest, and that of the sinker is intermediate. If the mantle viscosity η_0 is uniform (Figure 4.11a), the mass anomalies of the two boundary deflections are comparable to each other and (to within a sign) to that of the sinker. The geoid anomaly is then dominated by the contribution ΔG_1 of the surface deflection and is negative. If, however, the lower mantle has a significantly higher viscosity than the upper mantle (Figure 4.11b), then most of the sinker's weight is supported by the CMB deflection rather than by the surface deflection, which is correspondingly reduced in size. The negative geoid anomaly ΔG_1 of the surface deflection can then be counteracted by the positive anomaly ΔG_0 of the sinker, even though the latter is further away from the equipotential surface. The total geoid anomaly is therefore positive if η_L/η_U is large enough. In summary, there is a direct link between the mantle viscosity profile and the degree of correlation or anticorrelation of the geoid with the distribution of internal loads.

To formalize the foregoing qualitative discussion, a natural way to proceed is to replace the finite sinker of Figure 4.11 by a suitably defined 'unit' load. The resulting flow field (described by a poloidal scalar) and geoid anomaly will then be Green functions that can be convolved with a 3-D load distribution. However, it proves convenient to define the unit load not as a point force, but rather as a surface force concentrated at a single radius and whose amplitude is proportional to a spherical harmonic $Y_l^m(\theta, \phi)$ of degree l and order m. Because the Stokes equations

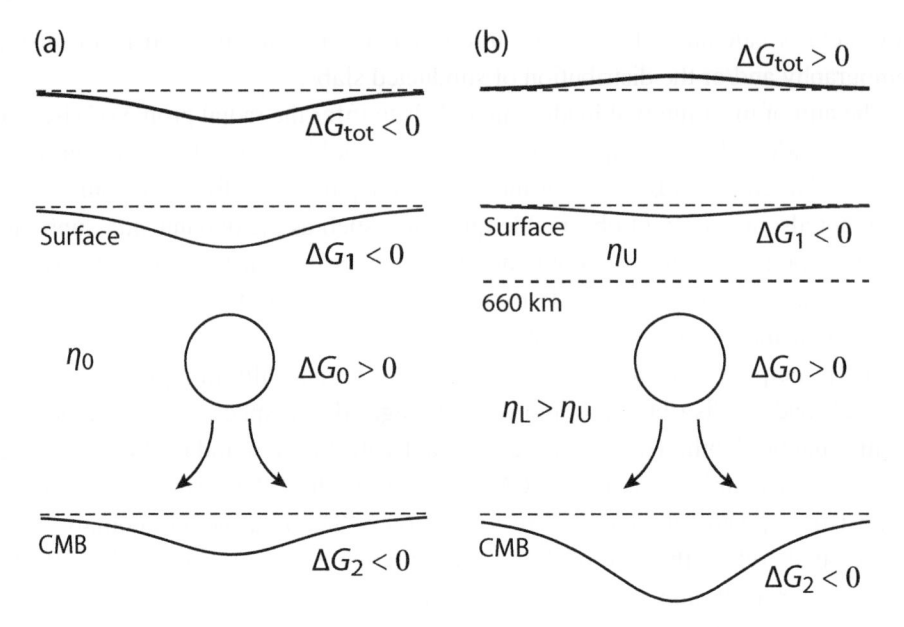

Figure 4.11 Boundary deflections (bottom) and geoid anomalies (top) generated by a dense spherical 'sinker' in the mid-mantle (circle). The thin black lines show the boundary deflections of the surface and the CMB, and the adjoining dashed lines indicate their undisturbed positions. The heavy black lines (top) show the total geoid anomaly ΔG_{tot}. The quantities ΔG_0, ΔG_1 and ΔG_2 are the contributions to the geoid anomaly of the sinker, the surface deflection, and the CMB deflection, respectively. (a) A mantle with a uniform viscosity η_0. (b) A two-layer mantle with viscosities η_U and $\eta_L > \eta_U$. Redrawn from Figures 1 and 2 of Hager (1984) by permission of John Wiley and Sons.

are separable in spherical coordinates, the Green function for the poloidal scalar satisfies an ordinary differential equation (ODE) in the radial coordinate r that can be solved analytically. This approach was adumbrated by Pekeris (1935), Runcorn (1967) and McKenzie (1977); formalized by Parsons and Daly (1983) for a plane layer with constant viscosity; and extended by Richards and Hager (1984), Ricard et al. (1984) and Forte and Peltier (1987) to spherical geometry with self-gravitation and radially variable viscosity. To illustrate the method, I sketch next the derivation by Forte and Peltier (1987) of the Green functions for the poloidal scalar and the surface gravitational potential for a self-gravitating mantle with uniform viscosity.

4.9.1 Wave-Domain Green Functions

Our starting point is the equations (4.1) with $n = 1$ that govern slow viscous flow in an incompressible, self-gravitating, Newtonian mantle. Let

$$\rho = \rho_0 + \hat{\rho}(r,\theta,\phi), \quad \sigma = \sigma_0(r) + \hat{\sigma}(r,\theta,\phi), \quad \chi = \chi_0(r) + \hat{\chi}(r,\theta,\phi), \quad (4.115)$$

4.9 Flow Driven by Internal Loads

where hatted quantities are perturbations of the field variables about a reference hydrostatic state denoted by a subscript 0. Substituting (4.115) into (4.1b) and (4.1d) and neglecting products of perturbation quantities, we obtain

$$0 = \nabla \cdot \hat{\sigma} + \mathbf{g}_0 \hat{\rho} - \rho_0 \nabla \hat{\chi}, \tag{4.116a}$$

$$\nabla^2 \hat{\chi} = 4\pi G \hat{\rho}, \tag{4.116b}$$

where $\mathbf{g}_0 = -\nabla \chi_0$. The second term on the RHS of (4.116a) represents the buoyancy force acting on the internal density anomalies, and the third term represents the additional force associated with the perturbations in the gravitational potential that they induce (self-gravitation). For consistency with other chapters of this book, the signs of all gravitational potentials (χ_0, $\hat{\chi}$, etc.) referred to later are opposite to those of Forte and Peltier (1987).

As noted in § 4.2, the flow driven by internal density anomalies in a fluid with constant or depth-dependent viscosity is purely poloidal. For the constant-viscosity case we are now considering, the poloidal scalar $\mathcal{P}(r, \theta, \phi)$ satisfies the first of eqns. (4.26), which remains valid in a self-gravitating mantle. To simplify the notation, the poloidal scalar and Green function are written without the superposed hat symbol even though they are perturbations relative to the background motionless hydrostatic state. We now substitute into the first equation of (4.26) the expansions

$$\mathcal{P} = \sum_{l=0}^{\infty} \sum_{m=-l}^{l} \mathcal{P}_l^m(r) Y_l^m(\theta, \phi), \quad \hat{\rho} = \sum_{l=0}^{\infty} \sum_{m=-l}^{l} \hat{\rho}_l^m(r) Y_l^m(\theta, \phi) \tag{4.117a,b}$$

where $Y_l^m(\theta, \phi)$ are surface spherical harmonics satisfying $\mathcal{B}^2 Y_l^m = -l(l+1) Y_l^m$ and $\mathcal{B}^2$ is defined by (4.25). We thereby find that $\mathcal{P}_l^m(r)$ satisfies

$$\mathcal{D}_l^4 \mathcal{P}_l^m(r) = \frac{g_0 \hat{\rho}_l^m}{\eta r}, \quad \mathcal{D}_l^2 = \frac{d^2}{dr^2} + \frac{2}{r} \frac{d}{dr} - \frac{l(l+1)}{r^2}, \tag{4.118}$$

where the gravitational acceleration g_0 has been assumed constant (Forte and Peltier, 1987). Now define a poloidal Green function $P_l(r, r')$ that satisfies

$$\mathcal{D}_l^4 P_l(r, r') = \delta(r - r'). \tag{4.119}$$

$P_l(r, r')$ represents the poloidal flow generated at the radius r by an infinitely thin density contrast of spherical harmonic degree l and unit amplitude located at a radius r'. The poloidal flow due to a distributed density anomaly $\hat{\rho}_l^m(r)$ is then obtained by convolving $\hat{\rho}_l^m(r)$ with the Green function, yielding

$$\mathcal{P}_l^m(r) = \frac{g_0}{\eta} \int_{a_2}^{a_1} \frac{\hat{\rho}_l^m(r')}{r'} P_l(r, r') dr', \tag{4.120}$$

where a_2 and a_1 are the inner and outer radii of the mantle, respectively.

At all radii $r \neq r'$, the Green function $P_l(r, r')$ satisfies the homogeneous form of (4.119), which has the general solution

$$P_l(r, r') = A_n r^l + B_n r^{-l-1} + C_n r^{l+2} + D_n r^{-l+1}, \tag{4.121}$$

where $A_n - D_n$ are undetermined constants with $n = 1$ for $r' < r \leq a_1$ and $n = 2$ for $a_2 \leq r < r'$. These eight constants are determined by the boundary conditions at $r = a_1$ and $r = a_2$ and by matching conditions at $r = r'$. The vanishing of the radial velocity at $r = a_1$ and $r = a_2$ requires

$$P_l(a_1, r') = P_l(a_2, r') = 0, \tag{4.122}$$

and the vanishing of the shear stress requires

$$\frac{d^2 P_l}{dr^2}(a_1, r') = \frac{d^2 P_l}{dr^2}(a_2, r') = 0. \tag{4.123}$$

Turning now to the matching conditions, we define $[A] \equiv A(r'_+) - A(r'_-)$ to be the jump in the quantity A across the radius $r = r'$, where r'_+ is just above $r = r'$ and r'_- is just below it. Continuity of the normal and tangential velocities and the shear stress requires

$$[P_l] = \left[\frac{dP_l}{dr}\right] = \left[\frac{d^2 P_l}{dr^2}\right] = 0. \tag{4.124}$$

However, the third derivative of P_l (related to the normal stress) is discontinuous at $r = r'$. By integrating (4.119) from r'_- to r'_+ and applying (4.124), we find

$$\left[\frac{d^3 P_l}{dr^3}\right] = 1. \tag{4.125}$$

Substitution of (4.121) into (4.122)–(4.125) yields eight equations for $A_n - D_n$, the solutions of which are

$$C_n = -\frac{B_n}{a_n^{2l+3}} = \frac{1}{2(2l+3)(2l+1)(r')^{l-1}} \frac{(a_1/a_n)^{2l+3} - (r'/a_2)^{2l+3}}{1 - (a_1/a_2)^{2l+3}} \tag{4.126a}$$

$$D_n = -a_n^{2l-1} A_n = \frac{a_n^{2l-1}}{2(4l^2 - 1)(r')^{l-3}} \frac{(a_1/a_n)^{2l-1} - (r'/a_2)^{2l-1}}{1 - (a_1/a_2)^{2l-1}}. \tag{4.126b}$$

The next step is to determine the gravitational potential anomaly. Because the flow induces deformations (dynamic topography) of Earth's surface and of the core-mantle boundary (CMB), the total potential anomaly is

$$\hat{\chi}_l^m(r) = (\hat{\chi}_0)_l^m(r) + (\hat{\chi}_1)_l^m(r) + (\hat{\chi}_2)_l^m(r), \tag{4.127}$$

4.9 Flow Driven by Internal Loads

where $\hat{\chi}_0$, $\hat{\chi}_1$ and $\hat{\chi}_2$ are the potentials associated with the internal load, the deformation $\hat{a}_1$ of Earth's surface and the deformation $\hat{a}_2$ of the CMB, respectively. To determine $\hat{\chi}_0$, we note that the general solution of (4.116b) is

$$\hat{\chi}(\mathbf{x}) = -G \int_V \frac{\hat{\rho}(\mathbf{x}')}{|\mathbf{x} - \mathbf{x}'|} dV' \tag{4.128}$$

where $\mathbf{x}$ is the 3-D position vector and the integral is over the whole mantle. We now invoke the expansion (Jackson, 1975, p. 102)

$$|\mathbf{x} - \mathbf{x}'|^{-1} = 4\pi \sum_{l=0}^{\infty} \sum_{m=-l}^{l} \frac{1}{2l+1} \frac{r_<^l}{r_>^{l+1}} Y_l^m(\theta, \phi) \overline{Y}_l^m(\theta', \phi') \tag{4.129}$$

where $r_< = \min(r, r')$ and $r_> = \max(r, r')$ and an overbar denotes complex conjugation. Now substitute (4.129) and (4.117b) into (4.128) and evaluate the angular integrals using the orthogonality relation for the spherical harmonics, viz.,

$$\int_0^{2\pi} d\phi \int_0^{\pi} d\theta \sin\theta\, Y_\lambda^\mu(\theta, \phi) \overline{Y}_l^m(\theta, \phi) = \delta_{\lambda l}\delta_{\mu m}. \tag{4.130}$$

The result is

$$(\hat{\chi}_0)_l^m(r) = -\frac{4\pi G}{2l+1} \int_{a_2}^{a_1} (r')^2 \frac{r_<^l}{r_>^{l+1}} \hat{\rho}_l^m(r') dr'. \tag{4.131}$$

Expressions for $\hat{\chi}_1$ and $\hat{\chi}_2$ can be obtained from (4.131) by replacing $\hat{\rho}_l^m(r')$ by $(\rho_0 - \rho_w)(\hat{a}_1)_l^m \delta(r' - a_1)$ and $(\rho_c - \rho_0)(\hat{a}_2)_l^m \delta(r' - a_2)$, respectively, where ρ_w is the density of seawater and ρ_c is the core density. The results are

$$(\hat{\chi}_1)_l^m(r) = -\frac{4\pi G a_1}{2l+1}(\rho_0 - \rho_w)\left(\frac{r}{a_1}\right)^l (\hat{a}_1)_l^m, \tag{4.132a}$$

$$(\hat{\chi}_2)_l^m(r) = -\frac{4\pi G a_2}{2l+1}(\rho_c - \rho_0)\left(\frac{a_2}{r}\right)^{l+1} (\hat{a}_2)_l^m. \tag{4.132b}$$

The boundary deflections $\hat{a}_n$ are determined from the condition that the normal stress be continuous across the deformed boundaries $r = a_n + \hat{a}_n$. Because the boundary deflections $\hat{a}_n$ are small, this condition can be linearized about the undisturbed reference radii $r = a_n$, following the procedure outlined in our later treatment of the Rayleigh–Taylor instability (§ 10.1). Expanded in spherical harmonics, the linearized condition is

$$-\hat{p}_l^m(a_n) + 2\eta \frac{d\hat{w}_l^m}{dr}(a_n) = -g_0 \Delta\rho_n (\hat{a}_n)_l^m, \tag{4.133}$$

where $\hat{p}_l^m$ is the nonhydrostatic pressure, $\hat{w}_l^m$ is the radial velocity, $\Delta\rho_1 = \rho_0 - \rho_w$ and $\Delta\rho_2 = \rho_0 - \rho_c$. The LHS of (4.133) is the perturbation of the radial normal stress $(\hat{\sigma}_{rr})_l^m$ evaluated at the undeformed boundaries $r = a_n$. An expression for $\hat{p}_l^m$ in terms of the poloidal scalar is obtained by integrating the $\mathbf{e}_\theta$-component of the momentum equation (4.116a) and expanding the result in spherical harmonics:

$$\hat{p}_l^m = -\eta \frac{d}{dr}\left(rD_l^2 \mathcal{P}_l^m\right) - \rho_0 \hat{\chi}_l^m. \tag{4.134}$$

Substituting (4.134) and $\hat{w}_l^m = -l(l+1)\mathcal{P}_l^m/r$ into (4.133) and applying the boundary conditions (4.122), we obtain

$$(\hat{a}_n)_l^m = \Delta\rho_n^{-1}\left[X_l^m(a_n) - \frac{\rho_0}{g_0}\hat{\chi}_l^m(a_n)\right], \tag{4.135a}$$

$$X_l^m(r) = \frac{\eta}{g_0}\left[-r\frac{d^3}{dr^3} + \frac{3l(l+1)}{r}\frac{d}{dr}\right]\mathcal{P}_l^m(r). \tag{4.135b}$$

Now by substituting (4.132) into (4.127) and using (4.135a), we obtain the following expression for the total gravitational potential:

$$\hat{\chi}_l^m(r) = (\hat{\chi}_0)_l^m(r) - \frac{4\pi a_1 G}{2l+1}\left(\frac{r}{a_1}\right)^l\left[X_l^m(a_1) - \frac{\rho_0}{g_0}\hat{\chi}_l^m(a_1)\right]$$
$$+ \frac{4\pi a_2 G}{2l+1}\left(\frac{a_2}{r}\right)^{l+1}\left[X_l^m(a_2) - \frac{\rho_0}{g_0}\hat{\chi}_l^m(a_2)\right]. \tag{4.136}$$

The boundary potentials $\hat{\chi}_l^m(a_1)$ and $\hat{\chi}_l^m(a_2)$ are determined by solving the coupled equations obtained by evaluating (4.136) at $r = a_1$ and $r = a_2$ and are then eliminated from (4.136). Next, the resulting equation for $\hat{\chi}_l^m(r)$ is rewritten as a convolution integral using (4.131), (4.135b), (4.120), (4.121) and (4.126). Finally, the result is evaluated at $r = a_1$ to obtain the surface potential

$$\hat{\chi}_l^m(a_1) = -\frac{4\pi a_1 G}{2l+1}\int_{a_2}^{a_1} G_l(r')\hat{\rho}_l^m(r')dr' \tag{4.137}$$

where the Green function or kernel G_l is

$$G_l(r) = \left(1 - K_1 + K_2 K_r \beta^{2l+1}\right)^{-1}\left(\frac{r}{a_1}\right)^{l+2}\left\{1 - K_r\beta\left(\frac{a_2}{r}\right)^{2l+1}\right.$$
$$\left. - E_1\left(1 - K_r\beta^{2l+2}\right)\left(\frac{a_1}{r}\right)^{l+2} - E_2(1 - K_r\beta)\left(\frac{a_2}{r}\right)^{l+2}\right\}, \tag{4.138a}$$

$$E_n = \frac{l(l+2)}{2l+1}\left(\frac{a_n}{r}\right)^{l-2}\frac{(r/a_1)^{2l-1} - (a_2/a_n)^{2l-1}}{1 - \beta^{2l-1}}$$
$$+ \frac{l(l-1)}{2l+1}\left(\frac{a_n}{r}\right)^l\frac{(r/a_1)^{2l+3} - (a_2/a_n)^{2l+3}}{1 - \beta^{2l+3}} \tag{4.138b}$$

$$K_n = \frac{4\pi a_n \rho_0 G}{(2l+1)g_0}, \quad K_r = \frac{K_1}{1 + K_2}, \quad \beta = \frac{a_2}{a_1}. \tag{4.138c}$$

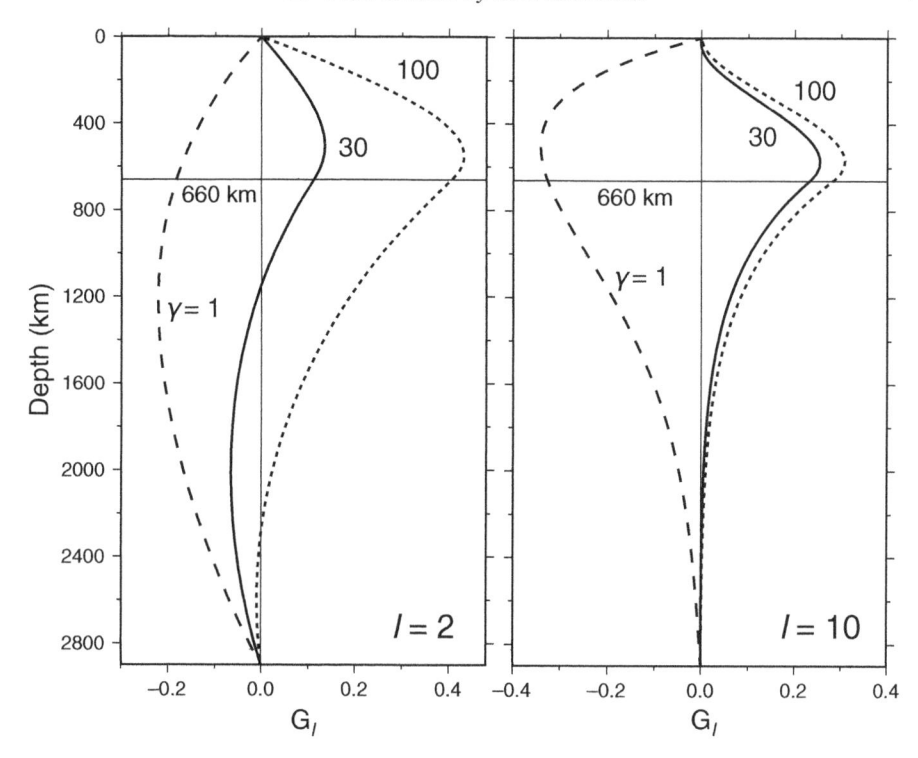

Figure 4.12 Gravitational potential kernels $G_l(r)$ for a two-layer mantle comprising an upper mantle with viscosity η and a lower mantle with viscosity $\gamma\eta$. Kernels are shown for spherical harmonic degrees $l = 2$ (left) and $l = 10$ (right) for three values of γ. See § 4.9.1 for a derivation of the kernels for an isoviscous ($\gamma = 1$) mantle. Figure redrawn from Figure 6 of Ricard (2015) by permission of Elsevier.

The dashed lines in Figure 4.12 show $G_l(r)$ for a self-gravitating mantle with uniform viscosity η. Kernels are shown for spherical harmonic degrees $l = 2$ (left) and $l = 10$ (right). The kernels are negative at all depths because the negative gravitational potential of the deformed upper surface exceeds the positive contribution of the internal mass anomaly itself. The maximum potential anomaly is produced by loads in the mid-mantle when $l = 2$ and in the upper mantle when $l = 10$.

It is straightforward but tedious to generalize the preceding derivation to a two-layer mantle comprising an upper layer $r > a_1 - z_{LU}$ with viscosity η_U and a lower layer $r < a_1 - z_{LU}$ with viscosity η_L (Forte and Peltier, 1987). As already discussed in connection with Figure 4.11, the general effect of a viscosity ratio $\eta_L/\eta_U \equiv \gamma > 1$ is to enhance the dynamic deflection of the CMB and reduce that of the upper surface. The reduced (negative) gravitational potential anomaly of the upper surface then counteracts less effectively the positive anomaly due to the (sinking) load. As a result, $G_l(r)$ increases at all depths relative to the kernels for $\gamma = 1$. The kernels for $z_{LU} = 660$ km are shown by the solid and dotted lines in

Figure 4.12 for $\gamma = 30$ and 100, respectively. For $\gamma = 30$, for example, $G_{10}(r) > 0$ for most depths, while $G_2(r)$ is positive in the upper mantle and negative in most of the lower mantle.

The analytical Green function approach outlined for a constant-viscosity mantle can in principle be extended to models comprising any number N of discrete layers with different viscosities. In practice, however, the rapidly increasing complexity of the analytical expressions limits the method to $N = 2$ (Forte and Peltier, 1987). One way to overcome this difficulty is to use a powerful semi-analytical technique known as the propagator matrix method. This is discussed next.

4.9.2 The Propagator-Matrix Method

An efficient approach to internal loading problems with multiple layers is the propagator–matrix method (PMM), whereby a flow solution is propagated from one layer interface to the next by simple matrix multiplication. The method is in fact applicable to any system of linear ODEs with constant coefficients of the form

$$\frac{d\mathbf{y}}{dz} = \mathbf{A}\mathbf{y} + \mathbf{b}(z), \tag{4.139}$$

where $\mathbf{y}(z)$ is a vector of dependent variables, $\mathbf{A}$ is a constant square matrix and $\mathbf{b}(z)$ is an inhomogeneous vector. The general solution of (4.139) is (Gantmacher, 1960, vol. I, p. 120, eqn. (53))

$$\mathbf{y}(z) = \mathbf{P}(z, z_0)\mathbf{y}(z_0) + \int_{z_0}^{z} \mathbf{P}(z, \zeta)\mathbf{b}(\zeta)d\zeta, \tag{4.140a}$$

$$\mathbf{P}(z, z_0) = \exp[\mathbf{A}(z - z_0)], \tag{4.140b}$$

where $\mathbf{y}(z_0)$ is the solution vector at the reference point $z = z_0$ and $\mathbf{P}(z, z_0)$ is the propagator matrix. The form of the solution (4.140) is identical to that for a scalar variable $y(z)$, except that the argument of the exponential in (4.140b) is now a matrix rather than a scalar quantity. Analytical expressions for functions of matrices such as (4.140b) are given by Gantmacher (1960, vol. I, pp. 95–110).

As an illustration, we determine the propagator matrix for a poloidal flow driven by internal density anomalies in a self-gravitating spherical shell with radially variable viscosity $\eta(r)$ and laterally averaged density $\rho(r)$ (Hager and O'Connell, 1981; Richards and Hager, 1984). As in § 4.9.1, we suppose that the driving density anomaly is $\hat{\rho}_l^m(r)Y_l^m(\theta, \phi)$, where Y_l^m is a surface spherical harmonic.

The first step is to transform the equations governing the flow in the mantle to the canonical form (4.139), where $\mathbf{A}$ – to repeat – is a constant matrix. The standard Stokes equations (4.1) for our model mantle do not have this form for three reasons: (i) the fluid properties $\eta(r)$ and $\rho(r)$ vary with radius, (ii) the expression for the

4.9 Flow Driven by Internal Loads

differential operator ∇ in spherical coordinates involves the factor r^{-1} and (iii) a system of first-order equations cannot be written in terms of the primitive variables u_i and p. Difficulty (i) is circumvented by dividing the mantle into N discrete layers $n = 1, 2, \ldots, N$, in each of which the viscosity η_n and the density ρ_n are constant. Difficulty (ii) is overcome by using a transformed radial variable $z = \ln(r/a_1)$, where a_1 is the outer radius of the mantle (Hager and O'Connell, 1979). Finally, one circumvents (iii) by using the independent variables

$$\mathbf{y} = \left[\hat{u}, \hat{v}, \frac{r\hat{\sigma}_{rr}}{\eta_0}, \frac{r\hat{\sigma}_{r\theta}}{\eta_0}, \frac{\rho_0 r \hat{\chi}}{\eta_0}, \frac{\rho_0 r^2 \partial_r \hat{\chi}}{\eta_0}\right]^T \quad (4.141)$$

where $\hat{u}$ and $\hat{v}$ are the $\mathbf{e}_r$- and $\mathbf{e}_\theta$-components, respectively, of the velocity, η_0 and ρ_0 are reference values of the viscosity and density, respectively, and the spherical harmonic dependence of each variable ($\hat{u}, \hat{\sigma}_{rr}, \hat{\chi}, \partial_r \hat{\chi} \propto Y_l^m$; $\hat{v}, \hat{\sigma}_{r\theta} \propto \partial_\theta Y_l^m$) has been suppressed for clarity. The Stokes equations within each layer n then take the form (4.139) with

$$\mathbf{A} = \begin{bmatrix} -2 & L & 0 & 0 & 0 & 0 \\ -1 & 1 & 0 & 1/\eta^* & 0 & 0 \\ 12\eta^* & -6L\eta^* & 1 & L & 0 & \rho^* \\ -6\eta^* & 2(2L-1)\eta^* & -1 & -2 & \rho^* & 0 \\ 0 & 0 & 0 & 0 & 1 & 1 \\ 0 & 0 & 0 & 0 & L & 0 \end{bmatrix} \quad (4.142a)$$

$$\mathbf{b} = [0, 0, r^2 g_0(r) \hat{\rho}(r)/\eta_0, 0, 0, 4\pi r^3 G \rho_0 \hat{\rho}(r)/\eta_0]^T, \quad (4.142b)$$

where $\hat{\rho} \equiv \hat{\rho}_l^m$, $L = l(l+1)$, $\eta^* = \eta_n/\eta_0$ and $\rho^* = \rho_n/\rho_0$. The signs of the elements A_{36}, A_{45} and b_6 in (4.142) are opposite to those of eqn. (A65) of Hager and O'Connell (1981), whose sign convention for the gravitational potential perturbation is opposite to the one adopted here. Further simplification is achieved by recasting the density anomaly $\hat{\rho}(r)$ as a sum of equivalent surface density contrasts ξ_n (units kg m^{-2}) localized at the midpoints r_n of the layers, according to

$$\hat{\rho}(r) = \sum_n \delta(r - r_n)\xi_n. \quad (4.143)$$

In practice, ξ_n is different in each layer, while η_n and ρ_n may be constant over several adjacent layers. For a group of M such adjacent layers bounded by the depths z and z_0, the solution (4.140a) now takes the approximate form

$$\mathbf{y}(z) = \mathbf{P}(z, z_0)\mathbf{y}(z_0) + \sum_{m=1}^{M} \mathbf{P}(z, z_m)\mathbf{b}_m, \quad (4.144a)$$

$$\mathbf{b}_m = [0, 0, r_m g_0(r_m)\xi_m/\eta_0, 0, 0, 4\pi r_m^2 G \rho_0 \xi_m/\eta_0]^T, \quad (4.144b)$$

where $z_m = \ln(r_m/a_1)$. To use (4.144), one must apply boundary conditions at the CMB and at Earth's surface, taking into account the dynamic topography of those boundaries (Richards and Hager, 1984, Appendix 1).

An analytical expression for $\mathbf{P}(z, z_0)$ can be written in terms of the minimal polynomial $\psi(\lambda)$ of $\mathbf{A}$. As a reminder, the minimal polynomial of a matrix is the annihilating polynomial of least degree with highest coefficient $= 1$. For the matrix (4.142a), $\psi(\lambda)$ is identical to the characteristic polynomial and is

$$\psi(\lambda) = \prod_{i=1}^{4} (\lambda - \lambda_i)^{m_i}, \tag{4.145a}$$

$$\lambda_1 = l+1, \quad \lambda_2 = -l, \quad \lambda_3 = l-1, \quad \lambda_4 = -l-2, \tag{4.145b}$$

$$m_1 = m_2 = 2, \quad m_3 = m_4 = 1, \tag{4.145c}$$

where λ_1-λ_4 are the four distinct eigenvalues of $\mathbf{A}$. Following Gantmacher (1960, vol. I, pp. 95–102),

$$\mathbf{P}(z, z_0) = \sum_{k=1}^{4} \sum_{j=1}^{m_k} \alpha_{kj} (\mathbf{A} - \lambda_k \mathbf{I})^{j-1} \Psi_k, \tag{4.146}$$

where

$$\alpha_{kj} = \frac{1}{(j-1)!} \frac{d^{j-1}}{d\lambda^{j-1}} \left[\frac{\exp \lambda(z - z_0)}{\psi_k(\lambda)} \right]_{\lambda = \lambda_k}, \tag{4.147}$$

$$\psi_k(\lambda) = \frac{\psi(\lambda)}{(\lambda - \lambda_k)^{m_k}}, \quad \Psi_k = \prod_{\substack{i=1 \\ i \neq k}}^{4} (\mathbf{A} - \lambda_i \mathbf{I})^{m_i} \tag{4.148}$$

and $\mathbf{I}$ is the identity matrix.

4.9.3 Inferences of Mantle Viscosity Structure

To summarize the results of § 4.9.1 and 4.9.2, there are three distinct methods for solving internal loading problems involving radially variable viscosity. The first is the analytical Green function method described in § 4.9.1. While the purely analytical character of that approach is appealing, its practical application is limited to models having no more than two layers with different viscosities. The second method is the PMM, using propagator matrices that are calculated either semi-analytically (§ 4.9.2) or analytically. The PMM offers an efficient way of solving problems involving many layers. However, it is less practical for problems in which the reference mantle density profile $\rho(r)$ and gravitational acceleration $g(r)$ are essentially continuous functions, given by a seismological reference model such as PREM (Dziewonski and Anderson, 1981). The main context in which this occurs is the Stokes equations for a compressible mantle (Forte and Peltier, 1991). Here a third method is used, wherein the solution is propagated from one viscosity interface to the next using a standard numerical method for the solution of coupled ODEs. These ODEs have the general form (4.139) except that $\mathbf{A}$ is a nonconstant matrix of which the elements are functions of r. The appearance of the functions $\rho(r)$ and $g(r)$ in the elements of $\mathbf{A}$ means that there is no point in making the transformation $z = \ln r/a_1$, as one does in the PMM.

The three methods just summarized have been widely used to infer the profile of mantle viscosity required by Earth's nonhydrostatic geoid. Here it is important to keep in mind the distinction between relative and absolute viscosity profiles. For a given distribution of internal loads, certain key geophysical observables such as dynamic topography and geoid anomalies do not depend on the absolute viscosity, but only on the profile of relative viscosity as a function of radius. This can be seen by noting that because velocities and strain rates are inversely proportional to the absolute viscosity, the stress (= viscosity times strain rate) is independent of it. Because the dynamic topography of the boundaries is proportional to the normal stress there, it must also be independent of the absolute viscosity. Finally, because the geoid anomaly depends only on the mass anomalies of the load and of the dynamic topography, it too is independent of the absolute viscosity. By contrast, observables that have units of velocity, such as surface plate speeds and rates of GIA, are sensitive to both absolute and relative viscosity.

The most common way of using internal loading calculations to infer the mantle viscosity profile is to start with a load function $\hat\rho(r, \theta, \phi)$ estimated from seismic tomography and/or the global distribution of subducted lithosphere, and then to determine by repeated forward modeling the parameters of a simple layered viscosity structure for which geoid anomalies predicted by formulae like (4.137) best match their observed counterparts on Earth. The earliest studies of this type assumed a simple two-layer viscosity structure (Hager, 1984; Hager et al., 1985; Forte and Peltier, 1987). The approach was subsequently generalized to multilayer (Hager and Richards, 1989; Ricard et al., 1989; King and Masters, 1992; Forte et al., 1993; Panasyuk and Hager, 2000) and quasi-continuous (Forte et al., 1991; Rudolph et al., 2015) viscosity profiles. Despite significant differences in detail, all these studies agree that the mantle viscosity increases by a factor of 10–150 from the upper to the lower mantle. A viscosity increase of this magnitude gives the best fit to the data because the sign of the geoid kernels as a function of depth matches well the sign of the observed correlation between the geoid and tomographically inferred density anomalies. However, the exact depth range over which this viscosity increase occurs remains controversial because inversion involves significant trade-offs between depth and viscosity ratio (Forte and Peltier, 1987). It is often assumed (e.g., Hager, 1984; Hager et al., 1985) that the principal viscosity increase should occur at 660 km, the depth of the ringwoodite-bridgmanite phase transition. However, several models suggest that the viscosity increase may rather take place in the depth range 800–1200 km (Forte and Peltier, 1987, 1991; Rudolph et al., 2015). A possible cause of such an increase is a progressive decrease in mineral hydration with increasing depth (Ohtani et al., 1997).

Exercises

4.1 A small spherical particle is freely suspended in a laminar flow in a cylindrical pipe with a parabolic velocity profile (Poiseuille flow). Does such a particle have a tendency to migrate sideways (toward or away from the wall of the pipe), assuming that it is not directly on the axis of the pipe?

4.2 (a) Apply the operator $\mathbf{e}_z \times \nabla$ to the momentum equation

$$-\nabla p + \eta \nabla^2 \mathbf{u} = g\rho(\mathbf{x})\mathbf{e}_z \qquad (4.149)$$

to show that the toroidal scalar satisfies the equation

$$\nabla_h^2 \nabla^2 \mathcal{T} = 0. \qquad (4.150)$$

(b) Apply the operator $\nabla \times (\mathbf{e}_z \times \nabla)$ to the same momentum equation to show that the poloidal scalar satisfies the equation

$$\nabla_h^2 \nabla^4 \mathcal{P} = \frac{g}{\eta} \nabla_h^2 \rho. \qquad (4.151)$$

4.3 The equations of conservation of momentum for a non-Newtonian corner flow can be written as

$$-\partial_r p + \frac{1}{r}\partial_\theta \tau = 0, \qquad (4.152a)$$

$$-\frac{1}{r}\partial_\theta p + \frac{2}{r}\tau + \partial_r \tau = 0, \qquad (4.152b)$$

where $\tau = 2\eta e_{r\theta}$, η is the viscosity (4.1f), and $e_{r\theta} \equiv U_0(F'' + F)/2r$ is the only nonzero component of the strain rate tensor. Starting from (4.152), derive (4.34).

4.4 The azimuthal velocity (4.52) for viscous eddies can be written as

$$v(r,0) = \Re\left[(a+ib)\frac{1}{r}\left(\frac{r}{r_0}\right)^{1+p_1+iq_1}\right] \qquad (4.153)$$

where $a + ib = -\lambda_1 A'[\cos(\lambda_1 - 2)\alpha - \cos\lambda_1\alpha]$. Show that (4.153) can be written in the form (4.53).

Exercises

4.5 Verify that the biharmonic equation is satisfied by the function $F = (y+ix)$ $f(x+iy)$ where $z = x + iy$ is a complex variable.

4.6 (a) Starting from (4.72), demonstrate (4.73). (b) Derive the expressions (4.75) for the components of the stress tensor.

4.7 Starting from the definition of the stress tensor, verify the expressions (4.88b) and (4.89b) for K_{ijk}.

4.8 Consider the 2-D flow due to a vertically acting line force located a distance d above the free-slip surface of a fluid half-space. Show that the integral of the normal stress over the boundary is equal and opposite to the strength of the line force.

4.9 [SM] Determine the poloidal scalar for the flow due to a vertical point force located at $z = z'$ in a horizontally infinite fluid layer $z \in [0, d]$ with free-slip boundaries and uniform viscosity η.

4.10 [SM] Use the singularity method to determine the disturbance flow generated by a solid sphere immersed in a linear flow $u_i^\infty = A_{ij}r_j$, where $\mathbf{A}$ is a constant traceless and (in general) nonsymmetric tensor. Make a contour plot of the Stokes streamfunction of the total flow for the axisymmetric case $A_{11} = A_{22} = -\dot{\epsilon}/2$, $A_{33} = \dot{\epsilon}$ and $A_{12} = A_{21} = A_{23} = A_{32} = A_{31} = A_{13} = 0$, where $\dot{\epsilon}$ is the rate of stretching along the r_3-direction. Hint: the disturbance flow can be expressed as the sum of flows due to a force dipole $G_{ijk}^{FD}(\mathbf{r})$ defined by (4.90b) and a source quadrupole $G_{ijk}^{SQ}(\mathbf{r}) = -\partial G_{ij}^{SD}(\mathbf{r})/\partial r_k$, where $G_{ij}^{SD}(\mathbf{r})$ is defined by (4.92).

4.11 Invert (4.112) to obtain (4.113).

4.12 [SM] Derive the matrix elements A_{1j}, A_{2j} and A_{3j} (first three rows) given by (4.142a).

4.13 Consider a horizontal plane layer of viscous fluid with constant reference density ρ_0, thickness d and infinite lateral extent. Flow in this layer is driven by a distribution of density anomalies (internal loads) $\hat{\rho}(\mathbf{x})$. The layer is bounded above ($z = d$) by an inviscid 'ocean' with density ρ_w, and below ($z = 0$) by an inviscid 'core' with density ρ_c. The gravitational acceleration g_0 is constant throughout the layer. Show that the total weight of the internal loads is equal and opposite to the total weight of the boundary deflections.

5
Elasticity and Viscoelasticity

Fluid convection with a period $\tau \sim 10^{15}$ s is the extreme limit of a spectrum of deformational processes in the mantle spanning an enormous range of time scales, including glacial isostatic adjustment (GIA; $\tau \sim 10^{11}$ s), the Chandler wobble ($\tau \sim 4 \times 10^7$ s) and elastic waves and free oscillations ($\tau \sim 1 - 10^3$ s). All these processes can be understood by regarding the mantle as a viscoelastic body that deforms as a fluid when $\tau \gg \tau_M$ and as an elastic solid when $\tau \ll \tau_M$, where τ_M is the Maxwell time of the material. τ_M is just the ratio of the viscosity of the material to its elastic shear modulus and is of the order of a few hundred years ($\sim 10^{10}$ s) in the mantle.

Because the theory of viscous flow is valid only at very long periods ($\tau \gg \tau_M$), other rheological models are required to understand phenomena with shorter periods. The two most commonly used models are the linear elastic solid (for short periods $\tau \ll \tau_M$) and the linear Maxwell solid (for intermediate periods $\tau \sim \tau_M$). I begin by reviewing the mathematical analogies (correspondence principles) among the viscous, elastic and viscoelastic models. I then turn to a classic geodynamic problem of elasticity theory: the deformation of an elastic plate (lithosphere) subject to a load. Finally, I discuss the problem of surface loading of a stratified viscoelastic sphere, the principal model used to study GIA.

5.1 Correspondence Principles

Stokes (1845) and Rayleigh (1945) demonstrated the existence of a mathematical correspondence (the Stokes–Rayleigh analogy) between small incompressible deformations of an elastic solid and slow flows of a viscous fluid. The constitutive law for a linear (Hookean) elastic solid is (Landau and Lifshitz, 1986)

$$\sigma_{ij} = K\varepsilon_{kk}\delta_{ij} + 2\mu\left(\varepsilon_{ij} - \frac{1}{3}\varepsilon_{kk}\delta_{ij}\right), \tag{5.1}$$

5.1 Correspondence Principles

where σ_{ij} is the stress tensor, $\varepsilon_{ij} \equiv (\partial_i \zeta_j + \partial_j \zeta_i)/2$ is the linearized strain tensor, ζ_i is the displacement vector, K is the bulk modulus and μ is the shear modulus. Alternatively, (5.1) can be written in terms of Young's modulus E and Poisson's ratio ν, which are related to K and μ by

$$E = \frac{9K\mu}{3K + \mu}, \quad \nu = \frac{3K - 2\mu}{2(3K + \mu)}. \tag{5.2}$$

An incompressible elastic solid corresponds to the limits $K/E \to \infty$, $\nu \to 1/2$ and an incompressible deformation to the limit $\varepsilon_{kk} \to 0$. As these limits are approached, however, the product $-K\varepsilon_{kk}$ tends to a finite value, the pressure p. Equation (5.1) then becomes

$$\sigma_{ij} = -p\delta_{ij} + 2\mu\varepsilon_{ij}, \tag{5.3}$$

which is identical to the constitutive relation (4.1e) for a viscous fluid if the shear modulus μ and the displacement ζ_i are replaced by the viscosity η and the velocity u_i, respectively. Consequently, the equations of (incompressible) Stokes flow can be obtained from those of (compressible) linear elasticity via the transformation

$$(\zeta_i, E, \nu) \to (u_i, 3\eta, 1/2). \tag{5.4}$$

However, the inverse transformation only works for the special case of incompressible elasticity, because the Stokes equations involve no parameter analogous to Poisson's ratio. The transformation (5.4) will be used in § 8 to obtain equations governing the slow deformation of thin viscous sheets.

The Stokes–Rayleigh analogy between linear elasticity and Stokes flow applies not only to the governing equations themselves, but also to solutions of BVPs for those equations. In that case, however, the analogy requires that the boundary conditions of the elastic and viscous problems be properly related. For example, imposed boundary velocities in the viscous problem must correspond to imposed boundary displacements in the elastic one. By contrast, elastic deformations and viscous flows driven by the same distribution of applied stresses are directly analogous because stress is invariant under the transformation (5.4).

A second useful correspondence principle relates problems in linear viscoelasticity and linear elasticity. Although this principle is valid for any linear viscoelastic body (Biot, 1954), its geodynamical application is usually limited to the special case of a linear Maxwell solid, for which the constitutive relation is (e.g., Peltier, 1974)

$$\dot{\sigma}_{ij} + \frac{\mu}{\eta}\left(\sigma_{ij} - \frac{1}{3}\sigma_{kk}\delta_{ij}\right) = 2\mu\left(\dot{\varepsilon}_{ij} - \frac{1}{3}\dot{\varepsilon}_{kk}\delta_{ij}\right) + K\dot{\varepsilon}_{kk}\delta_{ij} \tag{5.5}$$

where dots denote time derivatives. Transforming (5.5) into the frequency domain using a Laplace transform, we obtain

$$\overline{\sigma}_{ij} = K\overline{\varepsilon}_{kk}\delta_{ij} + 2\overline{\mu}(s)\left(\overline{\varepsilon}_{ij} - \frac{1}{3}\overline{\varepsilon}_{kk}\delta_{ij}\right), \qquad (5.6a)$$

$$\overline{\mu}(s) = \frac{\mu s}{s + \mu/\eta}, \qquad (5.6b)$$

where s is the complex frequency and overbars denote Laplace transforms of the quantities beneath. Equation (5.6) is identical in form to Hooke's law (5.1) for an elastic solid, but with a frequency-dependent shear modulus $\overline{\mu}$. Accordingly, any viscoelastic problem can be reduced to an equivalent elastic problem, which can be solved and then inverse-Laplace transformed back into the time domain to yield the solution of the original viscoelastic problem (Lee, 1955).

The simple form (5.5) of the Maxwell constitutive law applies only to infinitesimal deformations and must be generalized for situations involving finite strain and rotation of material elements. Large-transformation viscoelasticity is a specialized topic that is beyond the scope of this book. An illuminating discussion and references to the recent literature can be found in Schrank et al. (2017).

5.2 Loading of an Elastic Lithosphere

As the first model problem of this chapter, consider the two-dimensional deformation of a horizontal elastic plate by spatially periodic normal loads applied to its upper surface. Figure 5.1a shows the geometry of the problem. An incompressible elastic plate with thickness T, density ρ_m and shear modulus μ is overlain by oceanic crust with density ρ_c and mean thickness h. The elevation of the upper surface of the crust is $\Delta \cos kx$, and the deformation of the upper surface of the plate is $\xi \cos kx$, so that the thickness of the crust is $h + (\Delta - \xi)\cos kx$. The plate is underlain by fluid with the same density ρ_m, and the whole is submerged beneath seawater with mean depth d and density ρ_w. My treatment of this problem follows McKenzie and Bowin (1976) with some changes of notation.

By the Stokes–Rayleigh analogy, the equations of incompressible elasticity are identical to those of slow viscous flow *modulo* a time derivative. The displacement vector ζ_i within the plate thus satisfies

$$\partial_j \zeta_j = 0, \qquad (5.7a)$$

$$\partial_i p = \mu \nabla^2 \zeta_i \qquad (5.7b)$$

where p is the pressure. As for slow viscous flow, these equations can be reduced to the biharmonic equation

$$\nabla^4 \psi = 0 \qquad (5.8)$$

5.2 Loading of an Elastic Lithosphere

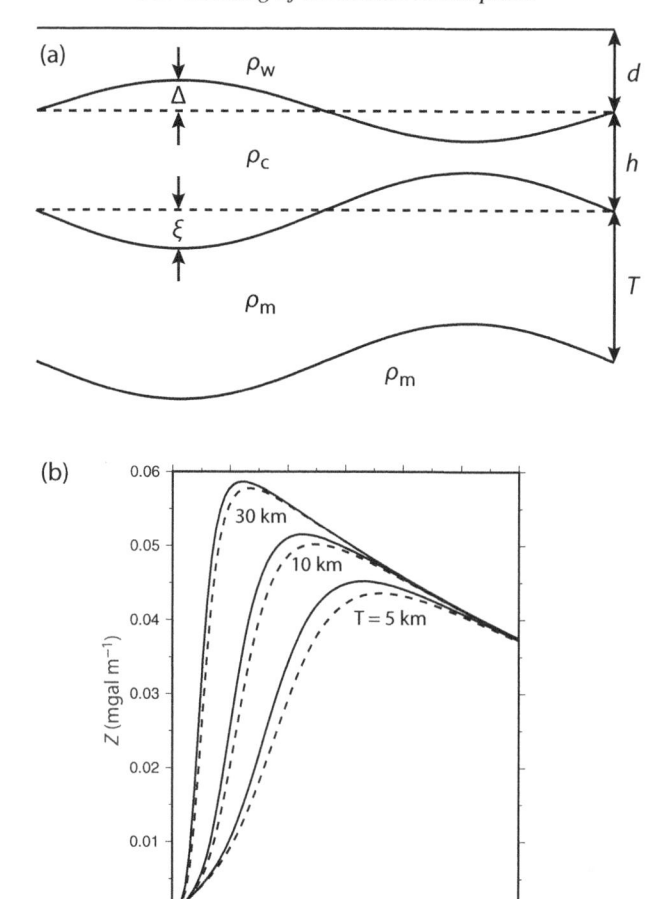

Figure 5.1 (a) Definition sketch for an elastic plate of thickness T deformed by an overlying crust with nonuniform thickness $h + (\Delta - \xi)\cos kx$. (b) Gravitational admittance for the model of part (a), for three values of the thickness T of the elastic plate. The solid curves are the admittances (5.17) for an incompressible plate (Poisson's ratio $\nu = 1/2$) and the dashed curves are for a plate with $\nu = 0.2$ (see Exercise 5.1). The admittances are calculated assuming $d = 5$ km, $h = 6$ km, $\rho_w = 1020$ kg m^{-3}, $\rho_c = 2650$ kg m^{-3}, $\rho_m = 3300$ kg m^{-3} and $\mu = 6.1 \times 10^{10}$ Pa. See § 5.2 for further discussion. Part (a) redrawn from Figure 10 of McKenzie and Bowin (1976) by permission of John Wiley and Sons.

where ψ is the elastic equivalent of the streamfunction and the components of the displacement are

$$(\zeta_1, \zeta_2, \zeta_3) = (\partial_z \psi, 0, -\partial_x \psi). \tag{5.9}$$

The general horizontally periodic solution of (5.8) is

$$\psi = \left[(A + Bkz)e^{kz} + (C + Dkz)e^{-kz}\right]\sin kx \tag{5.10}$$

where *A–D* are constants that are determined by the boundary conditions. The corresponding pressure is obtained by integrating the $i = 1$ component of (5.7b) and is

$$p = -2k^2\mu \left(Be^{kz} + De^{-kz}\right) \cos kx. \tag{5.11}$$

Two of the four required boundary conditions are that the tangential stress vanishes on both surfaces $z = \pm T/2$ of the plate, or

$$(\partial_x \zeta_3 + \partial_z \zeta_1)|_{z=\pm T/2} = 0. \tag{5.12a}$$

Moreover, the modified normal stress must vanish on the bottom of the plate, or

$$(-p + 2\mu \partial_z \zeta_3)|_{z=-T/2} = 0. \tag{5.12b}$$

Finally, the normal stress on the top of the plate is

$$(-p + 2\mu \partial_z \zeta_3)|_{z=T/2} = -[g\Delta(\rho_c - \rho_w) + g\xi(\rho_m - \rho_c)] \cos kx, \tag{5.12c}$$

where the minus sign on the right-hand side (RHS) is present because the load acts in the $-z$ direction. The preceding boundary conditions yield four coupled linear equations that can be readily solved for *A–D*. Once these constants are known, the ratio Δ/ξ is determined from the definition of ξ, viz.,

$$\xi \cos kx = \zeta_3|_{z=T/2}. \tag{5.13}$$

The result is

$$\frac{\Delta}{\xi} = -\frac{\rho_m - \rho_c}{\rho_c - \rho_w} - \frac{\mu F(k')}{gT(\rho_c - \rho_w)} \tag{5.14}$$

where

$$F(k') = \frac{2k'(\cosh 2k' - 2k'^2 - 1)}{\sinh 2k' + 2k'}, \quad k' = kT. \tag{5.15}$$

Equation (5.14) has a simple physical interpretation. In the limit of infinite wavelength ($k' = 0$), $F(k') = 0$ and only the first term on the RHS of (5.14) remains. The equation then becomes $\Delta(\rho_c - \rho_w) + \xi(\rho_m - \rho_c) = 0$, which is the condition for local (Airy) isostatic compensation of the topography $\Delta \cos kx$ by variations in crustal thickness. The second term on the RHS of (5.14) represents the nonlocal contribution to the isostatic compensation due to plate flexure.

The principal utility of the solution presented is for the interpretation of free-air gravity anomalies (McKenzie and Bowin, 1976). In particular, the solution allows us to write down an expression for the gravitational admittance $Z = \delta g/\Delta$, where

$\delta g \cos kx$ is the free-air gravity anomaly measured at the surface of the ocean, given by (McKenzie, 1967)

$$\delta g = 2\pi G(\rho_c - \rho_w)\Delta e^{-kd} + 2\pi G(\rho_m - \rho_c)\xi e^{-k(d+h)} \tag{5.16}$$

where G is the universal gravitational constant. Using (5.14) to eliminate ξ from (5.16), we obtain the admittance as

$$Z = 2\pi G(\rho_c - \rho_w)e^{-kd}\left\{1 - \left[1 + \frac{\mu F(k')}{gT(\rho_m - \rho_c)}\right]^{-1} e^{-kh}\right\}. \tag{5.17}$$

Figure 5.1b shows $Z(k)$ for $T = 5$ km, 10 km and 30 km. In the limit $kT \gg 1$, the admittance is $Z = 2\pi G(\rho_c - \rho_w)\exp(-kd)$, which corresponds to uncompensated topography with a gravity anomaly that is attenuated by a factor $\exp(-kd)$ due to upward continuation of the potential field across the depth of the ocean.

Admittance functions similar to (5.17) have been extensively used to interpret profiles of free-air gravity anomalies collected by seagoing research ships. The standard procedure is to take the Fourier transform of the gravity data, multiply it by the admittance calculated for some candidate value of the elastic thickness T and finally transform the result back into the space domain to obtain a predicted free-air gravity profile. The process is then repeated iteratively to find the value of T for which the predicted gravity best matches the observed. A classic study of this kind is Watts (1978), who determined $T = 25\pm 5$ km for the lithosphere beneath the Hawaiian Island chain. The admittance used by Watts (1978) was in fact calculated using thin-plate theory rather than the full thick-plate theory used earlier; Exercise 8.5 invites you to determine the thin-plate admittance.

5.3 Surface Loading of a Stratified Viscoelastic Sphere

The fundamental problem of GIA is to determine the response of a radially stratified Maxwell Earth to a time-dependent surface load $P(\theta, \phi, t)$ that represents the changing global distribution of glacial ice and meltwater. Formally, this problem can be solved by a mathematical convolution of the load distribution with a Green function that describes Earth's response to an impulsive point load applied at time t at a point $\mathbf{r}$ on Earth's surface and then immediately removed. Now an impulse function in the time domain corresponds to a constant in the frequency domain. According to the correspondence principle, therefore, our problem reduces to that of determining the deformation of a stratified elastic sphere by a static point load (Longman, 1962; Farrell, 1972; Peltier, 1974). My treatment of this problem follows Peltier (1989).

The general equations governing the deformation of a self-gravitating elastic sphere are (Backus, 1967)

$$0 = \nabla \cdot \hat{\sigma} + \mathbf{g}_0 \hat{\rho} - \rho_0 \nabla \hat{\chi} - \nabla(\rho_0 g_0 \hat{u}), \tag{5.18a}$$

$$\nabla^2 \hat{\chi} = 4\pi G \hat{\rho}, \tag{5.18b}$$

where $\hat{u}$ is the radial component of the displacement vector $\hat{\zeta} \equiv \hat{u}\mathbf{e}_r + \hat{v}\mathbf{e}_\theta + \hat{w}\mathbf{e}_\phi$ and the other symbols are the same as in the analogous equations (4.116) for slow viscous flow. There are three significant differences between the two sets of equations. First, the stress tensor $\hat{\sigma}$ in (5.18a) is related to the displacement vector $\hat{\zeta}$ by the elastic constitutive law (5.6) with a frequency-dependent shear modulus. Second, because deformation in the elastic case is driven by surface loading rather than by internal density anomalies, the perturbation density $\hat{\rho} \equiv -\nabla \cdot (\rho_0 \hat{\zeta})$ in (5.18) is determined entirely by the requirement of mass conservation in the deformed solid. Third, (5.18a) contains an additional term $-\nabla(\rho_0 g_0 \hat{u})$ that corrects for the fact that the strain tensor that appears in the elastic constitutive law is at a fixed material particle and not at a fixed point in space (Love, 1967; Backus, 1967, p. 96).

The reduction of the governing equations (5.18) to solvable form proceeds as for the viscous flow equations in § 4.9.2, but with a few significant differences. As in the viscous case, the goal is to reduce (5.18) to a sixth-order system of ordinary differential equations (ODEs) of the form

$$\frac{d\mathbf{y}}{dr} = \mathbf{A}\mathbf{y}, \tag{5.19}$$

where $\mathbf{y}$ is the unknown vector and $\mathbf{A}$ is a 6×6 matrix. Equation (5.19) contains no inhomogeneous vector $\mathbf{b}$ like the one in the viscous equations (4.139), because the loads in the elastic problem are applied at the boundaries rather than internally. Another important point is that the reference profiles of density $\rho_0(r)$ and the elastic moduli $\mu(r)$ and $K(r)$ that appear in the equations are essentially continuous functions, given *a priori* by a seismological reference model (e.g., PREM; Dziewonski and Anderson, 1981).

We turn now to the definition of the variables $\mathbf{y}$. By symmetry, the azimuthal deformation $\hat{w}$ is zero, and the components $\hat{u}$ and $\hat{v}$ can depend only on the radius r and the angular distance θ from the point load. The field variables can therefore be expanded as

$$\hat{\zeta} = \sum_{l=0}^{\infty} \left[\hat{u}_l(r,s) P_l(\cos\theta) \mathbf{e}_r + \hat{v}_l(r,s) \partial_\theta P_l(\cos\theta) \mathbf{e}_\theta \right], \tag{5.20a}$$

$$\hat{\chi} = \sum_{l=0}^{\infty} \hat{\chi}_l(r,s) P_l(\cos\theta). \tag{5.20b}$$

5.3 Surface Loading

where $P_l(\cos\theta)$ are the Legendre polynomials and s is the complex frequency. Now for each angular degree l, let

$$\mathbf{y} = \left[\hat{u}, \hat{v}, \hat{\sigma}_{rr}, \hat{\sigma}_{r\theta}, \hat{\chi}, \partial_r\hat{\chi} + (l+1)\hat{\chi}/r + 4\pi G\rho_0\hat{u}\right]^T \tag{5.21}$$

where the subscript l on each variable has been suppressed for simplicity. Apart from factors of r, the definition of y_6 and the replacement of velocities by displacements, (5.21) is identical to its viscous analog (4.141). With the choice (5.21), the matrix $\mathbf{A}$ in (5.19) is

$$\mathbf{A} = \begin{bmatrix}
-\dfrac{2\lambda}{\beta r} & \dfrac{L\lambda}{\beta r} & \dfrac{1}{\beta} & 0 & 0 & 0 \\
-r^{-1} & r^{-1} & 0 & \overline{\mu}^{-1} & 0 & 0 \\
\dfrac{4}{r}\left(\dfrac{\gamma}{r} - \rho_0 g_0\right) & -\dfrac{L}{r}\left(\dfrac{2\gamma}{r} - \rho_0 g_0\right) & -\dfrac{4\overline{\mu}}{\beta r} & \dfrac{L}{r} & -\dfrac{\rho_0 M}{r} & \rho_0 \\
\dfrac{1}{r}\left(\rho_0 g_0 - \dfrac{2\gamma}{r}\right) & \dfrac{L(\gamma + \overline{\mu}) - 2\overline{\mu}}{r^2} & -\dfrac{\lambda}{\beta r} & -\dfrac{3}{r} & \dfrac{\rho_0}{r} & 0 \\
-4\pi G\rho_0 & 0 & 0 & 0 & -\dfrac{M}{r} & 1 \\
-\dfrac{4\pi MG\rho_0}{r} & \dfrac{4\pi LG\rho_0}{r} & 0 & 0 & 0 & \dfrac{l-1}{r}
\end{bmatrix}$$
$$\tag{5.22}$$

where $L = l(l+1)$, $M = l+1$, $\beta = K + 4\overline{\mu}/3$, $\gamma = 3K\overline{\mu}/\beta$, $\lambda = K - 2\overline{\mu}/3$ and $\overline{\mu}$ is defined by (5.6b). Equation (5.22) corrects misprints in the matrix elements A_{43}, A_{51} and A_{66} in eqns. (16) of Peltier (1989) and the matrix elements A_{12}, A_{32}, A_{51} and A_{66} given in the appendix of Peltier (1974). Because ρ_0, K and $\overline{\mu}$ in the preceding expression for $\mathbf{A}$ are continuous functions of r, the transformation $z = \ln(r/a_1)$ used in § 4.9.2 has not been applied. The system (5.19) can be solved by a standard numerical method for two-point BVPs (e.g., shooting) or by an analytical propagator matrix technique designed to handle a large number of layers (Vermeersen and Sabadini, 1997). Inversion of the resulting solution back into the time domain and its convolution with a given time-dependent surface load function are then performed numerically. The use of this procedure to infer the mantle viscosity profile $\eta(r)$ is reviewed by Peltier (1989, 2004) and Mitrovica and Peltier (1995).

The first attempt to use GIA data to determine the viscosity of Earth's mantle was the classic study of Haskell (1935), who inferred a value $\eta = 10^{21}$ Pa s based on uplift rates in Fennoscandia (see the end of § 10.1 for a simple version of Haskell's model). Since that time, GIA-based inferences of the mantle viscosity profile have been greatly refined by the incorporation of more and better relative

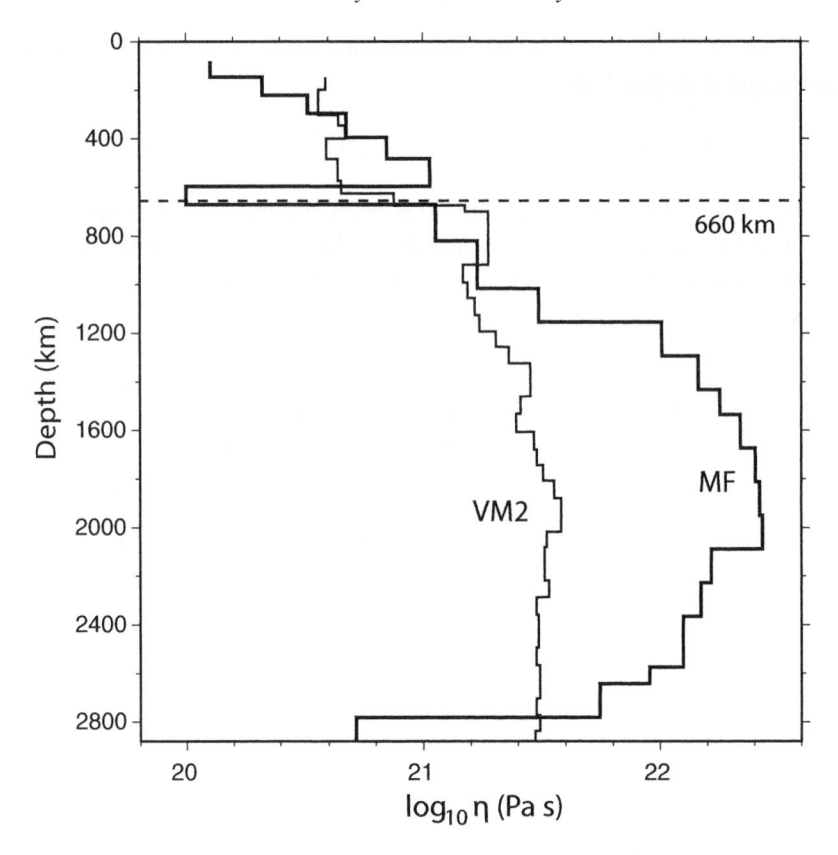

Figure 5.2 Radial viscosity profiles discussed in the text. Heavy line: model 'MF' of Mitrovica et al. (2015). Light line: model 'VM2' of Peltier (1996, 2004). Figure redrawn from Figure 2 of Mitrovica et al. (2015) by permission of the American Association for the Advancement of Science and Figure 1 of Peltier (2004) by permission of Annual Reviews, Inc.

sea-level data and the use of more sophisticated continuum mechanical models like the one described earlier. A widely cited modern example of a GIA-based viscosity profile is the model 'VM2' of Peltier (1996, 2004), which is shown by the thin line in Figure 5.2. Its main features are a nearly constant viscosity in the upper mantle, an increase by a factor ≈ 4.3 in the depth range 625-700 km and a subsequent slow increase in the upper half of the lower mantle.

Up until the mid-1990s, inferences of mantle viscosity profiles based on convection-related data such as the long-wavelength geoid (§ 4.9.3) and on GIA proceeded on parallel tracks with little interaction between them. The significant differences between the results of the two approaches were explained by supposing that GIA-based inferences reflected a transient viscosity, as opposed to the steady-state viscosity that applied at the much longer time scales characteristic of convection (e.g., Peltier et al., 1986). However, the need for transient viscosity was later obviated by the success of joint inversions of convection and GIA data, which showed that both could be explained by a single viscosity profile (Forte and Mitrovica, 1996; Kaufmann and Lambeck, 2000; Mitrovica and Forte, 2004). Figure 5.2 shows one such profile, the model 'MF' of Mitrovica and Forte (2004) and Mitrovica et al. (2015). Its main feature is an increase in viscosity by more than two orders of magnitude between the base of the lithosphere and 2100 km depth. Also evident are thin layers of strongly (by a factor ≈ 10) reduced viscosity at the base of the transition zone and in D''.

Exercises

5.1 [SM] Plane elastic deformations can be described by a so-called stress function $\chi(x, z)$ defined such that

$$\sigma_{xx} = \partial_{zz}^2 \chi, \quad \sigma_{xz} = -\partial_{xz}^2 \chi, \quad \sigma_{zz} = \partial_{xx}^2 \chi. \tag{5.23}$$

The stress function satisfies the biharmonic equation $\nabla^4 \chi = 0$ (Landau and Lifshitz, 1986, § 7). Use the stress function formulation to generalize the solution (5.14) for an incompressible elastic plate to the case of a compressible plate with Poisson's ratio $\nu \neq 1/2$.

5.2 [SM] Derive the matrix elements A_{1j}, A_{2j} and A_{3j} (first three rows) given by (5.22).

6
Boundary-Layer Theory

6.1 The Boundary-Layer Approximation

Many geophysical flows occur in layers or conduits whose length greatly exceeds their thickness: examples include thermal boundary layers (TBLs), subducted oceanic lithosphere, mantle plume conduits, and gravity currents of buoyant plume material spreading beneath the lithosphere. In all these cases, the gradients of the fluid velocity and/or temperature across the layer greatly exceed the gradients along it, a fact that can be exploited to simplify the governing equations substantially. The classic example of this approach is boundary-layer (BL) theory, which describes the flow in layers whose thickness is controlled by a balance of diffusion and advection of a transported quantity.

A boundary layer (BL) is defined as a thin region in a flow field, usually adjoining an interface or boundary, where the gradients of some quantity transported by the fluid (e.g., vorticity, heat, or a chemical species) are large relative to those elsewhere in the flow. Physically, BLs arise when the boundary acts as a source of the transported quantity, which is then prevented from diffusing far from the boundary by strong advection. Boundary layers thus occur when $UL/D \gg 1$, where U and L are characteristic velocity and length scales for the flow, and D is the diffusivity of the quantity in question. In classical boundary-layer theory, the transported quantity is vorticity, the relevant diffusivity is the kinematic viscosity ν and BLs form when the Reynolds number $Re \equiv UL/\nu \gg 1$. Such BLs do not occur in the mantle, where $Re \sim 10^{-20}$. However, because the thermal diffusivity $\kappa \sim 10^{-23}\nu$, the Péclet number $UL/\kappa \gg 1$ for typical mantle flows, implying that TBLs will be present.

Although BLs are 3-D structures in general, nearly all BL models used in geodynamics involve one of the three simple geometries shown in Figure 6.1: (a) 2-D flow, (b) an axisymmetric plume and (c) axisymmetric flow along a surface of revolution. Let x and y be the coordinates parallel to and normal to the boundary

6.1 The Boundary-Layer Approximation

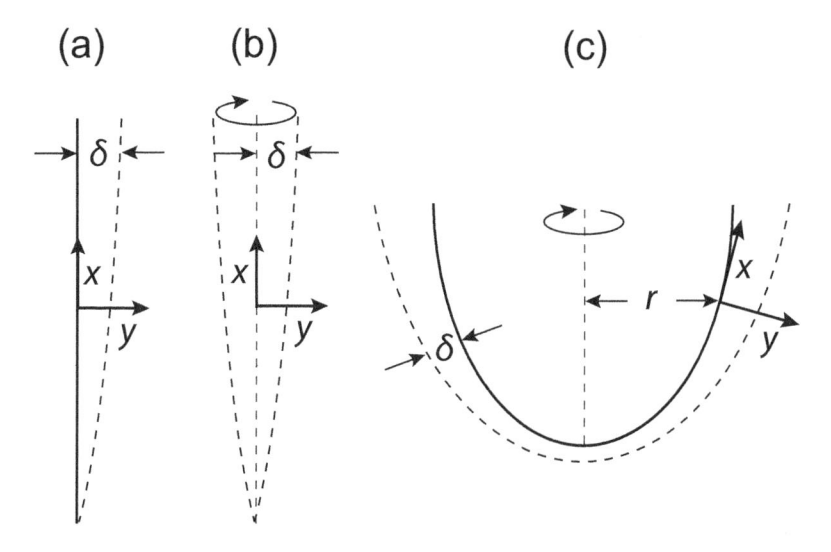

Figure 6.1 Canonical geometries for TBLs: (a) 2-D flow, (b) an axisymmetric plume and (c) axisymmetric flow along a surface of revolution. Impermeable boundaries are shown by heavy lines, axes of symmetry by light dashed lines and the edges of BLs by heavy dashed lines. The coordinates x and y are parallel to and normal to the boundary (or symmetry axis), respectively, $\delta(x)$ is the thickness of the BL and $r(x)$ is the radius of the surface of revolution. See § 6.1 for discussion.

(or symmetry axis), respectively, u and v be the corresponding velocity components and $\delta(x)$ be the thickness of the BL.

The fundamental hypothesis of BL theory is that gradients of the transported quantity (heat in this case) along the BL are much smaller than gradients across it. Diffusion of heat along the layer is therefore negligible. As an illustration, consider the simplest case of a thermal BL in steady flow with negligible viscous dissipation of energy and a constant thermal diffusivity. The BL forms of the continuity and energy equations are then

$$u_x + v_y = 0, \tag{6.1a}$$

$$uT_x + vT_y = \kappa T_{yy} \tag{6.1b}$$

for 2-D flow,

$$u_x + y^{-1}(yv)_y = 0, \tag{6.2a}$$

$$uT_x + vT_y = \kappa y^{-1}(yT_y)_y \tag{6.2b}$$

for an axisymmetric plume and

$$(ru)_x + rv_y = 0, \tag{6.3a}$$

$$uT_x + vT_y = \kappa T_{yy} \tag{6.3b}$$

for a surface of revolution. As already noted in § 2.3, the advection terms uT_x and vT_y are of the same order.

6.2 Solution of the Boundary-Layer Equations Using Variable Transformations

A powerful technique for solving problems involving BLs is the use of variable transformations. One of the most important of these is that of von Mises (1927), which transforms the 2-D BL equations into the classical heat conduction equation. A second useful transformation, due to Mangler (1948), relates the structure of an axisymmetric BL on a surface of revolution to that of a 2-D BL on a flat surface. After introducing both transformations, I will show how they can be used together to obtain a solution for the heat transfer from a hot sphere moving in a viscous fluid (Figure 1.1a).

6.2.1 Von Mises's Transformation

The essential trick involved in this transformation is to use the streamfunction ψ instead of y as the transverse coordinate in the BL. Consider for definiteness the 2-D BL equations (6.1) and denote the streamwise coordinate by a new symbol $s \equiv x$ for clarity. We transform the derivatives in (6.1b) using the chain rule as

$$\frac{\partial T}{\partial x} = \frac{\partial T}{\partial s}\frac{\partial s}{\partial x} + \frac{\partial T}{\partial \psi}\frac{\partial \psi}{\partial x} \equiv \frac{\partial T}{\partial s} - v\frac{\partial T}{\partial \psi}, \tag{6.4}$$

$$\frac{\partial T}{\partial y} = \frac{\partial T}{\partial s}\frac{\partial s}{\partial y} + \frac{\partial T}{\partial \psi}\frac{\partial \psi}{\partial y} \equiv u\frac{\partial T}{\partial \psi}, \tag{6.5}$$

where the streamfunction is defined according to the convention $(u, v) = (\psi_y, -\psi_x)$. Equation (6.1b) then becomes

$$\frac{\partial T}{\partial s} = \kappa \frac{\partial}{\partial \psi}\left(u\frac{\partial T}{\partial \psi}\right). \tag{6.6}$$

Equation (6.6) takes still simpler forms if the surface $y = \psi = 0$ is either free-slip or rigid. Near a free-slip surface, $u \approx U(s)$ is constant across the BL. Upon introducing a new downstream coordinate τ such that $d\tau = U(s)ds$, (6.6) becomes

$$\frac{\partial T}{\partial \tau} = \kappa \frac{\partial^2 T}{\partial \psi^2}, \tag{6.7}$$

which is just the classical equation for diffusion of heat in a medium with constant thermal diffusivity κ. The units of the time-like variable τ, however, are now those of diffusivity (m^2 s^{-1}) rather than of time. Near a rigid surface, $u = yf(s)$

and $\psi = y^2 f(s)/2$, implying $u \approx (2\psi f)^{1/2}$, where $f(s)$ is arbitrary. Substituting this result into (6.6) and introducing a new downstream variable τ such that $d\tau = (2f)^{1/2} ds$, we obtain

$$\frac{\partial T}{\partial \tau} = \kappa \frac{\partial}{\partial \psi} \left(\psi^{1/2} \frac{\partial T}{\partial \psi} \right), \tag{6.8}$$

which describes the diffusion of heat in a medium with a position-dependent thermal diffusivity $\kappa \psi^{1/2}$ as a function of a time-like variable τ having units m s$^{-1/2}$.

Of special interest are the self-similar solutions of (6.7) and (6.8) that exist when the wall temperature T_0 is constant ($= \Delta T$, say) and the upstream temperature profile is $T(0, \psi) = 0$. The solution of the free-slip surface equation (6.7) has the form $T = \Delta T F(\eta)$, where $\eta = \psi/\delta(\tau)$. Separating variables, we obtain $-F''/\eta F' = \delta \dot{\delta}/\kappa = \lambda^2$ (constant), which when solved subject to the conditions $F(0) - 1 = F(\infty) = \delta(0) = 0$ yields

$$T = \Delta T \operatorname{erfc}\left(\frac{\psi}{2\sqrt{\kappa \tau}}\right). \tag{6.9}$$

The solution of the rigid-surface equation (6.8) has the form $T = \Delta T F(\eta)$, where $\eta = \sqrt{\psi}/\delta(\tau)$. The form of η is chosen to simplify the ordinary differential equations (ODEs) satisfied by $F(\eta)$ and $\delta(\tau)$; the choice $\eta = \psi/\epsilon(\tau)$ with $\epsilon = \delta^2$ would also work. Separating variables, we find $-F''/4\eta^2 F' = \delta^2 \dot{\delta}/\kappa = \lambda^2$ (constant), which when solved subject to $F(0) - 1 = F(\infty) = \delta(0) = 0$ with $\lambda^2 = 1/3$ for convenience gives

$$\frac{T}{\Delta T} = 1 - \frac{3}{\Gamma(1/3)} \left(\frac{4}{9}\right)^{1/3} \int_0^{\eta} \exp\left(-\frac{4}{9} x^3\right) dx, \quad \eta = \frac{\psi^{1/2}}{(\kappa \tau)^{1/3}}, \tag{6.10}$$

where Γ is the gamma function. Later we shall see how the solution (6.9) can be applied to heat transfer from a moving sphere (§ 6.2.2) and to steady cellular convection (§ 10.4.3).

6.2.2 Mangler's Transformation

Mangler (1948) showed that the equations governing an axisymmetric BL on a surface of revolution with radius $r(x)$ (Figure 6.1) are related to the 2-D BL equations by the variable transformation

$$\bar{x} = \int_0^x (r/L)^2 dx, \quad \bar{y} = \frac{r}{L} y, \quad \bar{u} = u, \quad \bar{v} = \frac{L}{r}\left(v + \frac{r'}{r} y u\right), \quad \bar{\psi} = L^{-1}\psi, \tag{6.11}$$

where the variables with and without overbars are those of the 2-D and the axisymmetric flows, respectively, and L is an arbitrary constant length scale. The transformation (6.11), which applies equally to vorticity and thermal BLs, allows solutions of the 2-D BL equations to be transformed directly into solutions of the axisymmetric BL equations on a surface of revolution. To illustrate, we determine the heat flow from a sphere of radius a with a free-slip surface and excess temperature ΔT in a stream of viscous fluid with velocity $-U\mathbf{e}_z$ far from the sphere (Figure 1.1). The starting point is the Cartesian BL solution (6.9), which in terms of the barred variables is

$$T = \Delta T \operatorname{erfc}\left(\frac{\bar{\psi}}{2\sqrt{\kappa\tau}}\right), \quad \tau(\bar{x}) = \int_0^{\bar{x}} \bar{u} d\bar{x}. \tag{6.12}$$

For a sphere, $r(x) = a \sin(x/a) \equiv a \sin\theta$, and $L = a$ is the natural choice. The Stokes–Hadamard–Rybczynski solution (4.31) with $\gamma = 0$ gives $\psi \approx (1/2) U a y \sin^2\theta$ and $u \approx (1/2) U \sin\theta$, and (6.11) implies

$$\bar{\psi} = \frac{1}{2} U y \sin^2\theta, \quad \bar{x} = \frac{1}{2}x - \frac{a}{4}\sin\frac{2x}{a}. \tag{6.13}$$

The stretched downstream variable τ is therefore

$$\tau = \int_0^{\bar{x}} \bar{u} d\bar{x} = \int_0^x u \frac{d\bar{x}}{dx} dx = \frac{2}{3} aU(2 + \cos\theta)\sin^4\frac{\theta}{2}. \tag{6.14}$$

Substitution of (6.13) and (6.14) into (6.12) yields the temperature $T(y, \theta)$ everywhere in the BL, and the corresponding local Nusselt number is

$$\mathcal{N}(\theta) \equiv -\frac{a}{\Delta T}\frac{\partial T}{\partial y}(y=0) = \left(\frac{3}{2\pi}\right)^{1/2}\frac{1+\cos\theta}{(2+\cos\theta)^{1/2}}Pe^{1/2}, \tag{6.15}$$

where $Pe = Ua/\kappa$ is the Péclet number. Note that $\mathcal{N} = 0$ at the rear stagnation point $\theta = \pi$, which indicates that the dimensionless BL thickness $\delta/a \to \infty$ there. The BL approximation therefore breaks down in the vicinity of the rear stagnation point. However, this failure is only local and does not vitiate the solution over the rest of the range $0 \leq \theta \leq \pi$.

The result (6.15), together with our previous treatments of the hot sphere in § 2.1, § 2.2 and § 2.3, shows that BL theory represents a third stage in a hierarchy of techniques (dimensional analysis, scaling analysis, BL theory) that give progressively more detailed information about the structure of the solution in the asymptotic limit of negligible inertia ($Re \ll 1$) and $Pe \to \infty$. Table 6.1 summarizes the local Nusselt number $\mathcal{N}(\theta)$ for a free-slip sphere predicted by each of the three techniques.

Table 6.1. *Heat transfer from a hot sphere for Re $\ll$ 1: analytical predictions*

Technique	Local Nusselt number $\mathcal{N}$	Validity
Dimensional analysis	fct(Pe, θ)	Universal
Scaling analysis	$Pe^{1/2}$fct(θ)	$Pe \gg 1$
BL theory	$\left(\dfrac{3}{2\pi}\right)^{1/2} \dfrac{1+\cos\theta}{(2+\cos\theta)^{1/2}} Pe^{1/2}$	$Pe \gg 1, 0 \leq \theta < \pi$

6.3 The Method of Matched Asymptotic Expansions

As noted earlier (§ 4.8), the method of matched asymptotic expansions (MMAE) is a powerful method for solving problems in which the field variables exhibit distinct regions characterized by very different length scales. The method is particularly well suited for BLs, whose characteristic thickness δ is much smaller than the scale L of the flow outside the BL. As an illustration of the method, consider a simple axisymmetric stagnation-point flow model for the steady temperature distribution in a plume upwelling beneath a rigid lithosphere (Figure 6.2), in which fluid with temperature T_1 and upward vertical velocity $-w_1$ at a depth $z = d$ ascends towards a rigid surface with temperature $T = T_0$. If viscous dissipation of energy is negligible, the pressure is (nearly) hydrostatic, and all physical properties of the fluid are constant, then $T(\mathbf{x})$ satisfies

$$\mathbf{u} \cdot \nabla T - \frac{g\alpha}{c_p} \mathbf{u} \cdot \mathbf{e}_z T = \kappa \nabla^2 T, \tag{6.16}$$

where α is the thermal expansion coefficient and c_p is the heat capacity at constant pressure. The three terms in (6.16) represent advection of temperature gradients, adiabatic decompression and thermal diffusion, respectively.

Let $\mathbf{u} = u(r,z)\mathbf{e}_r + w(r,z)\mathbf{e}_z$. Because $w(r,d) \equiv -w_1$ is constant, $w = w(z)$, whence the continuity equation implies $u = -rw'(z)/2$. Equation (6.16) therefore admits a 1-D solution $T = T(z)$ that satisfies

$$wT' - \frac{g\alpha}{c_p} wT = \kappa T'', \tag{6.17}$$

where primes denote d/dz. By substituting $w = w(z)$ and $u = -rw'(z)/2$ into the constant-viscosity Stokes equations in cylindrical coordinates and solving the resulting equation for $w(z)$ subject to the boundary conditions shown in Figure 6.2, we find

$$\frac{w}{w_1} = 2\left(\frac{z}{d}\right)^3 - 3\left(\frac{z}{d}\right)^2. \tag{6.18}$$

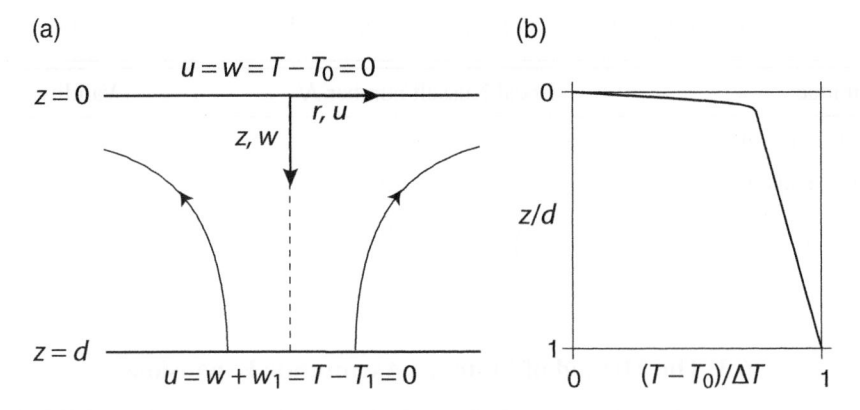

Figure 6.2 Model for the temperature distribution in a steady stagnation-point flow beneath a rigid lithosphere (§ 6.3). (a) The base of the lithosphere $z = 0$ is at temperature T_0, and a uniform upward vertical velocity $w_1 > 0$ is imposed at a depth $z = d$ where the temperature is T_1. The curved lines with arrows are streamlines. (b) Temperature as a function of depth for $\epsilon = 0.0001$, $\beta = 0.1$ and $\gamma = 2.0$.

Upon introducing dimensionless variables $\tilde{z} = z/d$ and $\tilde{T} = (T - T_0)/(T_1 - T_0) \equiv (T - T_0)/\Delta T$ and then immediately dropping the tildes, (6.16) together with (6.18) and the boundary conditions on T become

$$(2z^3 - 3z^2)\left[T' - \beta(T + \lambda)\right] = \epsilon T'', \tag{6.19a}$$

$$T(0) = T(1) - 1 = 0, \tag{6.19b}$$

where

$$\epsilon = \frac{\kappa}{dw_1} \equiv Pe^{-1}, \quad \beta = \frac{g\alpha d}{c_p}, \quad \lambda = \frac{T_0}{T_1 - T_0}. \tag{6.20}$$

We wish to solve (6.19) in the limit $\epsilon \to 0$ ($Pe \to \infty$), assuming for simplicity that $\beta = O(1)$ and $\lambda = O(1)$. However, note that ϵ appears in (6.19a) as the coefficient of the most highly differentiated term. We therefore cannot simply set $\epsilon = 0$ in (6.19a), because that would reduce the order of the ODE and make it impossible to satisfy all the boundary conditions. Equation (6.19) therefore constitutes a singular perturbation problem: the solution for small values of $\epsilon > 0$ is not a small perturbation of a solution for $\epsilon = 0$, which in any case does not exist. The resolution of this apparent paradox is that the solution exhibits a thin BL where the term $\epsilon T''$ in (6.19a) is important, no matter how small ϵ may be.

We therefore anticipate that the solution to (6.19) will comprise two distinct regions governed by different dynamics: an inner region (the BL) of dimensionless thickness $\delta \ll 1$, in which advection is balanced by diffusion, and an outer region,

6.3 The Method of Matched Asymptotic Expansions

where advection is balanced by adiabatic decompression. Consider the outer region first, and let the temperature there be $T = h(z, \epsilon)$. We seek a solution in the form of an asymptotic expansion

$$h = h_0(z) + \eta_1(\epsilon)h_1(z) + \eta_2(\epsilon)h_2(z) + \cdots \quad (6.21)$$

where $\eta_n(\epsilon)$ are (as yet unknown) gauge functions that form an asymptotic sequence such that $\lim_{\epsilon \to 0}(\eta_n/\eta_{n-1}) = 0$. The function $\eta_0(\epsilon) = 1$ because $T = O(1)$ in the outer region. Substituting (6.21) into (6.19a) and retaining only the lowest-order terms, we obtain

$$h_0' - \beta(h_0 + \lambda) = 0. \quad (6.22)$$

Because (6.22) is a first-order ODE, its solution can satisfy only one of the boundary conditions (6.19b). Evidently this must be the condition at $z = 1$, because $z = 0$ is the upper limit of a TBL that cannot be described by (6.22). The same conclusion can be reached in a more formal algorithmic way by assuming contrary to fact that the BL is at $z = 1$, solving (6.22) subject to the wrong boundary condition $T(0) = 0$ and finally realizing that the inner solution for the supposed BL is unphysical because it increases exponentially upward. The lowest-order terms of the (correct) boundary condition are $h_0(1) = 1$, and the solution of (6.22) that satisfies this is

$$h_0 = (1 + \lambda)\exp\beta(z - 1) - \lambda. \quad (6.23)$$

Turning now to the inner region, the first task is to determine which term on the left-hand side (LHS) of (6.19a) balances the right-hand side (RHS). Suppose (contrary to fact, as will soon appear) that $\beta(T+\lambda) \gg T'$ (decompression $\gg$ advection). Because $z \sim \delta$ and $z^2 \gg z^3$ in the BL, the balance $z^2\beta(T + \lambda) \sim \epsilon T''$ implies $\delta \sim \epsilon^{1/4}$. But then $T' \gg \beta(T + \lambda)$, which contradicts our original assumption. The correct balance must therefore be $z^2 T' \sim \epsilon T''$ (advection $\sim$ diffusion), which implies $\delta \sim \epsilon^{1/3}$ and $T' \gg \beta(T + \lambda)$, consistent with our original assumption. A more physical argument that leads to the same conclusion is to note that the vertical scale length over which adiabatic decompression is significant in the mantle ($\sim$ 1000 km) is much greater than a typical BL thickness.

Now that we know the thickness of the BL, we can proceed to determine its structure. To discern the thin BL distinctly, we use a sort of mathematical magnifying glass: a new stretched depth coordinate $\hat{z} = z/\delta \equiv z\epsilon^{-1/3}$ that is of order unity in the BL. Denoting the inner solution by $f(\hat{z}, \epsilon)$ and writing (6.19a) in terms of $\hat{z}$, we find

$$(2\epsilon\hat{z}^3 - 3\epsilon^{2/3}\hat{z}^2)\left[\epsilon^{-1/3}f' - \beta(f + \lambda)\right] = \epsilon^{1/3}f'', \quad (6.24)$$

where primes denote differentiation with respect to $\hat{z}$. We now seek a solution in the form

$$f = f_0(z) + \nu_1(\epsilon)f_1(z) + \nu_2(\epsilon)f_2(z) + \cdots \quad (6.25)$$

where $\nu_n(\epsilon)$ are unknown gauge functions. Substituting (6.25) into (6.24) and retaining only the lowest-order terms, we obtain

$$f_0'' + 3\hat{z}^2 f_0' = 0. \quad (6.26)$$

The solution of (6.26) that satisfies the boundary condition $f_0(0) = 0$ is

$$f_0 = A \int_0^{\hat{z}} \exp(-x^3)\mathrm{d}x \equiv \frac{A}{3}\gamma(1/3, \hat{z}^3), \quad (6.27)$$

where $\gamma(a, x) = \int_0^x \exp(-t)t^{a-1}\mathrm{d}t$ is the incomplete Gamma function and A is an unknown constant that must be determined by matching (6.27) to the outer solution (6.23).

The most rigorous way to do the matching is to use the so-called intermediate matching principle (IMP). In the present example, this would consist of rewriting both the inner and outer expansions in terms of an intermediate variable $z_{int} = \xi(\epsilon)z$ such that $\epsilon^{-1/3} \ll \xi \ll 1$, and then choosing the values of any unknown constants (A in this case) so that the two expressions agree. The IMP is explained in detail in the solution to Exercise 6.6. However, it is often possible to use a simpler matching principle, due to Prandtl, which states that the inner limit of the outer expansion must be equal to the outer limit of the inner expansion – roughly speaking, that the two expansions must match at the edge of the BL. For our problem, Prandtl's principle is

$$\lim_{\hat{z} \to \infty} f_0(\hat{z}) = \lim_{z \to 0} h_0(z), \quad \text{whence} \quad A = \frac{3[(1+\lambda)\exp(-\beta) - \lambda]}{\Gamma(1/3)} \quad (6.28)$$

where Γ is the Gamma function.

The last step is to construct a composite expansion that is valid both inside and outside the BL. This is just the sum of the inner and outer expansions less their shared common part $h_0(z \to 0) \equiv (1+\lambda)\exp(-\beta) - \lambda$, or

$$T = h_0(z) + \left[(1+\lambda)\exp(-\beta) - \lambda\right]\left[\frac{\gamma(1/3, z^3/\epsilon)}{\Gamma(1/3)} - 1\right]. \quad (6.29)$$

The composite expansion (6.29) is shown in Figure 6.2b for $\epsilon = 0.0001$, $\beta = 0.1$ and $\gamma = 2.0$.

The procedure just described is a first-order matching that retains only the first terms in the expansions (6.21) and (6.25). If desired, the matching can be carried out to higher order by working back and forth between the inner and outer expansions, determining the gauge functions $\eta_n(\epsilon)$ and $\nu_n(\epsilon)$ and matching at each step. Higher-order matching often requires the use of the more rigorous IMP; for examples, see Hinch (1991), Kevorkian and Cole (1996) or the somewhat less formal treatments of Nayfeh (1973) or Van Dyke (1975). For many problems, however, first-order matching suffices to reveal the essential structure and physical significance of the solution.

The MMAE has been used to solve a variety of geodynamically relevant problems involving BLs, mostly with constant viscosity. Umemura and Busse (1989) studied the axisymmetric convective flow in a cylindrical container of height d with free-slip boundaries, using the MMAE to match the interior flow to the central rising plume and the circumferential downwelling. They found that the vertical velocity in the plume is $w \sim (\kappa/d)(-\ln \epsilon)^{1/2} Ra^{2/3}$, where the dimensionless plume radius $\epsilon \equiv a/d$ satisfies $\epsilon(-\ln \epsilon)^{1/4} \sim Ra^{-1/6}$, $Ra \gg 1$ being the Rayleigh number. Whittaker and Lister (2006a) studied creeping plumes from a point heat source with buoyancy flux B on an impermeable plane boundary by matching the BL flow to an outer flow that sees the plume as a line distribution of Stokeslets. This problem will be discussed in more detail in the following section. Whittaker and Lister (2006b) studied the dynamics of a plume above a heated disk on a plane boundary and used the MMAE to match the flow within the plume to both an outer flow and a horizontal BL flow across the disk. The results show that the Nusselt number $Nu \sim Ra^{1/5}$ for a rigid boundary and $Nu \sim (Ra/\ln Ra)^{1/3}$ for a free-slip boundary, where $Ra \gg 1$ is the Rayleigh number defined using the disk radius. Finally, Morris (1982) used the MMAE to study the problem of heat transfer from a hot sphere moving in a fluid with temperature-dependent viscosity (§ 1.1). A simple scaling analysis of this situation is performed in § 6.5.2, and you are invited to apply the MMAE to the problem in Exercise 6.7.

6.4 A Plume from a Point Source of Buoyancy

A phenomenon of fundamental geodynamic importance is a thermal plume from a point source of buoyancy in an isoviscous fluid. This problem is of particular interest in the context of this book because its solution requires many of the concepts and techniques we have already seen, including singular solutions of the Stokes equations, slender-body theory, boundary-layer theory and matched asymptotic expansions. The following abbreviated discussion is based on Whittaker and Lister (2006a) (henceforth WL06) with some changes of notation.

Much of the complexity of the point-source plume problem is due to the need to avoid Stokes's paradox. If the fluid containing the point source is infinite in all directions then the Stokes equations have no solution because the velocity diverges logarithmically at great distances from the plume. It is therefore necessary to find

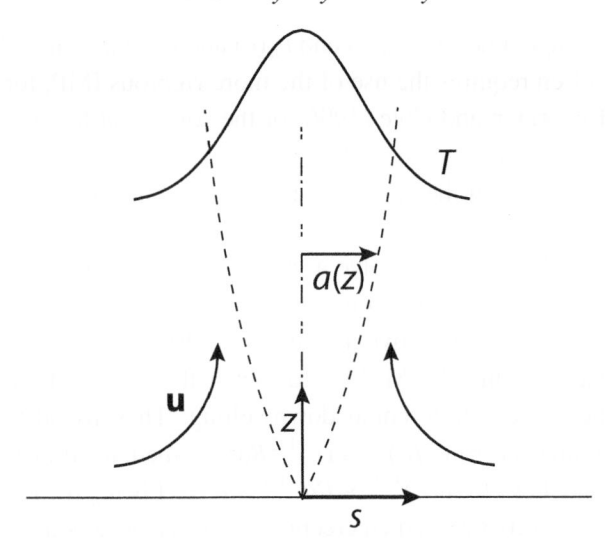

Figure 6.3 Definition sketch for a plume from a point source of buoyancy on a free-slip boundary, as discussed in § 6.4. Figure reproduced from Figure 1 of Whittaker and Lister (2006a) by permission of Cambridge University Press.

a means of removing this divergence. One way to do this is to include inertia (Worster, 1986), but this expedient is not geodynamically realistic. A more realistic approach is to suppose that the point source is located on an infinite horizontal boundary, which can be either free-slip or rigid. The buoyancy of the plume can then be balanced by normal stresses acting on the horizontal boundary, obviating the need for inertia. Such an approach also makes geodynamic sense because the CMB acts as a free-slip surface for plumes arising from the D'' layer.

Figure 6.3 shows a sketch of the plume. Cylindrical coordinates (s, z) are used to describe the plume, whose radius is $a(z)$. The temperature profile across the plume and the velocity in the fluid outside it are shown schematically.

The governing equations for the plume are

$$\nabla \cdot \mathbf{u} = 0, \tag{6.30a}$$

$$\nu \nabla^2 \mathbf{u} = \rho_0^{-1} \nabla p - b \mathbf{e}_z, \tag{6.30b}$$

$$\mathbf{u} \cdot \nabla b = \kappa \nabla^2 b, \tag{6.30c}$$

where $\mathbf{u}$ is the velocity, $b = \alpha g(T - T_0)$ is the buoyancy, ν is the kinematic viscosity, α is the thermal expansivity, g is the gravitational acceleration, T_0 is the far-field temperature, ρ_0 is the density at the temperature T_0, $p = p_{\text{tot}} + \rho g z$ is the modified pressure, p_{tot} is the total pressure, $\mathbf{e}_z$ is an upward-pointing vertical unit vector and κ is the thermal diffusivity. The far-field boundary conditions are $\mathbf{u} \to \mathbf{0}$ and

6.4 A Plume from a Point Source of Buoyancy

$b \to 0$. To maximize geodynamic realism we shall consider only a free-slip horizontal boundary. The final condition is that of conservation of buoyancy, which implies that the vertical buoyancy flux

$$B = 2\pi \int_0^\infty wbs\,ds \tag{6.31}$$

is independent of height, where w is the vertical velocity.

We begin by performing a scaling analysis of the governing equations. From the advection-diffusion equation (6.30c) and the conservation of buoyancy (6.31), we obtain

$$\frac{w}{z} \sim \frac{\kappa}{a^2}, \quad wba^2 \sim B \quad \text{whence} \quad b \sim \frac{B}{\kappa z}. \tag{6.32a-c}$$

An estimate of the vertical velocity w can be obtained using slender-body theory, modelling the plume as a vertical line distribution of Stokeslets with local density $F \sim ba^2$. The leading order scaling is obtained from (4.114b) for a rod moving longitudinally and is (Exercise 6.2)

$$w \sim \frac{F}{2\pi\nu} \ln\left(\frac{z}{a}\right). \tag{6.33}$$

Combining (6.32) and (6.33) and introducing a length scale $z_0 \propto (\nu\kappa^2/B)^{1/2}$, we obtain

$$w \sim \left(\frac{B}{\nu}\right)^{1/2} \left[\ln\left(\frac{z}{a}\right)\right]^{1/2}, \tag{6.34a}$$

$$F \sim (B\nu)^{1/2} \left[\ln\left(\frac{z}{a}\right)\right]^{-1/2}, \tag{6.34b}$$

$$a \sim (z_0 z)^{1/2} \left[\ln\left(\frac{z}{a}\right)\right]^{-1/4}. \tag{6.34c}$$

Finally, it can be shown (Exercise 6.3) that the vertical velocity inside the plume is dominated by a plug-flow component $w_0(z)$, which is much greater in magnitude than the radial variation $\tilde{w}(s,z)$ across the plume. The separation of scales $a \ll z$ and $\tilde{w} \ll w_0$ can then be exploited to derive inner and outer expansions for the plume and the surrounding fluid.

Consider first the inner solution. To describe it, we introduce a Stokes streamfunction $\psi(s,z)$, which is related to the velocity components by

$$u \equiv \mathbf{u} \cdot \mathbf{e}_s = -\frac{1}{s}\frac{\partial\psi}{\partial z}, \quad w \equiv \mathbf{u} \cdot \mathbf{e}_z = \frac{1}{s}\frac{\partial\psi}{\partial s}. \tag{6.35}$$

Now from the advection-diffusion balance (6.32a), we expect the plume radius to scale as $a = (4\kappa z/w_0)^{1/2}$, where the leading-order vertical velocity $w_0(z)$ is as yet unknown. We therefore introduce a scaled radial variable

$$\xi = s \left(\frac{w_0(z)}{4\kappa z} \right)^{1/2}. \tag{6.36}$$

Now using the scales (6.34), we write the Stokes streamfunction and buoyancy fields in terms of ξ and z as

$$\psi = 4\kappa z f(\xi; z), \quad b = \frac{B}{2\pi} \frac{h(\xi; z)}{4\kappa z} \tag{6.37a,b}$$

where the factor of $1/(2\pi)$ is added for later convenience. Now with the definitions (6.36) and (6.37a), the vertical velocity is

$$w = \frac{w_0}{\xi} \frac{\partial f}{\partial \xi}. \tag{6.38}$$

We therefore require $f'(\xi; z) \sim \xi$ at leading order, where a prime denotes differentiation with respect to ξ. This must hold as $z \to \infty$ with ξ fixed, which ensures an asymptotically uniform vertical flow in each cross-section of the plume.

The expressions (6.37) for ψ and b are now substituted into the momentum equation (6.30b) and (6.30c) and the buoyancy flux condition (6.31). After applying standard BL approximations to neglect the vertical derivatives in the Laplacians and the vertical gradient of the modified pressure, we obtain (Exercise 6.4)

$$\frac{1}{\xi} \frac{\partial}{\partial \xi} \left(\xi \frac{\partial}{\partial \xi} \left(\frac{1}{\xi} \frac{\partial f}{\partial \xi} \right) \right) = -\epsilon h, \tag{6.39a}$$

$$f'(-h + z h_z) - (f + z f_z) h' = \frac{1}{4} (\xi h')', \tag{6.39b}$$

$$\int_0^\infty h f' \, d\xi = 1, \tag{6.39c}$$

where subscripts z denote partial differentiation with respect to z and

$$\epsilon(z) = \frac{B}{2\pi \nu w_0^2} \ll 1. \tag{6.40}$$

The scalings (6.34) imply that $\epsilon(z) \sim (\ln z)^{-1}$.

WL06 seek solutions for f and h as expansions of the forms

$$f(\xi; z) = f_0(\xi) + \epsilon(z) f_1(\xi) + O(z\epsilon_z, \epsilon^2), \tag{6.41a}$$

$$h(\xi; z) = h_0(\xi) + \epsilon(z) h_1(\xi) + O(z\epsilon_z, \epsilon^2). \tag{6.41b}$$

Substituting (6.41) into (6.39) and gathering terms proportional to like powers of ϵ, we obtain equations for (f_0, h_0) and (f_1, h_1) that can be solved sequentially. The solutions for f_0, h_0 and f_1 are

6.4 A Plume from a Point Source of Buoyancy

$$f_0(\xi) = \frac{1}{2}\xi^2, \quad h_0(\xi) = 2\exp(-\xi^2), \tag{6.42a}$$

$$f_1(\xi) = \frac{1}{2}A\xi^2 - \frac{1}{4}\xi^2(2\ln\xi + Ei(\xi^2) + \gamma - 1) + \frac{1}{4}[\exp(-\xi^2) - 1], \tag{6.42b}$$

where $Ei(x) = \int_1^\infty t^{-1}e^{-xt}\,dt$ and $\gamma = 0.5772\ldots$ is Euler's constant. The constant A is the amplitude of an $O(\epsilon)$ correction to the vertical plug flow and can only be determined by matching with the outer solution. A closed-form solution for h_1 can also be obtained, but WL06 do not write it down due to its complexity. Substitution of (6.41a) into (6.38) now implies

$$w = w_0(z)\left[1 + \epsilon(z)\left(-\ln\xi - \frac{1}{2}Ei(\xi^2) + A - \frac{\gamma}{2}\right) + \cdots\right]. \tag{6.43}$$

We now turn to the outer solution, which is driven by the buoyancy distribution $b(\mathbf{x})$ of the inner solution. The basic idea is to express the outer velocity as a linear superposition of contributions from a distribution of point forces (Stokeslets) acting vertically. Thus we have

$$\mathbf{u}(\mathbf{x}) = \frac{1}{\nu}\int_V b(\mathbf{x}')\mathbf{e}_z \cdot \mathbf{G}^B(\mathbf{x}, \mathbf{x}')dV', \tag{6.44}$$

where the Green function $\mathbf{G}^B(\mathbf{x}, \mathbf{x}')$ satisfies the free-slip boundary conditions at $z = 0$ and is given by (4.93), (4.94) and (4.88a). To simplify (6.44), we now exploit the fact that the buoyancy is confined to a narrow region of radial extent $\sim a \ll z$. We may therefore replace (6.44) by its multipole expansion, valid outside the plume. Because the buoyancy distribution is confined radially but not vertically, we expand (6.44) in terms of the radial moments while leaving the vertical integrals intact. WL06 show that keeping only the first term in this expansion introduces a negligible error, whence we obtain

$$\mathbf{u}(\mathbf{x}) = \frac{1}{\nu}\int_0^\infty \mathcal{F}(z')\mathbf{e}_z \cdot \mathbf{G}^B(\mathbf{x}, z'\mathbf{e}_z)dz' \tag{6.45}$$

where

$$\mathcal{F}(z) = 2\pi\int_0^\infty b(s, z)s\,ds. \tag{6.46}$$

The explicit expression for $\mathcal{F}(z)$ is

$$\mathcal{F}(z) = \frac{B}{w_0}\int_0^\infty h(\xi; z)\xi\,d\xi = \frac{B}{w_0}\left[1 - \epsilon(z)\left(A - \frac{1}{2}\ln 2\right) + \cdots\right]. \tag{6.47}$$

The detailed calculation of the outer velocity field is complicated, and WL06 relegate it to their Appendix B. However, all we need for now is the asymptotic form of the outer solution as $s/z \to 0$ because this will allow us to perform the matching to the inner solution and thereby determine $w_0(z)$. Defining $K(s, z; z') = 8\pi G^B_{33}(\mathbf{x}, z'\mathbf{e}_z)$, we can write the vertical component of (6.45) as

$$8\pi \nu w(s, z) = \int_0^\infty \mathcal{F}(z') K(s, z; z') \, dz' \tag{6.48}$$

where

$$K(s, z; z') = \frac{s^2 + 2(z - z')^2}{(s^2 + (z - z')^2)^{3/2}} - \frac{s^2 + 2(z + z')^2}{(s^2 + (z + z')^2)^{3/2}}. \tag{6.49}$$

The first term on the RHS of (6.49) represents a Stokeslet on the plume axis and is of the same form as the integrand in slender-body theory. The second term represents the image Stokeslet, which cancels the far-field $O(z'^{-1})$ behaviour of the direct Stokeslet and thereby ensures convergence of the integrals.

Now from the estimate (6.34a) for w_0, we expect $\mathcal{F}$ to be a slowly varying logarithmic function of z'. We can therefore approximate $\mathcal{F}(z')$ in (6.48) by its value at $z' = z$, where the dominant contribution to the integral occurs. The factor $\mathcal{F}(z)$ can then be moved outside the integral, leaving an integral of K that can be evaluated analytically:

$$\int_0^\infty K(s, z; z') \, dz' = -4\ln\left(\frac{s}{z}\right) + 4\left(\ln 2 - \frac{1}{2}\right) + O\left(\frac{s^2}{z^2}\right). \tag{6.50}$$

Combining (6.47), (6.48) with $\mathcal{F}(z') \to \mathcal{F}(z)$ and (6.50), we find that

$$w \sim \frac{B}{2\pi \nu w_0(z)} \left[1 - \epsilon(z)\left(A - \frac{1}{2}\ln 2\right)\right]\left[-\ln\left(\frac{s}{z}\right) + \ln 2 - \frac{1}{2}\right] \tag{6.51}$$

at the inner edge of the outer solution, i.e., for $s \ll z$.

We are now in a position to match the inner (6.43) and outer (6.51) expressions for the vertical velocity. Here we do only the leading-order matching; WL06 may be consulted for the details of the $O(\epsilon)$ matching. Equating the leading-order terms in (6.43) and (6.51), we have

$$w_0 \sim -\frac{B}{2\pi \nu w_0} \ln\left(\frac{s}{z}\right). \tag{6.52}$$

Now (6.36) and (6.40) imply that

$$\frac{s}{z} = \xi\left(\frac{4\kappa}{zw_0}\right)^{1/2} = \xi \epsilon^{1/4}\left(\frac{z}{z_0}\right)^{-1/2}, \quad \text{where } z_0 = \left(\frac{32\pi \kappa^2 \nu}{B}\right)^{1/2}. \tag{6.53}$$

Substituting (6.53) into (6.52) and using (6.40), we find

$$w_0 = w_0 \epsilon \left[-\ln \xi - \frac{1}{4}\ln \epsilon + \frac{1}{2}\ln\left(\frac{z}{z_0}\right)\right]. \tag{6.54}$$

Now for fixed ξ, either the second or the third term in the square brackets is dominant as $\epsilon \to 0$. Assuming the third term to be the dominant one, we solve (6.54) to obtain $\epsilon \sim 2[\ln(z/z_0)]^{-1}$. The third term $(= \epsilon^{-1})$ then greatly exceeds the second, and so our assumption is consistent. Finally, we rearrange the definition (6.40) of ϵ to obtain the leading-order expression for the vertical velocity,

$$w_0(z) \sim \left(\frac{B \ln(z/z_0)}{4\pi \nu}\right)^{1/2}, \qquad (6.55)$$

which is consistent with the scaling estimates (6.34a) and (6.34c).

In closing, I emphasize that the parameter $\epsilon \sim 2[\ln(z/z_0)]^{-1}$ is only logarithmically small for $z \gg z_0$. In such problems, accuracy requires that one go beyond leading-order matching, as WL06 did by matching to $O(\epsilon)$. For reference, the resulting expression for w_0 is

$$w_0(z) \sim \left(\frac{B \ln(z/z_0)}{4\pi \nu}\right)^{1/2} \left(1 + \frac{\ln \ln(z/z_0) - \ln 2}{4 \ln(z/z_0)} + \cdots\right). \qquad (6.56)$$

6.5 Boundary Layers with Strongly Variable Viscosity

In a fluid with constant viscosity and infinite Prandtl number, TBLs are not accompanied by vorticity BLs, because the velocity field varies on a length scale that greatly exceeds the thickness of the TBL. As a result, the velocity within the TBL can be represented by the first (constant) term of the Taylor series expansion of the velocity profile. This greatly simplifies the task of solving the TBL equation. If, however, the viscosity depends on temperature, then the velocity varies across the TBL and is strongly coupled to the temperature field. Problems of this type are much harder to solve than their constant-viscosity counterparts.

Most studies of variable-viscosity BLs in the geodynamics literature focus on mantle plumes in either planar (Figure 6.1a) or axisymmetric (Figure 6.1b) geometry. The basic procedure is to supplement the BL equations (6.1) or (6.2) with a simplified BL form of the vertical component of the momentum equation in which derivatives along the BL are neglected relative to those normal to it. In physical terms, this equation simply states that the buoyancy force in the plume is balanced by the lateral gradient of the vertical shear stress τ, or

$$y^{-m} \partial_y (y^m \tau) + \rho g \alpha (T - T_\infty) = 0, \qquad (6.57)$$

where T_∞ is the temperature far from the plume, y is the coordinate normal to the plume and $m = 0$ or 1 for planar or axisymmetric geometries, respectively.

An important early study based on (6.57) in planar geometry was that of Yuen and Schubert (1976), who investigated the buoyant upwelling of a fluid with temperature-dependent non-Newtonian rheology adjacent to an isothermal free-slip boundary. The BL equations for this case admit a similarity transformation of the form

$$T = T_\infty + f(\eta), \quad \psi = x^{(n+2)/(n+3)} g(\eta), \quad \eta = yx^{-1/(n+3)}, \qquad (6.58)$$

where ψ is the streamfunction and n is the power-law exponent in the non-Newtonian constitutive law (4.1f). However, neither the planar geometry nor the isothermal boundary condition is realistic for Earth.

6.5.1 A Plume from a Point Source of Heat

A more realistic situation is a plume from a point source of heat in a fluid with temperature-dependent viscosity, which has been studied by several authors. Morris (1980) studied the case of a viscosity varying exponentially with temperature. Using a similarity transformation, he showed that the temperature on the plume axis decreases exponentially upward with a scale height $Q\gamma/12\pi k$, where Q is the total heat flux carried by the plume, γ^{-1} is the temperature change required to change the viscosity by a factor e and k is the thermal conductivity. Because Morris's solution only appeared in his PhD. thesis, it remained largely unknown to subsequent researchers. Loper and Stacey (1983) examined the case of a viscosity varying as the exponential of the inverse temperature, with a prescribed flux of hot material at the plume's base. Olson et al. (1993) determined an approximate solution for exponentially temperature-dependent viscosity in which the form of the temperature profile across the plume was assumed. Hauri et al. (1994) found a similarity solution for a plume with a complex superexponential temperature-, stress- and depth-dependent viscosity law. More recently, the problem of Morris (1980) was solved using a different method by Crosby and Lister (2014) (henceforth CL14). The following development is essentially that of CL14 but with a different nondimensionalization.

Because the problem is axisymmetric, the Stokes equations with temperature-dependent viscosity have no solution if the fluid is infinite in all directions, just as for the previous problem of an isoviscous plume from a point source. CL14's way of getting around this difficulty is conditioned by their use of a BL approximation for the flow in the plume. Because there is no plume-parallel diffusion of momentum in the BL approximation, the plume's buoyancy cannot be supported by normal stresses on a horizontal boundary, and so that resolution of Stokes's paradox is unavailable. Instead, CL14 choose to restrict the domain of their solution to a finite maximum radius where the plume's buoyancy is supported. The finitude of the

6.5 Boundary Layers with Strongly Variable Viscosity

solution domain also implies that the effects of inertia can be neglected. Moreover, as long as the maximum radius is well outside the low-viscosity core of the plume, its exact value has little influence on the solution of the BL equations in the core.

CL14 assume that the viscosity depends exponentially on temperature as

$$\eta = \eta_\infty \exp[-\gamma(T - T_\infty)], \tag{6.59}$$

where T_∞ is the background temperature far from the plume. The boundary-layer equations governing the plume are (6.2) and (6.57) with $m = 1$ and $\tau = \eta \partial_y u$, where (to recall the notation of § 6.1) y is the radial coordinate and u is the plume-parallel velocity. Finally, we define the heat flux

$$Q = 2\pi \rho_\infty c_p \int_0^{y_{max}} (T - T_\infty) u y \, dy \tag{6.60}$$

where ρ_∞ is the density at temperature $T = T_\infty$, c_p is the heat capacity at constant pressure and y_{max} is the maximum radius. Because there is negligible heat loss across $y = y_{max}$, Q is independent of height.

Now define dimensionless (hatted) variables

$$\hat{T} = \gamma(T - T_\infty), \quad \hat{r} = \frac{yRa^{1/4}}{L}, \quad \hat{z} = \frac{x}{L}, \quad \hat{u} = \frac{Lv}{\kappa Ra^{1/4}}, \quad \hat{w} = \frac{Lu}{\kappa Ra^{1/2}}, \tag{6.61}$$

where $L = Q\gamma/k$, $\kappa = k/\rho_\infty c_p$ is the thermal diffusivity and the Rayleigh number is

$$Ra = \frac{\rho_\infty \alpha g L^3}{\gamma \kappa \eta_\infty} \tag{6.62}$$

where α is the thermal expansivity. The notation for the dimensionless variables is chosen to correspond to the more familiar notation for mantle plumes: (r, u) in the radial direction and (z, w) in the vertical direction. Note also that Ra is defined using the rheological temperature scale γ^{-1}, which is the only temperature scale in the problem.

Rewriting the BL equations and the expression for the heat flux in terms of the dimensionless variables (6.61) and dropping the hats, we obtain

$$\frac{1}{r}\partial_r(ru) + \partial_z w = 0, \tag{6.63a}$$

$$\frac{1}{r}\partial_r\left(re^{-T}\partial_r w\right) + T = 0, \tag{6.63b}$$

$$u\partial_r T + w\partial_z T = \frac{1}{r}\partial_r(r\partial_r T), \tag{6.63c}$$

$$2\pi \int_0^{r_{max}} Twr \, dr = 1, \tag{6.63d}$$

114   Boundary-Layer Theory

which are free of parameters. In fact, the scales used to define the dimensionless variables (6.61) are precisely those required to make (6.63) parameter-free and can be determined by a simple scaling analysis of the governing equations (Exercise 6.5). Unless otherwise stated, all variables in the following derivation are dimensionless.

We now derive an asymptotic similarity solution for the fast-flowing core of a plume with a dimensionless centerline temperature $T_0(z) \gg 1$. Let $\phi(r, z) \equiv T_0(z) - T(r, z)$ be the deviation of the temperature from its value on the centerline. The viscosity increases as $\exp(\phi)$ as the temperature decreases, and so we expect the fast-flowing core of the plume to be confined to the region where $\phi = O(1)$. When the overall temperature contrast is large ($T_0 \gg 1$), the temperature in the core is comparable to the centerline temperature, and thus the core lies within a wider thermal halo where $\phi = O(T_0)$ (Loper and Stacey, 1983).

Let $\delta(z)$ be the radial length scale of the fast-flowing core and $W_0(z)$ be the vertical velocity scale, both of which must be determined. Define scaled variables

$$\zeta \equiv r/\delta(z) \quad \text{and} \quad W(r, z) \equiv w(r, z)/W_0(z). \tag{6.64}$$

Now rewrite the vertical force balance (6.63b) in terms of ζ and W, using $T_0 \gg 1$ to neglect the temperature deviation ϕ in the buoyancy force $T = T_0 - \phi$. The result is

$$\zeta^{-1} \partial_\zeta (\zeta e^\phi \partial_\zeta W) \sim -\frac{T_0 e^{T_0} \delta^2}{W_0}. \tag{6.65}$$

Here and henceforth, the notation $\sim$ means 'asymptotically equal to in the limit $T_0 \gg 1$'. Assuming that the vertical heat flux is dominated by the fast-flowing core and neglecting the effect of the temperature deviation ϕ in (6.63d), we obtain

$$\int_0^\infty W\zeta \, d\zeta \sim \frac{1}{2\pi W_0 \delta^2 T_0}. \tag{6.66}$$

The upper limit of integration $r_{max}/\delta \gg 1$ in (6.66) has been set to ∞ because r_{max} lies well into the region of exponentially larger viscosity.

Now because (6.65) and (6.66) are written in terms of scaled variables, the RHSs of these equations must be of order unity. We set them equal to -1 and 1, respectively, which amounts to choosing precise definitions of $\delta(z)$ and $W_0(z)$. The result is

$$\delta(z) \equiv (2\pi T_0^2 e^{T_0})^{-1/4} \quad \text{and} \quad W_0(z) \equiv (e^{T_0}/2\pi)^{1/2}. \tag{6.67a,b}$$

Because $T_0(z)$ decreases with height, (6.67) shows that the core radius δ increases and the vertical velocity W_0 decreases. The scaled velocity and temperature

6.5 Boundary Layers with Strongly Variable Viscosity

deviation can now be assumed to be self-similar functions $W(\zeta)$ and $\phi(\zeta)$ of the scaled radial variable ζ, which satisfy

$$\frac{1}{\zeta}\frac{d}{d\zeta}\left(\zeta e^{\phi}\frac{dW}{d\zeta}\right) \sim -1 \quad \text{and} \quad \int_0^{\infty} W\zeta \, d\zeta \sim 1. \tag{6.68a,b}$$

In addition to (6.68), the temperature and velocity fields must satisfy the energy equation (6.63c). Written in terms of the scaled variables (6.64), (6.63c) takes the form

$$\frac{1}{\zeta}\frac{d}{d\zeta}\left(\zeta\frac{d\phi}{d\zeta}\right) = -W_0\delta^2\frac{dT_0}{dz}W + \delta\left(u - W_0\frac{d\delta}{dz}\zeta W\right)\frac{d\phi}{d\zeta}. \tag{6.69}$$

The preceding equation involves the radial velocity u, which we must now determine. The first step is to relate the vertical derivatives of $W_0(z)$ and $\delta(z)$ to the derivative of the centerline temperature $T_0(z)$ by differentiating (6.67). This yields

$$\frac{1}{W_0}\frac{dW_0}{dz} = \frac{1}{2}\frac{dT_0}{dz} \quad \text{and} \quad \frac{1}{\delta}\frac{d\delta}{dz} = -\frac{1}{4}\frac{dT_0}{dz} - \frac{1}{2T_0}\frac{dT_0}{dz} \sim -\frac{1}{4}\frac{dT_0}{dz} \tag{6.70}$$

where $T_0 \gg 1$ has been used in the last step. Next we evaluate the vertical derivative of $w(r,z) = W_0(z)W(\zeta = r/\delta)$ and use (6.70) to obtain

$$\partial_z w = \frac{dW_0}{dz}W(\zeta) - W_0\frac{\zeta}{\delta}\frac{d\delta}{dz}\frac{dW}{d\zeta} \sim -\frac{W_0}{\delta}\frac{d\delta}{dz}\frac{1}{\zeta}\frac{d}{d\zeta}(\zeta^2 W). \tag{6.71}$$

Now the continuity equation (6.63a) yields

$$\partial_z w = -r^{-1}\partial_r(ru) = -\frac{1}{\delta\zeta}\partial_\zeta(\zeta u). \tag{6.72}$$

Combining (6.72) with (6.71), we obtain

$$u \sim W_0(z)\frac{d\delta}{dz}\zeta W(\zeta). \tag{6.73}$$

Note that $u > 0$ because the plume core decelerates as it rises and hence increases in width.

Returning to the energy equation (6.69), we note that the last term on the RHS cancels exactly owing to (6.73). The cancellation occurs because the streamlines and the contours of constant ϕ are parallel to leading order, so that advection of gradients of the temperature deviation ϕ is negligible. Effecting the cancellation and using (6.67), we can simplify (6.69) to

$$\frac{1}{\zeta W}\frac{d}{d\zeta}\left(\zeta\frac{d\phi}{d\zeta}\right) \sim -\frac{1}{2\pi T_0}\frac{dT_0}{dz}. \tag{6.74}$$

Now the LHS of (6.74) is a function of ζ, while the RHS is a function of z. The two sides can therefore only be equal to one another if each is equal to a constant, β say, whence

$$\frac{1}{\zeta}\frac{d}{d\zeta}\left(\zeta\frac{d\phi}{d\zeta}\right) = \beta W \quad \text{and} \quad \frac{dT_0}{dz} = -2\pi\beta T_0. \tag{6.75a,b}$$

Now (6.68a) and (6.75a) are a fourth-order system of coupled ordinary differential equations for $W(\zeta)$ and $\phi(\zeta)$. Three of the necessary boundary conditions come from regularity (smoothness) at the origin and from the definition of ϕ as the deviation of the temperature from its centerline value, which require $\phi'(0) = W'(0) = \phi(0) = 0$. The fourth boundary condition is $W(\infty) \approx 0$, which essentially matches the self-similar solution $W(\zeta)$ in the fast-flowing core to an outer solution where the vertical velocity is much smaller due to the exponentially larger viscosity.

While the equations (6.68a) and (6.75a) have a solution for any value of β, the heat flux constraint (6.68b) is satisfied only for $\beta = 6$. The analytical solutions for W and ϕ for this value of β are

$$W(\zeta) \sim \frac{1}{2}\left(1 + \frac{\zeta^2}{4}\right)^{-2} \quad \text{and} \quad \phi(\zeta) \sim 3\ln\left(1 + \frac{\zeta^2}{4}\right). \tag{6.76}$$

With $\beta = 6$, (6.75b) implies that the centerline temperature decays exponentially with height as

$$T_0(z) \sim T_0(0)\exp(-12\pi z). \tag{6.77}$$

The width scale δ and the vertical velocity scale W_0 are then given in terms of T_0 by (6.67).

Reverting to dimensional variables, (6.77) takes the form

$$T_0(z) \sim T_0(0)\exp\left(-\frac{12\pi k}{\gamma Q}z\right). \tag{6.78}$$

The scale height for the exponential decay of the centerline temperature is $\gamma Q/12\pi k$. To estimate this scale height for the lowermost mantle, consider the strongest of all mantle plumes, the Hawaiian plume. The buoyancy flux of a plume may be defined as

$$B = 2\pi\rho_0\alpha\int_0^\infty w(T - T_\infty)r\,dr, \tag{6.79}$$

and is ≈ 3000 kg s^{-1} for the Hawaiian plume (Ribe and Christensen, 1999). Comparing (6.79) with (6.60), we see that $Q = c_p B/\alpha$. Using characteristic upper-mantle values $\alpha = 3.5 \times 10^{-5}$ K^{-1} and $c_p = 1000$ J kg^{-1} K^{-1}, we find $Q \approx 8.5 \times 10^{10}$ W. It is appropriate here to use upper-mantle values of α and c_p because

6.5 Boundary Layers with Strongly Variable Viscosity

B and Q are independent of height. A typical value of the thermal conductivity for the lowermost mantle is $k \approx 11$ W m^{-1} K^{-1} (Ohta et al., 2012, 2017). To estimate γ for the lowermost mantle, we suppose that the viscosity obeys an Arrhenius law

$$\eta = \eta_\infty \exp\left[\frac{H}{R}\left(\frac{1}{T} - \frac{1}{T_\infty}\right)\right] \tag{6.80}$$

where H is the activation enthalpy and R is the universal gas constant. The quantity γ can then be estimated as

$$\gamma = -\frac{1}{\eta}\frac{\partial \eta}{\partial T} \equiv \frac{H}{RT^2}. \tag{6.81}$$

An appropriate average value of H is 4.5×10^5 J mol^{-1} (Yamazaki and Karato, 2001, Figure 2B), and a representative value of T is 3000 K. With $R = 8.3$ J mol^{-1} K^{-1}, we find $\gamma \approx 0.006$ K^{-1}, which gives finally $Q\gamma/(12\pi k) \approx 1230$ km. Even a strong plume like the Hawaiian plume will therefore be unable to traverse the whole mantle without a significant decrease of its centerline temperature.

6.5.2 Heat Transfer from a Hot Sphere Moving in a Fluid with Temperature-Dependent Viscosity

While plumes involve free or buoyancy-driven convection, variable-viscosity BLs can also arise in situations where a large-scale background flow or wind is imposed externally (forced convection). An example is the previously introduced stagnation-point flow model for the heat transfer from a hot sphere moving in a fluid with strongly temperature-dependent viscosity (§ 1.1). The geometry is shown in Figure 1.1a, where the BL thickness δ is now to be interpreted as the thickness of the deformation layer in which the viscosity is much smaller than the viscosity far from the sphere. The essential dynamics of this situation can be understood by means of a simple scaling analysis (Morris, 1982). Let (u_0, v_0) and (u_1, v_1) be the velocities inside and outside the BL, respectively, and let η_0 and η_1 be the corresponding viscosities. Global conservation of mass requires that

$$aU \approx \delta u_0 + au_1, \tag{6.82}$$

which states that the total rate at which fluid volume is displaced by the sphere is equal to the sum of the volume fluxes through the deformation layer and outside it. Now because the deformation layer is thin, the lateral pressure gradient inside it is of the same order as the pressure gradient outside. Conservation of momentum thus requires that

$$\eta_0 \frac{u_0}{\delta^2} \approx \frac{1}{a}\frac{\partial p}{\partial \theta} \approx \eta_1 \frac{u_1}{a^2}. \tag{6.83}$$

Equations (6.82) and (6.83) together determine u_0 and u_1 as

$$u_0 \approx \frac{(a/\delta)U}{1+\Delta}, \quad u_1 \approx \frac{\Delta}{1+\Delta}U \quad \text{where} \quad \Delta = \frac{\eta_0}{\eta_1}\left(\frac{a}{\delta}\right)^3. \tag{6.84}$$

The drag on the sphere is $D \approx (p-p_0)\pi a^2$, where p_0 is the pressure on the equator. From the second equality in (6.83) it follows that the drag is

$$D \approx \frac{\eta_1 aU}{1+\Delta^{-1}}. \tag{6.85}$$

The drag has two limits. If $\Delta \gg 1$, then most of the volume flux around the sphere is carried by the fluid outside the deformation layer, and the drag $D \approx \eta_1 aU$ is given to order of magnitude by Stokes's law (Stokes limit). If however $\Delta \ll 1$, then most of the volume flux is carried by the deformation layer, and $D \approx \eta_0 (a/\delta)^3 aU$ (lubrication limit). The determination of δ requires consideration of the energy equation, which is beyond our scope here. Asymptotic solutions to the full governing equations have been obtained using the MMAE both for squeezing flow in cylindrical geometry (Figure 1.1b) and for spherical geometry (Figure 1.1a) by Morris (1982) and Ansari and Morris (1985). Exercise 6.7 invites you to apply the MMAE to the squeezing flow problem of Morris (1982).

The scaling laws resulting from the analysis have been applied to estimate the distance through which a hot diapir can rise by the lubrication mechanism before losing so much heat that it solidifies (Morris, 1982). The key point is that the presence of a low-viscosity lubrication layer increases not only the diapir's ascent speed, but also the rate of heat loss from it because the diapir must soften the whole column of fluid ahead of it to move. As a result, a single diapir can only rise through a distance of the order of its radius before solidifying. However, Morris (1982) suggested that a train of diapirs rising one after the other could create a preheated pathway allowing magma to be transported vertically over long distances.

Exercises

6.1 Consider a solid sphere with radius a and uniform temperature T_0 moving at speed U in a fluid with constant viscosity η and thermal diffusivity κ whose temperature is T_∞ far from the sphere. The Péclet number $Ua/\kappa \gg 1$. Working in spherical coordinates, use the Von Mises transformation to determine the temperature within the TBL and thence the Nusselt number. The solution involves a similarity transformation.

6.2 Show that (4.114b) implies (6.33).

6.3 Starting from the Stokes equation (6.30b), show that the variation $\tilde{w}$ of the vertical velocity across the plume is much less than the axial velocity $w_0(z)$ when $z \gg z_0$.

6.4 Starting from the governing equations (6.30b) and (6.30c) and the integral constraint (6.31), derive the BL equations (6.39).

6.5 Perform a scaling analysis of the governing equations to determine the scales used to define the dimensionless variables (6.61).

6.6 Consider the two-point boundary problem

$$\epsilon y'' + y' + y = 0, \tag{6.86a}$$

$$y(0) = y(1) - e^{-1} = 0. \tag{6.86b}$$

Use the MMAE to determine a two-term composite expansion of the solution to this problem, and compare your result for $\epsilon = 0.05$ with the exact analytical solution

$$y = \frac{\exp\left(\frac{1-2\epsilon-x}{2\epsilon}\right) \sinh \frac{\sqrt{1-4\epsilon}}{2\epsilon} x}{\sinh \frac{\sqrt{1-4\epsilon}}{2\epsilon}}. \tag{6.87}$$

Note: the solution to this exercise introduces the rigorous IMP.

6.7 Consider further the stagnation flow with temperature-dependent viscosity discussed in § 6.5.2 and sketched in Figure 1.1. The problem admits a solution of the form

$$(u, w) = \left(\frac{r}{2} f'(z), -f(z)\right), \quad T = T(z), \tag{6.88}$$

where $f(z)$ satisfies

$$[\nu(T) f'']'' = 0 \tag{6.89}$$

and primes denote d/dz. Use the MMAE guided by the scaling analysis of § 6.5.2 to determine a one-term composite expansion of the solution of (6.89). Hints: (1) assume that the temperature varies linearly in the deformation layer as $T = T_0 - (Nu \Delta T/a) z$, where Nu is an unknown Nusselt number that can be determined later by solving the energy equation; (2) match the constant and linear terms in the inner and outer expansions for f and the shear stress $\nu(T) f''(z)$.

6.8 Ansari and Morris (1985) performed laboratory experiments on the motion of a hot sphere in a fluid with strongly temperature-dependent viscosity. The apparatus consisted of an electrically heated brass sphere of radius a connected by a vertical rod to a platform on which a mass M could be placed to drive the sphere's downward motion through a container of corn syrup held at a fixed temperature $T_\infty = 0°$ C. The temperature T_0 of the sphere was controlled by adjusting manually the power supplied to it. Suppose that you wish to repeat the experiment for $T_0 = 50°$ C and $a = 2$ cm, the radius of the larger of the two spheres used by Ansari and Morris (1985). Using the fluid properties and experimental data reported in that paper, estimate the mass M required to drive the sphere with a velocity $U = 0.05$ cm s^{-1}.

7

Long-Wave Theories, 1: Lubrication Theory and Related Techniques

Long-wave theories comprise a variety of loosely related approaches whose goal is to describe the flow or elastic deformation within layers whose thickness is much less than their characteristic lateral dimensions. These theories can be roughly divided into two classes depending on which component of the stress tensor is dominant within the layer. For definiteness, let s be a coordinate in the layer-parallel direction, and n be a coordinate normal to the layer. Theories of the first, or 'lubrication', class apply to flows dominated by the shear stress σ_{sn} acting on layer-parallel planes. This situation typically arises when the layer adjoins one or more boundaries or walls on which the no-slip condition applies. The second, or 'shell', class comprises theories for either viscous or elastic shells that are not attached to a wall, but are instead able to deform freely by a combination of stretching and bending. In such objects the dominant stress component is the fiber stress σ_{ss}. Theories of the lubrication class (together with some closely related theories) are treated in this chapter, while theories of the shell class are the subject of the next chapter. Both types of theory differ from boundary-layer theory (Chapter 6) in that they do not involve an explicit transport (advection-diffusion) equation.

A fundamental idea underlying many long-wave theories is the distinction between small amplitude and small slope. To take a simple example, if a 2-D fluid layer has thickness $h_0 + \Delta h \sin kx$, its surface has small amplitude if $\Delta h/h_0 \ll 1$ and small slope (or long wavelength) if $k\Delta h \ll 1$. Long-wave theories often exploit the fact that small slope does not imply small amplitude to derive equations governing the nonlinear (finite-amplitude) evolution of long-wave disturbances in a layer. Examples to be treated in this chapter and the next include the spreading of viscous gravity currents and the large-amplitude deformation of 2-D viscous shells.

7.1 Lubrication Theory

Shear-dominated viscous flows in thin layers are common in geodynamics: examples include lava flows and the spreading of buoyant plume material beneath the lithosphere. Such flows are described by a simplified form of the Navier–Stokes equations, called the lubrication equations because of their importance in the design of industrial lubrication bearings.

To develop the theory, I use the simple idealized 2-D model shown in Figure 7.1a. A layer of viscous fluid with density ρ, dynamic viscosity η and thickness h is bounded below by a rigid surface. Flow in the layer is driven by a horizontal velocity $u = U \sin kx$ applied to the upper surface, the wavelength $L \equiv 2\pi/k$ of which greatly exceeds the layer depth. The vertical velocity on the upper surface is zero.

The Navier–Stokes equations governing the flow in the layer are

$$u_x + w_z = 0, \tag{7.1a}$$

$$\rho(uu_x + wu_z) = -p_x + \eta(u_{xx} + u_{zz}), \tag{7.1b}$$

$$\rho(uw_x + ww_z) = -p_z + \eta(w_{xx} + w_{zz}), \tag{7.1c}$$

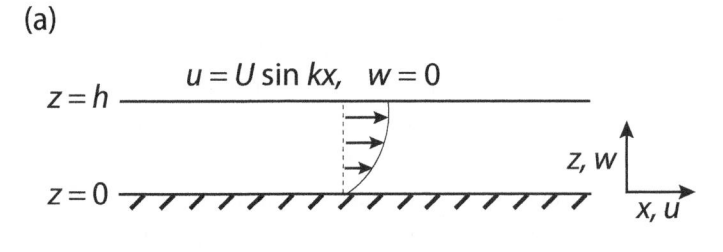

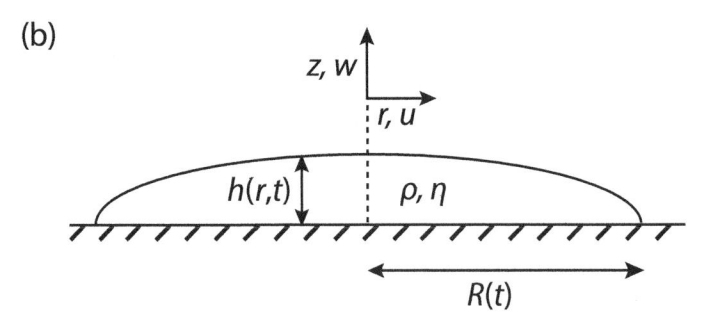

Figure 7.1 Models used to derive the equations of lubrication theory (§ 7.1). (a) A layer of viscous fluid of depth h, driven by a sinusoidally varying horizontal velocity $U \sin kx$ applied to its upper surface, where the vertical velocity is zero. The wavelength of the applied velocity is much greater than the layer thickness. (b) An axisymmetric gravity current with density ρ, viscosity η and constant volume V.

7.1 Lubrication Theory

where subscripts denote partial derivatives. In writing (7.1) it has been assumed that the acceleration $\partial \mathbf{u}/\partial t$ of the fluid is negligible, which is nearly always the case in very viscous lubrication flows.

Now define dimensionless (hatted) variables

$$\hat{x} = \frac{x}{L}, \quad \hat{z} = \frac{z}{h}, \quad \hat{u} = \frac{u}{U}, \quad \hat{w} = \frac{w}{W}, \quad \hat{p} = \frac{p}{P}. \tag{7.2}$$

In writing (7.2), we have used the obvious scales for x, z and u, while leaving the less-obvious scales W and P undefined for the moment. The dimensionless form of the continuity equation is

$$\frac{U}{L}\hat{u}_{\hat{x}} + \frac{W}{h}\hat{w}_{\hat{z}} = 0. \tag{7.3}$$

Requiring the two terms in the preceding equation to balance gives

$$W = U\frac{h}{L}, \tag{7.4}$$

which implies that the velocity perpendicular to the layer is much smaller than that parallel to the layer. The dimensionless form of (7.1b) is now

$$\epsilon \frac{Uh}{\nu}(\hat{u}\hat{u}_{\hat{x}} + \hat{w}\hat{u}_{\hat{z}}) = -\frac{Ph^2}{\eta UL}\hat{p}_{\hat{x}} + (\epsilon^2 \hat{u}_{\hat{x}\hat{x}} + \hat{u}_{\hat{z}\hat{z}}), \tag{7.5}$$

where $\epsilon = h/L \ll 1$. The Reynolds number Uh/ν is generally small in lubrication flows, and the modified Reynolds number $\epsilon Uh/\nu$ is smaller still, so that the left-hand side (LHS) of (7.5) can be neglected. The term $\epsilon^2 \hat{u}_{\hat{x}\hat{x}}$ is also negligible. Requiring the remaining two terms to balance implies that the pressure scale is

$$P = \frac{\eta UL}{h^2}, \tag{7.6}$$

whereupon (7.5) reduces to

$$\hat{p}_{\hat{x}} = \hat{u}_{\hat{z}\hat{z}}. \tag{7.7}$$

Finally, the dimensionless form of (7.1c) is

$$\epsilon^3 \frac{Uh}{\nu}(\hat{u}\hat{w}_{\hat{x}} + \hat{w}\hat{w}_{\hat{z}}) = -\hat{p}_{\hat{z}} + (\epsilon^4 \hat{w}_{\hat{x}\hat{x}} + \epsilon^2 \hat{w}_{\hat{z}\hat{z}}). \tag{7.8}$$

Neglecting all terms proportional to a power of ϵ, one obtains simply

$$\hat{p}_{\hat{z}} = 0. \tag{7.9}$$

Gathering the equations (7.3), (7.7) and (7.9) and transforming them back to dimensional form, we obtain the lubrication equations

$$u_x + w_z = 0, \qquad (7.10a)$$

$$p_x = \eta u_{zz}, \qquad (7.10b)$$

$$p_z = 0. \qquad (7.10c)$$

We now solve the equations (7.10) for the problem of Figure 7.1a. Equation (7.10c) implies that $p = p(x)$, i.e., that the pressure is constant across the layer. Equation (7.10b) can therefore be integrated twice with respect to z subject to the boundary conditions $u(z=0) = 0$ and $u(z=h) = U \sin kx$, yielding

$$u = \frac{p_x}{2\eta} z(z-h) + \frac{zU}{h} \sin kx. \qquad (7.11)$$

Equation (7.11) states that the horizontal velocity is the sum of a plane Poiseuille flow driven by the (still unknown) pressure gradient p_x and a Couette flow driven by the motion of the upper boundary. Next, substitute (7.11) into (7.10a) and integrate the result once with respect to z subject to the boundary condition $w(z=0) = 0$ to obtain

$$w = -\frac{p_{xx}}{12\eta} z^2(2z - 3h) - \frac{z^2 kU}{2h} \cos kx. \qquad (7.12)$$

Finally, the remaining boundary condition $w(z = h) = 0$ yields a differential equation for p whose periodic solution is

$$p = -\frac{6\eta U}{kh^2} \cos kx. \qquad (7.13)$$

Substituting (7.13) into (7.11) and (7.12) yields

$$u = \frac{U}{h^2}(3z - 2h)z \sin kx, \quad w = \frac{kU}{h^2}(h - z)z^2 \cos kx. \qquad (7.14)$$

The pressure is in phase with the vertical velocity and out of phase by $\pi/2$ with the horizontal velocity.

We now turn to the problem of a viscous gravity current of constant volume V spreading on a rigid surface, sketched in Figure 7.1b (Huppert, 1982). In the limit of long times, $h/R \ll 1$, so that the problem can be solved using lubrication theory. However, the lubrication equations in the form (7.10) are not appropriate for this task because they contain no term representing the Archimedean buoyancy force that drives the spreading of the gravity current. This difficulty can be easily remedied by adding to the right-hand side (RHS) of (7.10c) a term $-\rho g$ that represents the hydrostatic variation of the pressure across the current. Let r be the (horizontal) radial coordinate measured from the center of the current, and

7.1 Lubrication Theory

let $u(z, r, t)$ be the radial component of the velocity. In cylindrical coordinates, the lubrication equations take the forms

$$r^{-1}(ru)_r + w_z = 0, \tag{7.15a}$$

$$p_r = \eta u_{zz}, \tag{7.15b}$$

$$p_z = -\rho g. \tag{7.15c}$$

Integrating (7.15c) subject to $p(z = h) = 0$, we obtain

$$p = \rho g(h - z). \tag{7.16}$$

Next, we substitute (7.16) into (7.15b) and integrate subject to the no-slip condition on the plate ($u(0, r, t) = 0$) and the condition of vanishing traction at $z = h$. Now because the current's upper surface is nearly horizontal, the traction there $\approx \eta(u_z + w_x)$. However, $w_x \sim (h/R)^2 u_z$ in the lubrication approximation, so the condition of vanishing traction is simply $u_z(h, r, t) = 0$. The profile of radial velocity across the current is therefore

$$u = \frac{\rho g h_r}{2\eta}(z^2 - 2hz), \tag{7.17}$$

where $\rho g h_r$ is the radial gradient of the hydrostatic pressure that drives the flow. Next, the continuity equation (7.15a) is integrated across the current subject to the impermeability condition $w(0, r, t) = 0$ to obtain

$$0 = w(h, r, t) + r^{-1} \int_0^h (ru)_r \, dz. \tag{7.18}$$

We now simplify (7.18) by using the kinematic surface condition to eliminate $w(h, r, t) \equiv h_t + u(h, r, t)h_r$; taking ∂_r outside the integral using the standard formula for the derivative of an integral with a variable limit; and evaluating the integral using (7.17). We thereby find that $h(r, t)$ satisfies the nonlinear diffusion equation

$$h_t = 4\sigma r^{-1}(rh^3 h_r)_r, \quad \sigma = \frac{\rho g}{12\eta}, \tag{7.19}$$

where σ is the 'spreadability'. Conservation of the current's volume requires

$$2\pi \int_0^R rh \, dr = V. \tag{7.20}$$

Important insight into the solution of (7.19) can be obained by a simple scaling analysis. Let $h_0(t)$ be the maximum thickness of the current. Then the constant-volume constraint (7.20) scales as

$$R^2 h_0 \sim V. \tag{7.21}$$

Furthermore, the diffusion equation (7.19) scales as

$$\dot{h}_0 \sim -\frac{\sigma h_0^4}{R^2} \tag{7.22}$$

where the time derivative $\dot{h}_0$ has been retained explicitly. Eliminating $\dot{h}_0$ between (7.21) and (7.22), we obtain

$$\dot{R} \sim \frac{\sigma V^3}{R^7}. \tag{7.23}$$

Integration of (7.23) subject to $R(0) = 0$ yields

$$R \sim (\sigma V^3 t)^{1/8}, \tag{7.24}$$

which shows that the radius of the current increases as the one-eighth power of the time.

The power-law form of (7.24) suggests that for long times, the current achieves a universal self-similar shape that retains no memory of the initial shape $h(r, 0)$. We are thereby led to seek a formal similarity solution of (7.19) and (7.20). Now self-similarity requires that h depend on the normalized radius $\zeta \equiv r/R(t) \in [0, 1]$, and conservation of volume requires $hR^2 \sim V$. We therefore seek a solution of the form

$$h = \frac{V}{R^2} H\left(\frac{r}{R}\right). \tag{7.25}$$

Upon substituting (7.25) into (7.19) and then separating variables in the now-familiar way (§ 3.3), we obtain

$$\frac{R^7 \dot{R}}{4\sigma V^3} = -\frac{(\zeta H^3 H')'}{(\zeta^2 H)'}. \tag{7.26}$$

The only way a function of t (the LHS of (7.26)) can be equal to a function of ζ (RHS) is if both are equal to a constant, C say. Solving the two resulting equations subject to the conditions $H'(0) = H(1) = R(0) = 0$, we obtain

$$H = \left[\frac{3C}{2}(1 - \zeta^2)\right]^{1/3}, \quad R = (32C\sigma V^3 t)^{1/8}. \tag{7.27}$$

Note that the expression (7.27) for $R(t)$ is consistent with the result (7.24) of the scaling analysis. Finally, the constant C is determined from the constant-volume constraint, which in terms of the similarity variables is

$$2\pi \int_0^1 \zeta H \, d\zeta = 1. \tag{7.28}$$

Substituting into (7.28) the expression (7.27) for H, we find

$$C = \frac{128}{81\pi^3}. \tag{7.29}$$

Huppert (1982) carried out laboratory experiments to test (7.27) and (7.29) and obtained remarkably good agreement.

The solution (7.27) is a special case of a more general class of similarity solutions of (7.19) studied by Gratton and Minotti (1990) using a phase-plane formalism. Rather than work with the single partial differential equation (PDE) (7.19), Gratton and Minotti (1990) wrote down two coupled PDEs for the thickness h and the mean horizontal velocity v, and then used a similarity transformation to reduce them to coupled ordinary differential equations (ODEs). The solutions of these ODEs can be represented as segments of integral curves on a phase plane that connect singular points representing different boundary conditions such as sources, sinks and current fronts. Gratton and Minotti (1990) give an exhaustive catalog of the solutions thus found, including a novel similarity solution of the second kind (see § 3.3) for the evolution of an axisymmetric gravity current surrounding a circular hole.

Viscous gravity-current theory has been further developed in several geodynamically useful directions. Lister and Kerr (1989b) extended the theory to 2-D and axisymmetric currents spreading at a fluid–fluid interface. Several authors have generalized the theory to non-Newtonian gravity currents with power-law rheology (e.g., Gratton et al., 1999; Sayag and Worster, 2013). The theory has also been extended to currents with temperature-dependent viscosity and buoyancy (Bercovici, 1994; Bercovici and Lin, 1996; Vasilyev et al., 2001). Finally, Michaut and Bercovici (2009) proposed a theory for two-phase gravity currents consisting of a low-viscosity fluid phase in a highly viscous porous matrix (see Chapter 9).

Applications of gravity current theory to mantle dynamics are numerous and diverse. Perhaps the most common application is to the interaction of ascending mantle plume heads and/or tails with the lithosphere (Griffiths and Campbell, 1991; Sleep, 1996; Nyblade and Sleep, 2003; Mittelstaedt and Ito, 2005; Ramalho et al., 2010). Kerr and Lister (1987) and Bercovici et al. (1993) invoked gravity currents to model the spread of subducted lithosphere along the 660 km discontinuity. Kemp and Stevenson (1996) proposed a model for subduction initiation in which mantle material wells up through a rift and spreads as a gravity current over old ocean floor. Bercovici and Long (2014) envisioned a 2-D gravity current as a model for a continent accumulating above a subduction zone.

In the next section, I discuss an extension of gravity-current theory to currents spreading on moving surfaces, which have been widely used to model the interaction of mantle plumes with a moving or rifting lithosphere.

7.2 Plume–Plate and Plume–Ridge Interaction Models

The geometry of these models was introduced in § 1.3 and is sketched in Figure 1.3b and c. Motion of a plate (or plates) with a horizontal velocity $\mathbf{U}_0(x, y)$ generates an ambient flow $\mathbf{U}(x, y, z)$ in the mantle below, which is assumed to have uniform density ρ and viscosity η_m. The plume conduit is represented as a volume source of strength Q fixed at $(x, y) = (0, 0)$ that emits buoyant fluid with density $\rho - \Delta\rho$ and viscosity η_p. This fluid spreads laterally beneath the lithosphere to form a thin pool whose thickness $h(x, y, t)$ is governed by a balance of buoyancy-driven spreading and advection by the ambient mantle flow. Strictly speaking, the solution of this problem requires the simultaneous determination of the flow in both fluids subject to the usual matching conditions on velocity and stress at their interface. However, useful results can be obtained via a simpler approach in which the mantle flow is assumed to be unaffected by the flow in the plume layer.

Consider first the case of a plume interacting with an intact (nonrifted) lithosphere moving at constant speed U relative to the hotspot (Figure 1.3b). Olson (1990) derived the lubrication equation governing $h(x, y, t)$ assuming a Newtonian rheology for the plume pool. That equation was generalized to a non-Newtonian (shear thinning) rheology by Asaadi et al. (2011), whose derivation we follow here.

The lubrication equations governing the flow in the pool are

$$\partial_\alpha u_\alpha + \partial_z w = 0, \tag{7.30a}$$

$$\partial_z(\eta \partial_z u_\alpha) = \partial_\alpha p, \tag{7.30b}$$

$$\partial_z p = -g\Delta\rho, \tag{7.30c}$$

where $\alpha = 1$ or 2, u_1 and u_2 are the horizontal velocity components, w is the vertical velocity (positive downward), z increases downward from the base of the lithosphere, and p is the pressure relative to the pressure outside the pool. To begin, we assume that the rheology of the pool has the power-law form

$$\dot\epsilon = D\tau^n, \quad \dot\epsilon = \left|\frac{\partial \mathbf{u}}{\partial z}\right|, \tag{7.31}$$

where $\dot\epsilon$ is the absolute value of the horizontal shear rate, τ is the absolute value of the horizontal shear stress, n is the power-law index and D is a rheological constant that reduces to the inverse of the viscosity if the fluid is Newtonian ($n = 1$). Equation (7.31) implies that the viscosity $\eta \equiv \tau/\dot\epsilon$ is

$$\eta = D^{-\frac{1}{n}} \left|\frac{\partial \mathbf{u}}{\partial z}\right|^{\frac{1-n}{n}}. \tag{7.32}$$

Next, we integrate the lubrication equation (7.30c) subject to the condition $p(z = h) = 0$ to obtain

7.2 Plume–Plate and Plume–Ridge Interaction Models

$$p = g\Delta\rho(h - z). \tag{7.33}$$

Substituting (7.33) into (7.30b), we obtain

$$\partial_z(\eta \partial_z u_\alpha) = g\Delta\rho \partial_\alpha h \tag{7.34}$$

where the viscosity is given by (7.32). The solution of (7.34) that satisfies the boundary conditions $u_\alpha(z=0) = U\delta_{1\alpha}$ and $\partial_z u_\alpha(z=h) = 0$ is

$$u_\alpha = U\delta_{1\alpha} - \frac{D}{n+1}(g\Delta\rho)^n |\nabla h|^{n-1}[h^{n+1} - (h-z)^{n+1}]\partial_\alpha h. \tag{7.35}$$

The boundary condition $\partial_z u_\alpha(z=h) = 0$ (vanishing shear stress at the bottom of the pool) is justified by Huppert (1982), Appendix B. Now conservation of mass over the whole thickness of the pool requires

$$\partial_t h + \partial_\alpha \int_0^h u_\alpha dz = \frac{Q}{\pi a^2}\exp\left(-\frac{r^2}{a^2}\right) \tag{7.36}$$

where $r^2 = x^2 + y^2$ and the last term represents a fluid source of volumetric strength Q distributed in the form of a Gaussian of radius a centered at the hotspot location $(x, y) = (0, 0)$. Substitution of (7.35) into (7.36) yields

$$\partial_t h + U\partial_x h = \sigma \nabla \cdot \left[|\nabla h|^{n-1}\nabla(h^{n+3})\right] + \frac{Q}{\pi a^2}\exp\left(-\frac{r^2}{a^2}\right), \tag{7.37a}$$

$$\sigma = \frac{D(g\Delta\rho)^n}{(n+2)(n+3)}. \tag{7.37b}$$

From left to right, the four terms in (7.37a) represent the rate of change of the pool's thickness, downstream advection of the pool, buoyancy-driven lateral spreading of the pool and localized injection of plume fluid into the pool, respectively.

The scales for the thickness h and half-width L of the pool can now be found via a scaling analysis of (7.37a). Balancing the advection term and the gravitational spreading term, we find

$$\frac{Uh}{L} \sim \sigma \frac{h^{2n+2}}{L^{n+1}}. \tag{7.38}$$

In addition, conservation of downstream volume flux requires

$$UhL \sim Q. \tag{7.39}$$

Solving (7.38) and (7.39) for L and h, we obtain the scales

$$L \sim \left(\frac{\sigma Q^{2n+1}}{U^{2n+2}}\right)^{\frac{1}{3n+1}} \equiv L_0, \quad h \sim \left(\frac{Q^n}{\sigma U^{n-1}}\right)^{\frac{1}{3n+1}} \equiv h_0. \tag{7.40}$$

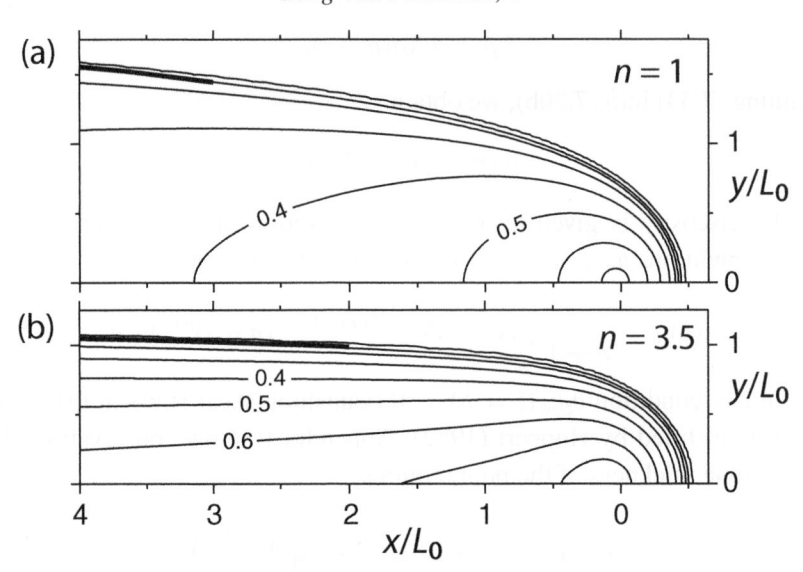

Figure 7.2 Steady-state solutions of the lubrication equation (7.37a) for (a) $n = 1$ and (b) $n = 3.5$. The contour interval is $0.1h_0$, where h_0 is defined by (7.40). The thick black lines show the similarity solution (7.44b) for the half-width of the pool. See § 7.2 for discussion.

Equation (7.37a) can be solved numerically using an explicit finite-difference method (Asaadi et al., 2011). The results are shown in Figure 7.2 for (a) a Newtonian rheology $n = 1$, and (b) a non-Newtonian shear-thinning rheology with $n = 3.5$. The non-Newtonian pool widens downstream more slowly than the Newtonian one.

Further insight is provided by an analytical similarity solution of the steady-state form of (7.37a) that is valid far downstream ($x \gg L_0$) from the plume source. We anticipate that at these distances, the pool thickness h will vary more strongly in the direction normal to the plate motion than parallel to it, so that all x-derivatives in the gravitational spreading term can be neglected. In addition, the fluid supply term is exponentially small for $x \gg L_0$, and Q appears instead in an integral relation expressing conservation of the downstream volume flux. Equation (7.37a) then simplifies to

$$U\partial_x h = \sigma(n+3)\partial_y \left(h^{n+2}|\partial_y h|^{n-1}\partial_y h\right) \quad (7.41a)$$

$$U\int_{-L}^{L} h\,dy = Q, \quad (7.41b)$$

where $L(x)$ is now defined precisely as the half-width of the pool.

7.2 Plume–Plate and Plume–Ridge Interaction Models

Equations (7.41) admit a similarity solution of the form

$$h = \frac{Q}{UL(x)} H(\zeta), \qquad \zeta = \frac{y}{L(x)}. \tag{7.42}$$

Substitution of (7.42) into (7.41) and applying the now-familiar procedure of separation of variables yields ordinary differential equations for $L(x)$ and $H(\zeta)$ that can be solved subject to the boundary conditions

$$0 = L(x_0) = H'(0) = H(1), \tag{7.43}$$

where x_0 is the virtual position of the hotspot as seen by the solution far downstream and the prime indicates $d/d\zeta$. The condition $H'(0) = 0$ states that the lateral slope of the pool is zero along the central symmetry axis, and $H(1) = 0$ states that the thickness of the pool is zero at its edge. The explicit solutions are

$$H(\zeta) = C_1 \left(1 - |\zeta|^{\frac{1+n}{n}}\right)^{\frac{n}{2n+1}}, \tag{7.44a}$$

$$L(x) = C_2 L_0 \left(\frac{x - x_0}{L_0}\right)^{\frac{1}{3n+2}}, \tag{7.44b}$$

$$C_1 = \frac{\Gamma\left(\frac{5n^2+5n+1}{(n+1)(2n+1)}\right)}{2\Gamma\left(\frac{2n+1}{n+1}\right)\Gamma\left(\frac{3n+1}{2n+1}\right)}, \tag{7.44c}$$

$$C_2 = \left[(n+3)(3n+2)\left(\frac{n+1}{2n+1}\right)^n C_1^{2n+1}\right]^{\frac{1}{3n+2}}. \tag{7.44d}$$

The virtual origin x_0 can be determined by fitting the numerically determined edges of the pools in Figure 7.2 to the expression (7.44b), yielding $x_0/L_0 = 0.753$ for $n = 1$ and 0.272 for $n = 3.5$. The corresponding similarity forms $L(x)$ from (7.44b) are shown as thick black lines in Figure 7.2. Equation (7.44b) shows that the half-width of the plume pool increases with downstream distance as $L \sim (x - x_0)^{1/5}$ for $n = 1$ and $L \sim (x - x_0)^{0.08}$ for $n = 3.5$. The difference between the two cases is substantial and suggests that the shape of hotspot swells – the Hawaiian swell especially – can be used as a 'rheometer' to determine whether the material compensating the swell is deforming as a Newtonian or a non-Newtonian fluid. Asaadi et al. (2011) concluded that a non-Newtonian rheology with $n = 3.5$ provides a better fit to the shape of the Hawaiian swell than a Newtonian rheology with $n = 1$.

To close this section, we turn to the interaction of a plume with an oceanic spreading ridge, the geometry of which is shown in Figure 1.3c. Ribe et al. (1995) studied this situation using a modified version of the model of Olson (1990) in which the uniform mantle flow $\mathbf{U} = U\mathbf{e}_x$ is replaced by a corner flow driven by surface plates diverging with a half-spreading rate U. Let the steady-state thickness of the pool of plume fluid be $h(x, y)$, and suppose that the source (hotspot) is directly below the ridge. The lubrication equation for this situation is similar to (7.37a), but with a modified LHS that describes how the plume pool is advected by the corner flow. The modified form of the term $U\partial_x h$ on the LHS of (7.37a) is

$$\nabla \cdot \int_0^h \mathbf{u} dz = \partial_x \int_0^h u dz = \partial_x \int_0^h \frac{\partial \psi}{\partial z} dz = \partial_x [\psi(x, h)] \qquad (7.45a)$$

where

$$\psi = \frac{2U}{\pi} z \tan^{-1} \frac{x}{z} \qquad (7.45b)$$

is the streamfunction for the corner flow driven by two infinitely thin plates moving as shown in Figure (1.3). Replacing the term $U\partial_x h$ in (7.37a) by (7.45) and setting $n = 1$ for a Newtonian rheology and $\partial_t h = 0$ for steady flow, we obtain the modified lubrication equation

$$\frac{2U}{\pi} \partial_x \left[h \tan^{-1} \frac{x}{h} \right] = \sigma \nabla_h^2 h^4 + \frac{Q}{\pi a^2} \exp\left(-\frac{x^2 + y^2}{a^2}\right). \qquad (7.46)$$

Figure 7.3a shows the shape of the plume pool for $\sigma Q/U^2 = 100$, determined by solving (7.46) numerically. The minimum width W of the pool is along the ridge ($x = 0$) and was called the 'waist width' by Feighner and Richards (1995). The scales for the thickness h of the pool and for the waist width can be obtained via a scaling analysis of (7.46). Directly beneath the ridge at $x = 0$, $\partial_x [h \tan^{-1}(x/h)] = 1$. The balance of advection (LHS of (7.46)) and gravitational spreading thus implies $U \sim \sigma h^4/W^2$. In addition, conservation of volume flux requires $Q \sim UWh$. Solving these two relations for h and W gives

$$h \sim \left(\frac{Q^2}{\sigma U}\right)^{1/6} \equiv h_0, \qquad (7.47a)$$

$$W \sim \left(\frac{\sigma Q^4}{U^5}\right)^{1/6} \equiv \left(\frac{Q}{U}\right)^{1/2} \left(\frac{\sigma Q}{U^2}\right)^{1/6} \equiv L_0 \left(\frac{\sigma Q}{U^2}\right)^{1/6}. \qquad (7.47b)$$

We now examine the range of validity of the scaling law (7.47b). Dimensional analysis using Buckingham's Π-theorem on the list of parameters W, σ, Q and U shows that the general form of the scaling law is $W = L_0 \text{fct}(\sigma Q/U^2)$, where L_0 is defined by (7.47b). Figure 7.3b shows W/L_0 as a function of $\sigma Q/U^2$, obtained by solving the lubrication equation (7.46) numerically. The scaling law (7.47b) is seen to be valid for $\sigma Q/U^2 \gg 1$. This result illustrates a typical feature of scaling laws having a power-law form: they are often valid only for very large or very small values of some dimensionless parameter.

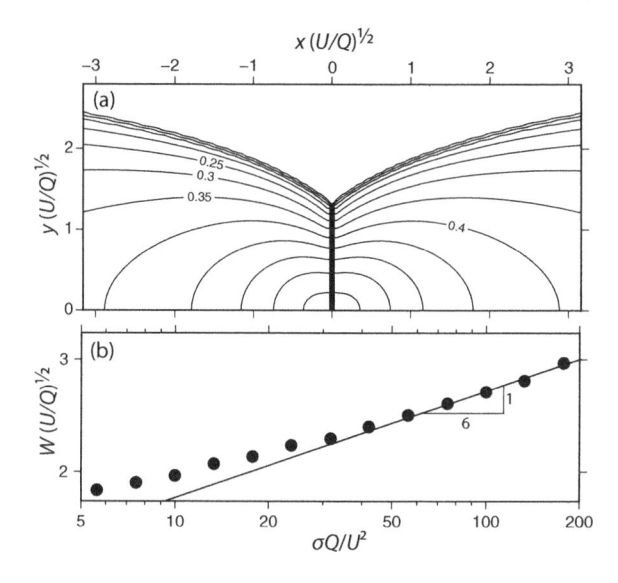

Figure 7.3 Lubrication theory model for the interaction of a ridge-centered plume with a mid-ocean ridge directly above it (§ 7.2). (a) Steady-state shape of the plume pool for $\sigma Q/U^2 = 100$, determined by solving (7.46) numerically. Only the upper half of the pool is shown. The contour interval is $0.05h_0$. The heavy black line is one-half the 'waist width' W. (b) Dimensionless waist width $W(U/Q)^{1/2}$ as a function of $\sigma Q/U^2$.

7.3 Long-Wave Analysis of Thermal Boundary-Layer Instability

In convecting systems such as Earth's mantle, plumes arise as instabilities of horizontal thermal boundary layers (TBLs). A long-wave model for this process has been proposed by Lemery et al. (2000) (henceforth LRS00), based on two assumptions: that the characteristic horizontal scale of the flow greatly exceeds the boundary layer (BL) thickness, and that the horizontal velocity of the fluid is approximately constant across the BL. The latter assumption excludes hot BLs within which the (temperature-dependent) viscosity is much lower than outside and makes the model most applicable to plume formation in cold BLs. The two assumptions allow the coupled 3-D dynamics of the BL and the fluid outside it to be reduced to 2-D equations for the lateral velocity at the edge of the BL and a temperature moment that describes the distribution of buoyancy within it.

The assumed constancy of horizontal velocity across the BL implies that the flow is dominated by the stress components σ_{11}, σ_{12} and σ_{22} associated with horizontal compression/stretching and shear. Taken alone, this fact suggests that the theory we are about to derive belongs in the 'shell' class treated in the next chapter. However, the problem nevertheless involves the nonlinear evolution of the thickness of a layer

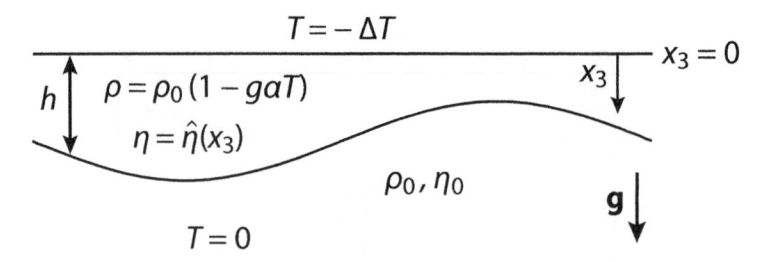

Figure 7.4 Definition sketch of the model problem for long-wave analysis of buoyant instability of a cold TBL (§ 7.3).

adjoining an impermeable (and free-slip) surface that prevents it from bending, and so it is appropriate to discuss it here.

The domain of the model is a fluid half-space bounded by a cold free-slip surface $x_3 = 0$ held at temperature $-\Delta T$ relative to the fluid far from it (Figure 7.4) The BL occupies the depth interval $x_3 < h(x_1, x_2, t)$, where x_α are Cartesian coordinates parallel to the BL and t is time. In the following, hatted and unhatted variables are those in the BL and in the interior, respectively, and an argument in parentheses indicates a value of x_3. The (constant) viscosity of the interior fluid is η_0, and the viscosity within the BL is $\hat{\eta}(x_3)$.

The starting point is the momentum equation within the BL, viz. $-\partial_i \hat{p} + \partial_j \hat{\tau}_{ij} = \rho_0 \alpha g \hat{T} \delta_{i3}$, where $\hat{\tau}_{ij}$ is the deviatoric part of the stress tensor. Taking the curl of this equation, applying the continuity equation $\hat{\tau}_{33} = -\hat{\tau}_{\gamma\gamma}$, and noting that $(\partial_{11}^2 \hat{\tau}_{\alpha 3}, \partial_{12}^2 \hat{\tau}_{\alpha 3}, \partial_{22}^2 \hat{\tau}_{\alpha 3}) \ll \partial_{33} \hat{\tau}_{\alpha 3}$ in the long-wavelength approximation, we obtain

$$\partial_{33}^2 \hat{\tau}_{\alpha 3} + \partial_3 \hat{A}_\alpha = -\rho_0 \alpha g \partial_\alpha \hat{T}, \quad \hat{A}_\alpha = \partial_\alpha \hat{\tau}_{\gamma\gamma} + \partial_\gamma \hat{\tau}_{\alpha\gamma}. \tag{7.48}$$

Physically, (7.48) are the lateral ($\alpha = 1$ or 2) components of the vorticity equation. Now multiply (7.48) by x_3, integrate across the BL from $x_3 = 0$ to $x_3 = h$ and take lateral derivatives outside the integral signs by neglecting the small lateral variation of the upper limit $h(x_1, x_2, t)$. The result is

$$-\hat{\tau}_{\alpha 3}(h) + h\partial_3 \hat{\tau}_{\alpha 3}(h) + h\hat{A}_\alpha(h) - \langle \hat{A}_\alpha \rangle = \rho_0 \alpha g \partial_\alpha M \tag{7.49}$$

where $\langle \rangle = \int_0^h dx_3$ and

$$M = -\langle x_3 \hat{T} \rangle \tag{7.50}$$

is the temperature moment. Now $h\partial_3 \hat{\tau}_{\alpha 3}(h) \ll \hat{\tau}_{\alpha 3}(h)$ in the long-wave limit, and continuity of shear stress at $x_3 = h$ requires $\hat{\tau}_{\alpha 3}(h) = \tau_{\alpha 3}(h)$. But because the interior fluid sees the BL as a skin with zero thickness, $\tau_{\alpha 3}(h) \approx \tau_{\alpha 3}(0) \approx \eta_0 \partial_3 u_\alpha(0)$. Moreover, the lateral velocity components are constant across the layer to lowest

7.3 Long-Wave Analysis of Boundary-Layer Instability

order and must match those in the interior fluid, requiring $\hat{u}_\alpha = u_\alpha(0)$ and $\hat{\tau}_{\alpha\beta} = \hat{\eta}(x_3)[\partial_\alpha u_\beta(0) + \partial_\beta u_\alpha(0)]$. Substituting these expressions into (7.49), we find

$$-\partial_3 \mathbf{u} - \nabla_h \cdot (\sigma \mathbf{T}) - 3\nabla_h(\sigma \nabla_h \cdot \mathbf{u}) = \frac{\rho_0 g\alpha}{\eta_0} \nabla_h M, \qquad (7.51)$$

where

$$\sigma = \left\langle \frac{\hat{\eta}(x_3) - \eta_0}{\eta_0} \right\rangle \equiv \frac{1}{\eta_0} \int_0^h \left(\hat{\eta}(x_3) - \eta_0\right) dx_3 \qquad (7.52)$$

is a quantity with units of length that measure the excess viscosity of the BL, $\mathbf{u}$ is the horizontal velocity vector, $\mathbf{T}$ is a modified strain rate tensor with components $T_{11} = -T_{22} = \partial_1 u_1 - \partial_2 u_2$ and $T_{12} = T_{21} = \partial_1 u_2 + \partial_2 u_1$, and all terms are evaluated at $x_3 = 0$. (7.51) is an effective boundary condition that represents the influence of the BL on the interior fluid. It shows that the BL acts like an extensible skin with (depth-integrated) shear viscosity $\sigma \eta_0$ and compressional viscosity $3\sigma \eta_0$ that applies a shear stress proportional to $\nabla_h M$ to the interior fluid.

The next step is to determine explicitly the flow in the interior fluid that is driven by lateral variations in M. This is done by using the Fourier transform to solve the Stokes equations in the halfspace $x_3 \geq 0$ subject to (7.51). Evaluating the resulting solution at $x_3 = 0$, we obtain the closure relationship (LRS00, eqn. 2.41)

$$\bar{\mathbf{u}}(\mathbf{k}) = \frac{i\mathbf{k}}{2k(1+2\sigma k)} \frac{\rho_0 g\alpha}{\eta_0} \bar{M}(\mathbf{k}) \qquad (7.53)$$

where $\mathbf{k}$ is the horizontal wavevector, $k = |\mathbf{k}|$ and overbars denote the Fourier transform. Equation (7.53) is only valid if σ does not vary laterally. Finally, an evolution equation for M is obtained by taking the first moment of the energy equation, yielding

$$\partial_t M + \mathbf{u} \cdot \nabla_h M + 2M \nabla_h \cdot \mathbf{u} = \kappa \left(\nabla_h^2 M + \Delta T\right). \qquad (7.54)$$

Equations (7.53) and (7.54) are three equations for $u_\alpha(x_1, x_2, t)$ and $M(x_1, x_2, t)$ that can be solved numerically subject to periodic boundary conditions in the lateral directions for a specified initial condition $M(x_1, x_2, 0)$.

An important special case of the preceding equations is the Rayleigh-Taylor instability (RTI) of a layer with density $\rho_0 + \Delta\rho$ and viscosity $\eta_1 = \gamma \eta_0$, obtained by the transformation

$$M \to h^2 \Delta\rho/2\rho_0\alpha, \quad \sigma \to h(\gamma - 1), \quad \kappa \to 0. \qquad (7.55)$$

A linear stability analysis of (7.53) and (7.54) can now be performed by setting $h = h_0 + \tilde{h} \exp(i\mathbf{k} \cdot \mathbf{x}) \exp(st)$ and $\mathbf{u} = \tilde{\mathbf{u}} \exp(i\mathbf{k} \cdot \mathbf{x}) \exp(st)$ and linearizing in the perturbations $\tilde{h}$ and $\tilde{\mathbf{u}}$. The resulting growth rate is $s/s_1 = \epsilon\gamma/2[1 + 2\epsilon(\gamma - 1)]$ where $\epsilon = h_0 k$ and $s_1 = g\Delta\rho h_0/\eta_1$. This agrees with the exact analytical expression

(3.21) if $\gamma \gg \epsilon$. Equations (7.51)–(7.54) are therefore valid as long as the BL is not too much less viscous than the interior fluid.

The preceding equations also describe the finite-amplitude evolution of the RTI of a dense viscous layer over a passive half-space (Canright and Morris, 1993, henceforth CM93). Because the half-space is effectively inviscid ($\gamma \to \infty$), the closure law (7.53) is not meaningful, and the relevant equations are (7.51) and (7.54). Rewriting these using (7.55) and noting that the first term in (7.51) (shear stress applied by the inviscid fluid) is negligible, we obtain

$$\nabla_h \left(\frac{g\Delta\rho}{4\eta_1} h^2 + h\nabla_h \cdot \mathbf{u} \right) + \nabla_h \cdot (h\mathbf{E}) = 0, \qquad (7.56)$$

$$\partial_t h + \nabla_h \cdot (h\mathbf{u}) = 0, \qquad (7.57)$$

where $\mathbf{E}$ is the standard 2-D strain rate tensor. Equations (7.56) and (7.57) are just the dimensional forms of eqns. (3.8) and (3.7), respectively, of CM93.

A remarkable feature of the equations (7.53)–(7.54) and (7.56)–(7.57) is the existence of solutions in which M or h becomes infinite at a finite time t^*, corresponding to the runaway escape of the plume from its source layer. Even though the flow in the vicinity of the developing singularity violates the original long-wave assumption of the model, this failure is only local and does not impair the model's global validity. Several examples are discussed by CM93, who obtained solutions of (7.56)–(7.57), which show that the amplitude of disturbances of a layer with initial thickness h_0 evolves at long times as $h \propto (t^* - t)^{-1}$, where $t^* = C\eta_1/h_0 g \Delta\rho$ and the value of C depends on the form of the initial disturbance. The generalization of this result to power-law fluids is discussed in § 10.1.

7.4 Effective Boundary Conditions from Thin-Layer Flows

The uppermost and lowermost portions of Earth's mantle comprise relatively thin layers, which are rheologically distinct from the rest of the mantle: the D'' layer, the lithosphere and (possibly) a low-viscosity channel beneath the lithosphere. Because such layers are thin, thin-layer theory can be used to reduce their dynamics to an equivalent boundary condition on the adjacent convective flow. There are two distinct limiting cases, according to whether the thin layer has a much higher or much lower viscosity than the adjacent convecting fluid.

7.4.1 Interaction of Convection with a Passive Lithosphere

We first consider the case of a high-viscosity layer in the form of a rheologically distinct and passive (nonbuoyant) lithosphere. A sketch of the model geometry

7.4 Effective Boundary Conditions

(a)

$x_3 = 0$

$\hat{\eta}(x_1, x_2)$

$x_3 = -h$

$\bar{\eta}(x_1, x_2)$

(b)

$y = d$

$y = 0$

$\hat{\eta}$

η

τ

$y = -D$

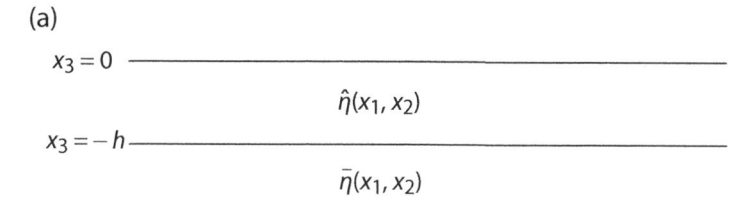

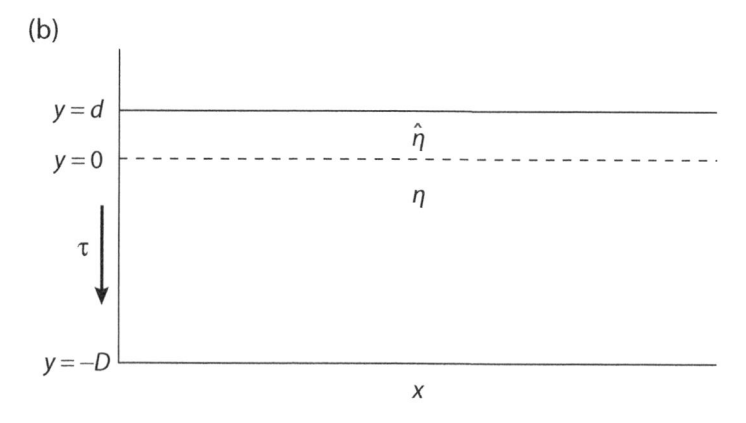

x

Figure 7.5 Models for effective boundary conditions from thin-layer flows. (a) A high-viscosity lithosphere with thickness h and viscosity $\hat{\eta}(x_1, x_2)$ overlying a mantle with viscosity $\bar{\eta}(x_1, x_2)$. See § 7.4.1 for discussion. (b) A low-viscosity channel of thickness d and viscosity $\hat{\eta}$ overlying a layer with thickness D and viscosity η. Flow is driven by a shear stress τ applied at $x = 0$. See § 7.4.2 for discussion. Part (b) redrawn from Figure 1 of Morris (2008) by permission of AIP Publishing LLC.

is shown in Figure 7.5. For simplicity, consider a flat fluid sheet with constant thickness h and laterally variable viscosity $\hat{\eta}(x_1, x_2)$, overlying a mantle whose viscosity just below the sheet is $\bar{\eta}(x_1, x_2) \ll \hat{\eta}$. In the following, superposed hats and bars denote quantities within the sheet and in the mantle just below it, respectively.

Because $\hat{\eta} \gg \bar{\eta}$, it is consistent to assume that the flow within the sheet is the sum of a dominant in-plane (membrane) component plus a small shear component that is required to balance the shear stress applied by the convective flow beneath. The dominance of membrane flow means that the lateral velocities $\hat{u}_\alpha$ and the pressure $\hat{p}$ are independent of the depth x_3 to lowest order, where $x_3 = 0$ and $x_3 = -h$ are the upper and lower surfaces of the sheet, respectively. Because the sheet is passive, the lateral force balance within it is $\partial_\alpha \hat{\sigma}_{\alpha\beta} + \partial_3 \hat{\sigma}_{\beta 3} = 0$. Integrating this across the sheet subject to the free-surface condition $\hat{\sigma}_{\beta 3}|_{z=0} = 0$, we obtain

$$h\partial_\alpha[-\hat{p}\delta_{\alpha\beta} + 2\hat{\eta}\hat{e}_{\alpha\beta}] - \hat{\sigma}_{\beta 3}|_{z=-h} = 0, \qquad (7.58)$$

where $e_{\alpha\beta} = (\partial_\alpha u_\beta + \partial_\beta u_\alpha)/2$ is the strain rate tensor. Now continuity of the velocity and stress at $x_3 = -h$ requires $\hat{u}_i = \bar{u}_i$, $\hat{e}_{\alpha\beta} = \bar{e}_{\alpha\beta}$, $\hat{\sigma}_{\beta 3}|_{z=-h} = \bar{\sigma}_{\beta 3}$ and $-\hat{p} + 2\hat{\eta}\hat{e}_{33} \equiv -\hat{p} - 2\hat{\eta}\bar{e}_{\lambda\lambda} = \bar{\sigma}_{33}$, where the continuity equation $\hat{e}_{33} = -\hat{e}_{\lambda\lambda}$ has been used. Substituting these relations into (7.58), we obtain an effective boundary condition that involves only mantle (barred) variables and the known viscosity $\hat{\eta}$ of the sheet:

$$h^{-1}\bar{\sigma}_{\beta 3} = 2\partial_\alpha \left[\hat{\eta}\left(\bar{e}_{\alpha\beta} + \bar{e}_{\lambda\lambda}\delta_{\alpha\beta}\right)\right] + \partial_\beta \bar{\sigma}_{33}. \tag{7.59}$$

Equation (7.59) remains valid even if the sheet has a power-law rheology (4.1f), because $\hat{I} \approx (\hat{e}_{\alpha\beta}\hat{e}_{\alpha\beta} + \hat{e}_{\lambda\lambda}^2)^{1/2}$ can be written in terms of the mantle variables using the matching condition $\hat{e}_{\alpha\beta} = \bar{e}_{\alpha\beta}$.

In reality, the sheet thickness is governed by the conservation law $\partial_t h = -\partial_\alpha(h\hat{u}_\alpha)$ and will therefore not remain constant in general. However, this effect is ignored in most applications. Weinstein and Olson (1992) used the 1-D version of (7.59) with a power-law rheology for $\hat{\eta}$ to study the conditions under which a highly non-Newtonian sheet above a vigorously convecting Newtonian fluid exhibits plate-like behavior. Ribe (1992) used the spherical-coordinate analog of (7.59) to investigate the generation of a toroidal component of mantle flow by lateral viscosity variations in the lithosphere.

7.4.2 Influence of a Low-Viscosity Channel

At the opposite extreme from the case just treated, the influence of thin horizontal low-viscosity channels on thermal convection in a more viscous fluid has been studied by Busse et al. (2006) and Lenardic et al. (2006), using a generalization of the boundary-layer analysis of Turcotte (1967). The motivation of this work was to understand better the physical factors that might give rise to large aspect-ratio (width/depth) convection cells whose existence in Earth's mantle seems to be indicated by the large lateral extent of plates such as the Pacific plate. Busse et al. (2006) considered the case of two symmetric low-viscosity channels adjoining the upper and lower boundaries of the convecting region and found that the presence of these channels increases the aspect ratio of the convection cells that maximizes the heat transport. Subsequently, Morris (2008) studied a simpler isothermal model (Figure 7.5b) in which a semi-infinite horizontal layer of thickness D and viscosity η is overlain by a channel of thickness d and viscosity $\hat{\eta}$, where $d/D \ll 1$ and $\hat{\eta}/\eta \ll 1$. Flow in this two-layer system is driven by a constant shear stress τ applied to the end $x = 0$ of the layers, modelling the influence of a subducting slab. The following discussion is based on that of Morris (2008) with some changes of notation.

Morris's (2008) study proceeds by reducing the effect of the low-viscosity channel to an equivalent boundary condition on the flow in the more viscous sublayer

7.4 Effective Boundary Conditions

beneath. The first step is to write down the equations of lubrication theory that govern the flow in the thin channel:

$$\hat{p}_x = \hat{\eta}\hat{u}_{yy}, \tag{7.60a}$$
$$\hat{p}_y = 0, \tag{7.60b}$$
$$\hat{u}_x + \hat{v}_y = 0. \tag{7.60c}$$

Here and henceforth all variables in the channel are denoted by hats, variables in the sublayer are unhatted and partial derivatives are denoted by subscripts. Now introduce a streamfunction

$$\hat{u} = -\hat{\psi}_y, \quad \hat{v} = \hat{\psi}_x. \tag{7.61}$$

Equation (7.60c) is then satisfied identically, and (7.60a) takes the form

$$\hat{p}_x = -\hat{\eta}\hat{\psi}_{yyy}. \tag{7.62}$$

Now we assume that the normal velocity and the shear traction are zero at the upper surface $y = d$ of the channel, or $\hat{v}(x, d) = \hat{u}_y(x, d) = 0$. We also assume that the viscosity contrast is sufficiently large that the channel sees the sublayer as a no-slip surface, or $\hat{u}(x, 0) = 0$. In terms of the streamfunction, these three boundary conditions are

$$\hat{\psi}(x, d) = \hat{\psi}_{yy}(x, d) = \hat{\psi}_y(x, 0) = 0. \tag{7.63}$$

Now (7.60b) implies that $\hat{p} = \hat{p}(x)$. Equation (7.62) can then be integrated subject to the boundary conditions (7.63) to obtain

$$\hat{\psi} = -\frac{d^3}{6\hat{\eta}}\hat{p}'(x)\left[2 - 3\left(\frac{y}{d}\right)^2 + \left(\frac{y}{d}\right)^3\right]. \tag{7.64}$$

At the viscosity interface $y = 0$, (7.64) gives

$$\hat{\psi}(x, 0) = -\frac{d^3}{3\hat{\eta}}\hat{p}'(x). \tag{7.65}$$

Continuity of vertical velocity across the viscosity interface requires $\hat{\psi}_x(x, 0) = \psi_x(x, 0)$. Integrating this equation with respect to x and noting that $\hat{\psi}(0, 0) = \psi(0, 0)$ because the sidewall $x = 0$ is impermeable, we find $\hat{\psi}(x, 0) = \psi(x, 0)$. Equation (7.65) can therefore be rewritten as

$$\psi(x, 0) = -\frac{d^3}{3\hat{\eta}}\hat{p}'(x). \tag{7.66}$$

Now continuity of normal stress across the viscosity interface requires

$$-p(x, 0) + 2\eta v_y(x, 0) = -\hat{p}(x). \tag{7.67}$$

As usual in lubrication theory, the deviatoric normal stress $2\hat{\eta}\hat{v}_y(x,0)$ is small relative to the pressure and has been neglected. Differentiating (7.67), we obtain

$$-p_x(x,0) + 2\eta v_{xy}(x,0) = -\hat{p}'(x). \tag{7.68}$$

Now use (7.66) to eliminate $\hat{p}'(x)$ from (7.68), and use the momentum equation $p_x = \eta \nabla^2 u$ to replace $p_x(x,0)$. Writing the result in terms of the streamfunction in the sublayer, we obtain

$$\frac{1}{3}\psi_{yyy}(x,0) + \psi_{xxy}(x,0) = \frac{\hat{\eta}}{\eta d^3}\psi(x,0). \tag{7.69}$$

Equation (7.69) is a coupling condition that describes the interaction between the channel and the more viscous sublayer, entirely in terms of the sublayer streamfunction.

The next step is to write down the boundary-value problem (BVP) for the flow in the sublayer. The streamfunction satisfies the biharmonic equation subject to the boundary conditions

$$\psi(x,-D) = \psi_{yy}(x,-D) = 0, \tag{7.70a}$$

$$\psi_{yy}(x,0) - \psi_{xx}(x,0) = 0, \tag{7.70b}$$

$$\frac{1}{3}\psi_{yyy}(x,0) + \psi_{xxy}(x,0) = \frac{\hat{\eta}}{\eta d^3}\psi(x,0), \tag{7.70c}$$

$$\psi(0,y) = 0, \tag{7.70d}$$

$$\eta \psi_{xx}(0,y) = \tau. \tag{7.70e}$$

Conditions (7.70a) state that the bottom boundary of the sublayer is impermeable and free of shear traction. Condition (7.70b) states that the sublayer sees the viscosity interface as an effectively traction-free surface, which is the case because the viscosity of the channel is much less than that of the sublayer. Condition (7.70c) is the coupling condition. Condition (7.70d) states that the sidewall $x = 0$ is impermeable. Finally, condition (7.70e) represents the driving tangential stress applied at $x = 0$.

To nondimensionalize the BVP, define dimensionless (starred) variables

$$(x,y) = D(x^*, y^*), \quad \psi = \frac{\tau D^2}{\eta}\psi^*. \tag{7.71}$$

Substituting these definitions into the biharmonic equation and the boundary conditions (7.70) and immediately dropping the stars, we obtain the dimensionless BVP

7.4 Effective Boundary Conditions

$$\nabla^4 \psi = 0, \tag{7.72a}$$

$$\psi(x,-1) = \psi_{yy}(x,-1) = 0, \tag{7.72b}$$

$$\psi_{yy}(x,0) - \psi_{xx}(x,0) = 0, \tag{7.72c}$$

$$\frac{1}{3}\psi_{yyy}(x,0) + \psi_{xxy}(x,0) = \epsilon\psi(x,0), \quad \epsilon = \frac{\hat{\eta}}{\eta}\left(\frac{D}{d}\right)^3, \tag{7.72d}$$

$$\psi(0,y) = \psi_{xx}(0,y) - 1 = 0. \tag{7.72e}$$

The sole dimensionless parameter that appears in the preceding BVP is ϵ, which is the effective hydrodynamic resistance of the low-viscosity channel. Small values of ϵ correspond to a channel with either a very low viscosity or a relatively large thickness (assuming, however, that $D/d \gg 1$). Both factors make it easier for the channel to carry a significant portion of the volume flux driven by the applied shear stress. Because $\hat{\eta}/\eta \ll 1$ and $(D/d)^3 \gg 1$, the product of the two factors ($=\epsilon$) may be either large or small. The BVP (7.72) is valid for $d/D \to 0$ with ϵ fixed (Morris, 2008).

The formal solution of the BVP (7.72) is the subject of Exercise 7.3. The resulting streamlines for the flow in the sublayer are shown in Figure 7.6 for $\epsilon = 50$ and 0.1. The horizontal length scale ℓ of the sublayer flow increases as the hydrodynamic resistance ϵ of the channel decreases, scaling as $\ell = 2D/\sqrt{3\epsilon}$ for small values of ϵ. Morris (2008) showed that this increase of the cell length is a consequence of Helmholtz's minimum dissipation principle, which is applicable because the viscosity is a specified function of position. Let u_s and u_c be the characteristic scales for the horizontal velocities in the sublayer and the channel, respectively. The total dissipation rate Φ (per unit length in the third dimension) comprises three contributions: a part Φ_e due to the sublayer flow near the driven end of the layer, and which is independent of ℓ; a part $\Phi_s \sim \eta(u_s/\ell)^2 \ell D$ due to the core (i.e., away from the end) of the sublayer flow; and a part $\Phi_c \sim \hat{\eta}(u_c/d)^2 \ell d$ due to the flow

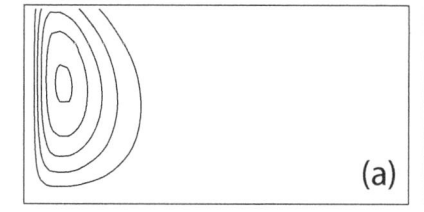

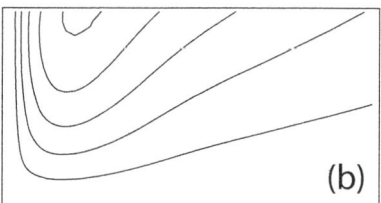

Figure 7.6 Streamlines of the flow in an end-driven layer adjoining a low-viscosity channel, calculated from the solution of the BVP (7.72) for (a) $\epsilon = 50$ and (b) $\epsilon = 0.1$. The contours in each panel are for $\psi/\psi_{max} = 0.2, 0.4, 0.6, 0.8$ and 0.99. The streamlines are closed by the flow in the thin channel (not shown). Redrawn from Figure 2 of Morris (2008) by permission of AIP Publishing LLC.

in the channel. Now because the net flow across any vertical plane normal to $\mathbf{e}_x$ is zero, $u_c d \sim u_s D$. Combining the preceding relations, we obtain

$$\frac{\Phi}{\eta u_s^2} = a_e + a_s \frac{D}{\ell} + a_c \frac{\epsilon \ell}{D} \tag{7.73}$$

where a_e, a_s and a_c are numerical constants. Minimization of (7.73) with respect to ℓ shows that the minimum dissipation occurs for $\ell = D\epsilon^{-1/2}(a_s/a_c)^{1/2}$, in agreement with the previously quoted result $\ell = 2D/\sqrt{3\epsilon}$.

7.5 Conduit Solitary Waves

Another flow that can be studied using a long-wave approximation is the motion of finite-amplitude solitary waves in a cylindrical conduit of low-viscosity fluid embedded in a fluid of higher viscosity (Olson and Christensen, 1986; Scott et al., 1986). These waves are examples of kinematic waves (Whitham, 1974), which occur when there is a functional relation between the density or amplitude of the medium and the flux of a conserved quantity like mass; other examples include the propagation of a pulse of wastewater down a gutter and the flow of traffic on a crowded highway.

To illustrate the basic ideas, we follow here the derivation of Olson and Christensen (1986). Consider an infinite vertical conduit with a circular cross-section of radius $R(z,t)$ and area $A(z,t)$, where z is the height and t is time. The conduit contains fluid with density ρ_L and viscosity η_L and is embedded in an infinite fluid with density $\rho_M \equiv \rho_L + \Delta\rho$ and viscosity $\eta_M \gg \eta_L$. Conservation of mass in the conduit requires

$$A_t + Q_z = 0, \tag{7.74}$$

where $Q(z,t)$ is the volume flux and subscripts indicate partial derivatives. Because $\eta_M \gg \eta_L$, the fluid in the conduit sees the wall as a no-slip surface. The volume flux is therefore given by Poiseuille's law

$$Q = -\frac{A^2}{8\pi\eta_L} P_z, \tag{7.75}$$

where P is the nonhydrostatic pressure in the conduit. P is determined by the requirement that the normal stress σ_{rr} in the radial (r) direction be continuous at $r = R$. Now the pressure in the matrix is hydrostatic, and the deviatoric component of σ_{rr} in the conduit is negligible relative to that in the matrix because $\eta_M \gg \eta_L$. Continuity of σ_{rr} therefore requires $\rho_L g z - P = \rho_M g z + 2\eta_M u_r(R)$. However, because the flow in the matrix is dominantly radial, the continuity equation requires $u \propto r^{-1}$, whence $u_r(R) = -u(R)/R = -A_t/2A$ and

7.5 Conduit Solitary Waves

$$P = -\Delta\rho g z + \eta_M A^{-1} A_t. \tag{7.76}$$

Substituting (7.76) into (7.75) and using (7.74), we obtain

$$Q = \frac{A^2}{8\pi\eta_L}\left[\Delta\rho g + \eta_M(A^{-1}Q_z)_z\right]. \tag{7.77}$$

The undisturbed background state is a steady Poiseuille flow with volume flux Q_0 in a conduit of constant radius A_0, for which (7.77) reduces to

$$Q_0 = \frac{g\Delta\rho}{8\pi\eta_L}A_0^2. \tag{7.78}$$

To make the coupled equations (7.74) and (7.77) dimensionless, we introduce length and time scales

$$L = \left(\frac{\eta_M A_0}{8\pi\eta_L}\right)^{1/2}, \quad T = \frac{1}{g\Delta\rho}\left(\frac{8\pi\eta_L\eta_M}{A_0}\right)^{1/2}. \tag{7.79}$$

Now define dimensionless (starred) variables

$$Q = Q_0 Q^*, \quad A = A_0 A^*, \quad z = Lz^*, \quad t = Tt^*. \tag{7.80}$$

Writing (7.74) and (7.77) in terms of the dimensionless variables and immediately dropping the stars, we obtain

$$Q = A^2(1 + (A^{-1}Q_z)_z), \tag{7.81a}$$

$$A_t + Q_z = 0. \tag{7.81b}$$

Reducing (7.81) to a single equation for A, we find

$$A_t + (A^2)_z - (A^2(A^{-1}A_t)_z)_z = 0, \tag{7.82}$$

which is sometimes called the 'conduit equation'.

Equations (7.81) admit finite-amplitude travelling wave solutions such that

$$A = A(z - ct) = A(y), \quad Q = Q(z - ct) = Q(y), \tag{7.83}$$

where c is the dimensionless wave speed in units of L/T. Substituting (7.83) into (7.81b) and integrating once with respect to y, we obtain

$$Q = Q_1 + cA \tag{7.84}$$

where Q_1 is a constant. Substitution of (7.83) and (7.84) into (7.81a) yields

$$c\frac{\mathrm{d}}{\mathrm{d}y}\left(\frac{1}{A}\frac{\mathrm{d}A}{\mathrm{d}y}\right) = \frac{c}{A} + \frac{Q_1}{A^2} - 1. \tag{7.85}$$

Multiplying each term in (7.85) by a suitable integrating factor (Exercise 7.5) converts the equation to an exact differential that can be integrated once, yielding

$$c\left(\frac{dA}{dy}\right)^2 = c_1 A^2 - 2A^2 \ln A - 2cA - Q_1 \tag{7.86}$$

where c_1 is a constant.

The dispersion relation for waves of finite amplitude can be found by requiring that dA/dy vanish at the points where the amplitude is minimum (A_{min}) and maximum (A_{max}). Evaluating (7.86) at these amplitudes and eliminating c_1 from the resulting expressions yields the implicit dispersion relation

$$\left(\frac{A_{min}}{A_{max}}\right)^2 = \frac{2A_{min}^2 \ln A_{min} + 2cA_{min} + Q_1}{2A_{max}^2 \ln A_{max} + 2cA_{max} + Q_1} \tag{7.87}$$

A particularly interesting case is that of an isolated solitary wave that propagates without change of shape along an otherwise uniform conduit. The constant Q_1 is then fixed by the conditions

$$A_{min}, Q \to 1 \quad \text{as } y \to \pm\infty, \tag{7.88}$$

whence (7.84) becomes

$$Q_1 = 1 - c. \tag{7.89}$$

Equation (7.89) corrects a misprint in eqn. (24) of Olson and Christensen (1986). The dispersion relation (7.87) now takes the simplified form

$$c = \frac{2A_{max}^2 \ln A_{max} - A_{max}^2 + 1}{(A_{max} - 1)^2}. \tag{7.90}$$

Figure 7.7 shows $c(A_{max})$ as given by (7.90). In the large-amplitude limit $A_{max} \gg 1$, the wavespeed is

$$c = 2\ln A_{max} - 1, \tag{7.91}$$

which is shown by the dashed line in Figure 7.7. The speed of the wave thus increases with its amplitude. Also in the large-amplitude limit, the solution of (7.86) is

$$A = A_{max} \exp(-y^2/2c), \tag{7.92}$$

implying that the disturbance has a Gaussian profile.

Some further remarks on the mathematical analysis of the conduit equation (7.82) will be found in § 9.4.2, where that equation is shown to be a special case of a more general family of equations for the transport of melt in a deformable porous matrix. The conduit equation has also been generalized to the case of a viscoelastic external fluid by Grimshaw et al. (1992). Another generalization (Richardson et al., 1996; Hewitt and Fowler, 2009) replaces the single-phase external fluid by a deformable melt-saturated porous medium (§ 9.2). Variations of pressure due to the passage of solitary waves then pump melt into and out of the conduit.

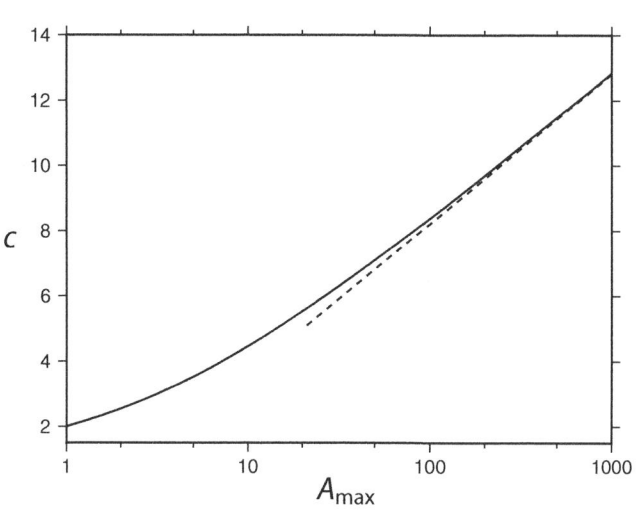

Figure 7.7 Dispersion relation (7.90) for the speed of an isolated solitary wave in a low-viscosity channel. The quantities c and A_{max} are dimensionless as described in the text. The dashed line shows the large-amplitude limit (7.91).

Geodynamic applications of conduit solitary waves include the suggestion of Whitehead and Helfrich (1988) that such waves might transport deep mantle material rapidly to the surface with little diffusion or contamination. Schubert et al. (1989) subsequently showed numerically that solitary waves can also propagate along the conduits of thermal plumes in a fluid with temperature-dependent viscosity. Such waves have been invoked to explain the origin of V-shaped topographic ridges on the ocean floor south of Iceland (Albers and Christensen, 2001; Ito, 2001) and of quasi-periodic variations of volcanic flux along the Hawaiian–Emperor Island/seamount chain (Van Ark and Lin, 2004; Vidal and Bonneville, 2004; Gavrilov and Boiko, 2012).

Exercises

7.1 A layer of viscous fluid with thickness d, viscosity η and density ρ rests on a rigid surface $z = -d$ in a field of gravity. At time $t = 0$, the upper surface of the fluid has the form $\zeta = \zeta_0 \sin kx$ where $\zeta_0 \ll d$ (small amplitude) and $\zeta_0 \ll 2\pi/k$ (small slope). The surface topography decays thereafter as $\zeta = \zeta_0 \sin kx \exp st$, where $s < 0$ is the decay rate.

(a) For arbitrary wavenumber k, the solution of the problem yields the decay rate

$$s = -\frac{g}{2k\nu}\left[\frac{\sinh 2kd - 2kd}{\cosh 2kd + 1 + 2k^2d^2}\right]. \quad (7.93)$$

Determine the limiting forms of the preceding expression when $kd \gg 1$ and $kd \ll 1$, and interpret the results physically.

(b) Use lubrication theory to derive a nonlinear partial differential equation for $\zeta(x,t)$. Linearize and solve this equation, and show that the resulting decay rate is the same as the result from part (a) for $kd \ll 1$.

7.2 Consider the subduction model shown in Figure 4.3 in the limit of small wedge angle $\alpha \ll 1$. Use lubrication theory to determine the pressure in the wedge, and compare your result with the exact solution (4.37), (4.42) in the limit $\alpha \to 0$.

7.3 [SM] Solve the BVP (7.72) using a sine transform, which is an appropriate method because only derivatives of even order in x appear. The sine transform (denoted by an overbar) and its inverse are defined as

$$\overline{F}(k) = \sqrt{\frac{2}{\pi}} \int_0^\infty F(x) \sin kx \, dx, \qquad (7.94a)$$

$$F(x) = \sqrt{\frac{2}{\pi}} \int_0^\infty \overline{F}(k) \sin xk \, dk. \qquad (7.94b)$$

7.4 Consider further the problem of flow in an end-driven layer adjoining a thin low-viscosity channel (Figure 7.5b). Using the theory of complex variables, Morris (2008) showed that for $\epsilon \to 0$ and $x \to \infty$ the dimensionless sublayer streamfunction is

$$\psi = -2Be^{-qx}[y \cos qy - (\cot q + q^{-1}) \sin qy], \qquad (7.95a)$$

$$B = \frac{(q - 2\sin q)\tan(q/2)}{2q^2(3 - 2q\cot q) + 3q \sin 2q}, \qquad (7.95b)$$

$$q = \frac{1}{2}\sqrt{3\epsilon}. \qquad (7.95c)$$

The dimensionless length scale over which the solution decays in the x-direction is therefore $2/\sqrt{3\epsilon}$. Now determine the analogous result for a layer without a low-viscosity channel by solving the BVP for an end-driven isoviscous layer, expressing the solution as a superposition of separable solutions. Compare the decay scales for the solutions with and without a low-viscosity channel.

7.5 Starting from (7.85), derive (7.86).

8
Long-Wave Theories, 2: Shells, Plates and Sheets

A central problem in geodynamics is to determine the response of the lithosphere to applied loads such as seamounts, plate boundary forces (ridge push, slab pull, etc.) and tractions imposed by underlying mantle convection. Such problems can be solved effectively using thin-shell theory, a branch of applied mechanics concerned with the behaviour of freely deformable bodies whose thickness h is much smaller than both their typical lateral dimension L and their typical radius of curvature R. This condition is evidently satisfied for Earth's lithosphere, for which $h \approx 100$ km, $L \approx 1000$–10000 km and $R \approx 6370$ km. More generally, thin-shell theory applies to layers in which the stress is dominated by the components $\sigma_{\alpha\beta}$, where α and β are coordinate directions parallel to the layer. This is the case when the layer is not attached to a stiff substrate and is free to deform by a combination of stretching and bending.

The basic idea of thin-shell theory is to exploit the smallness of h/L and h/R to reduce the full 3-D dynamic equations to equivalent 2-D equations for the dynamics of the shell's midsurface. Let the 3-D Cartesian coordinates of any point on this surface be $\mathbf{x}_0(\theta_1, \theta_2)$, where θ_α are coordinates on the midsurface itself. In the most general formulations of thin-shell theory (e.g., Niordson, 1985), θ_α are allowed to be arbitrary and nonorthogonal. Such a formulation is indispensable for thin viscous sheets that can experience large deformations, because material coordinate lines that follow the directions of principal curvature at a given time will not do so at later times. If the deformation is small, however, it makes sense to define θ_α as orthogonal lines-of-curvature coordinates whose isolines are parallel to the two directions of principal curvature of the midsurface at each point. This less elegant and more limited formulation is the one that has been used in most geodynamic applications of thin-shell theory.

Following common practice, I shall use the word *shell* for elastic shells, and the word *plate* for elastic shells that are flat. Viscous shells and plates will both be called *sheets*. The theories of shells and sheets are closely related: the equations governing sheets can be obtained from those governing shells by applying the Stokes–Rayleigh transformation (5.4). However, the reverse is not true in general because (incompressible) sheets have no property analogous to Poisson's ratio. The equations for sheets can be transformed into those for shells only for the special case of incompressible shells.

In light of the preceding remarks, I begin this chapter with a presentation of thin viscous-sheet theory in general nonorthogonal coordinates. Next, I present the equations of elastic thin-shell theory in lines-of-curvature coordinates that follow the directions of principal curvature on the shell's midsurface. Finally, I discuss several geodynamic applications and show how each represents a special case of the full lines-of-curvature theory.

8.1 Theory of Thin Viscous Sheets in General Coordinates

8.1.1 Differential Geometry of Thin Sheets

Because θ_1 and θ_2 are nonorthogonal coordinates, the formalism of differential geometry and general tensor calculus is required to describe the shape of the sheet. The following development is based on Ribe (2002) and uses as far as possible the notation of Green and Zerna (1992) (henceforth GZ), whose chapters 1 and 10 may be consulted for more detail. Latin indices range over the values 1, 2 and 3; Greek indices range over the values 1 and 2 only; and the summation convention for repeated indices (subscript plus superscript pairs) is assumed. For simplicity, the sheet's thickness T is assumed constant.

Figure 8.1a shows a segment of a curved sheet with thickness T, indicating the isolines of the general coordinates θ_1 and θ_2 on the midsurface (dashed lines). Let $\mathbf{r}_0(\theta_1, \theta_2)$ be the position vector of a point on the midsurface relative to an arbitrary origin, and let $\mathbf{a}_3(\theta_1, \theta_2)$ be a unit vector normal to the midsurface. Then the position vector of an arbitrary point in the sheet is

$$\mathbf{r}(\theta_1, \theta_2, z) = \mathbf{r}_0(\theta_1, \theta_2) + z\mathbf{a}_3(\theta_1, \theta_2). \tag{8.1}$$

The vectors $\mathbf{r}_0$ and $\mathbf{a}_3$ are also functions of time in general. However, because inertia has been assumed negligible, time plays the role of a mere parameter, and so the time argument is suppressed here and henceforth.

Because the coordinates θ_1 and θ_2 may be nonorthogonal, one must distinguish between covariant and contravariant components of vectors and tensors defined on the midsurface. The covariant midsurface base vectors are just the partial derivatives of the position vector, viz.,

$$\mathbf{a}_\alpha = \mathbf{r}_{0,\alpha} \tag{8.2}$$

where a comma denotes partial differentiation with respect to θ_α. While both $\mathbf{a}_1$ and $\mathbf{a}_2$ are tangent to the midsurface, they are neither orthogonal nor unitary in general.

8.1 Theory of Thin Viscous Sheets

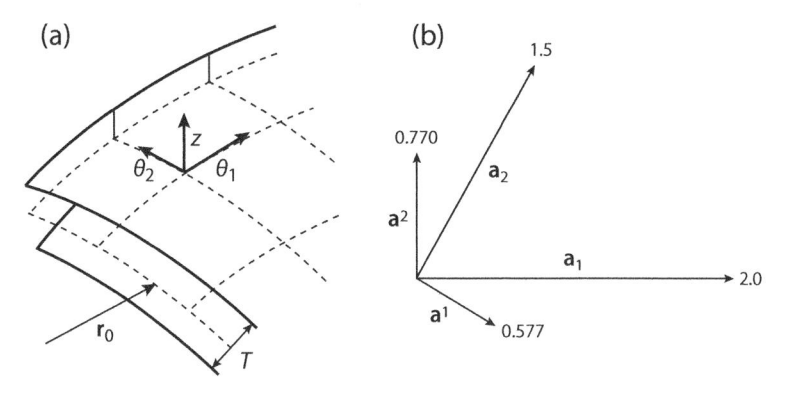

Figure 8.1 (a) Section of a curved viscous sheet with constant thickness T. The isolines of the coordinates θ_1 and θ_2 on the midsurface are indicated by dashed lines. The coordinate normal to the midsurface is $z \in [-T/2, T/2]$. (b) Illustration of the covariant (subscripts) and contravariant (superscripts) base vectors on the midsurface, which satisfy $\mathbf{a}^\alpha \cdot \mathbf{a}_\beta = \delta^\alpha_\beta$. The length of each vector is indicated at its end. See § 8.1.1 for discussion.

The contravariant base vectors $\mathbf{a}^\alpha$ are the reciprocals of the covariant ones, which means that they satisfy

$$\mathbf{a}^\alpha \cdot \mathbf{a}_\beta = \delta^\alpha_\beta, \tag{8.3}$$

where δ^α_β is the Kronecker delta. The quantities

$$a_{\alpha\beta} = \mathbf{a}_\alpha \cdot \mathbf{a}_\beta, \qquad a^{\alpha\beta} = \mathbf{a}^\alpha \cdot \mathbf{a}^\beta \tag{8.4}$$

are respectively the covariant and contravariant components of the (symmetric) metric tensor of the midsurface.

The second fundamental tensor of the midsurface is the curvature tensor, with covariant components

$$b_{\alpha\beta} = -\mathbf{a}_\alpha \cdot \mathbf{a}_{3,\beta} = -\mathbf{a}_\beta \cdot \mathbf{a}_{3,\alpha}. \tag{8.5}$$

Equation (8.5) implies that the curvature tensor is proportional to the rate of change of the normal vector as one moves along the midsurface. The covariant components of the curvature tensor are related to the mixed components b^α_β and contravariant components $b^{\alpha\beta}$ by the formulae

$$b^\alpha_\beta = a^{\alpha\lambda} b_{\beta\lambda} = a_{\beta\lambda} b^{\alpha\lambda}, \qquad b^{\alpha\beta} = a^{\alpha\lambda} b^\beta_\lambda, \qquad b_{\alpha\beta} = a_{\alpha\lambda} b^\lambda_\beta, \tag{8.6}$$

illustrating the general rule that inner multiplication of a surface vector or tensor by $a^{\alpha\beta}$ (or $a_{\alpha\beta}$) is equivalent to raising (or lowering) the appropriate index. The invariant quantities

$$H = \frac{1}{2} b^\alpha_\alpha, \qquad G = b^1_1 b^2_2 - b^1_2 b^2_1 \tag{8.7}$$

are, respectively, the mean curvature and the Gaussian curvature of the midsurface. A third invariant quantity that is not in common use but will prove useful later is the curvature modulus

$$K = (4H^2 - 2G)^{1/2} = (b_\beta^\alpha b_\alpha^\beta)^{1/2}. \tag{8.8}$$

The mutually perpendicular principal directions of the curvature tensor are those with respect to which the midsurface is not twisted, i.e., along which adjacent normals to the midsurface are coplanar. The principal values k_1 and k_2 of b_β^α are called the principal curvatures of the midsurface. In terms of these, the three invariant quantities defined above are

$$H = \frac{k_1 + k_2}{2}, \quad G = k_1 k_2, \quad K = (k_1^2 + k_2^2)^{1/2}. \tag{8.9}$$

Because the sheet is curved, the base vectors $\mathbf{r}_{,\alpha}$ at points off the midsurface are not identical to the base vectors $\mathbf{r}_{0,\alpha}$ on the midsurface. The covariant base vectors $\mathbf{g}_i$ and the contravariant base vectors $\mathbf{g}^i$ at an arbitrary point in the sheet are

$$\mathbf{g}_\alpha = \mathbf{r}_{,\alpha} = \mu_\alpha^\lambda \mathbf{a}_\lambda, \quad \mathbf{g}^\alpha = h^{-1}\left(\mu_\rho^\rho \delta_\lambda^\alpha - \mu_\lambda^\alpha\right) \mathbf{a}^\lambda, \quad \mathbf{g}_3 = \mathbf{g}^3 = \mathbf{a}_3, \tag{8.10}$$

where

$$\mu_\alpha^\beta = \delta_\alpha^\beta - z b_\alpha^\beta \tag{8.11}$$

and

$$h = 1 - 2Hz + Gz^2 \tag{8.12}$$

is the ratio of an element of surface area at a distance z from the midsurface to the corresponding area on the midsurface itself (Figure 8.2). The covariant and contravariant components of the metric tensor at an arbitrary point in the sheet are

$$g_{ij} = \mathbf{g}_i \cdot \mathbf{g}_j, \qquad g^{ij} = \mathbf{g}^i \cdot \mathbf{g}^j. \tag{8.13}$$

A final important notion from differential geometry is that of covariant differentiation. To illustrate, consider the partial derivative $\mathbf{v}_{,i}$ of a vector $\mathbf{v}$ with respect to the coordinate θ_i. Depending on whether $\mathbf{v}$ is expressed with respect to the covariant base vectors $\mathbf{g}_k$ or the contravariant base vectors $\mathbf{g}^k$, $\mathbf{v}_{,i}$ can be written in two ways:

$$\mathbf{v}_{,i} = (v^k \mathbf{g}_k)_{,i} = v^k_{,i} \mathbf{g}_k + v^k \mathbf{g}_{k,i} = (v^k_{,i} + \Gamma^k_{ij} v^j) \mathbf{g}_k = v^k|_i \mathbf{g}_k, \tag{8.14a}$$

$$\mathbf{v}_{,i} = (v_k \mathbf{g}^k)_{,i} = v_{k,i} \mathbf{g}^k + v_k \mathbf{g}^k_{,i} = (v_{k,i} - \Gamma^j_{ik} v_j) \mathbf{g}^k = v_k|_i \mathbf{g}^k, \tag{8.14b}$$

where $\Gamma^k_{ij} = \mathbf{g}^k \cdot \mathbf{g}_{i,j} = -\mathbf{g}_i \cdot \mathbf{g}^k_{,j}$ is a so-called Christoffel symbol of the second kind. The quantities $v^k|_i$ and $v_k|_i$ are called the covariant derivatives of the vector components v^k and v_k. As (8.14) shows, the covariant derivative of a vector

8.1 Theory of Thin Viscous Sheets

$$dS(1 - 2Hz + Gz^2)$$

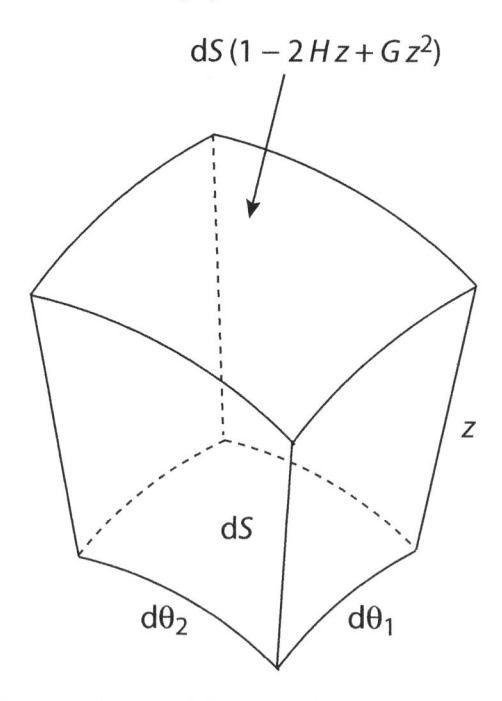

Figure 8.2 An element of a curved sheet. An elemental area dS on the midsurface corresponds to an area $dS(1 - 2Hz + Gz^2)$ at a distance z from the midsurface, where H and G are the mean curvature and Gaussian curvature, respectively, of the midsurface. See § 8.1.1 for discussion.

(or tensor) component is just the corresponding component of the partial derivative of the vector (or tensor) itself. As indicated by the occurrence of $\mathbf{g}_{k,i}$ and $\mathbf{g}^k_{,i}$ in (8.14), the covariant derivative takes into account the variation of the base vectors from point to point. The covariant derivative of any invariant quantity (such as a scalar) is just the usual partial derivative. Moreover, the covariant derivatives of all components of the metric tensor are zero (Exercise 8.3). The covariant derivative may therefore be thought of as a derivative that follows the changing metric of a surface. Expressions for the covariant derivatives of the components of vectors and tensors are given in chapter 1 of GZ. In the sequel, we shall be primarily concerned with the covariant derivatives of components of vectors and tensors defined on the sheet's midsurface, for which the indices i and k in (8.14) range over the values 1 and 2 only.

8.1.2 Governing Equations

In the theory of thin sheets, the fundamental dynamic quantities are the stress resultants and bending moments, defined as (weighted) integrals of the stresses across the sheet. For these integrals to be meaningful, the stresses must be expressed both per unit area of a single reference surface and relative to base vectors that do not vary across the sheet. The natural choice for this purpose is the midsurface and its intrinsic base vectors $\mathbf{a}_i$. The stress tensor referred to the midsurface in this way is (GZ, p. 375)

$$\sigma^{i\lambda} = h\mu_\alpha^\lambda \tau^{i\alpha}, \quad \sigma^{i3} = h\tau^{i3}, \tag{8.15}$$

where τ^{ij} is the stress tensor per unit *local* area and relative to the *local* basis $\mathbf{g}_i$. The tensor σ^{ij} is not symmetric, precisely because the base vectors to which it is referred are not the local ones.

The equations of equilibrium in terms of σ^{ij} are (GZ, p. 379)

$$\sigma^{\alpha\beta}|_\alpha - b_\alpha^\beta \sigma^{\alpha 3} + \sigma^{3\beta}_{,3} = -h\rho f^\beta, \tag{8.16a}$$

$$\sigma^{\alpha 3}|_\alpha + b_{\alpha\beta} \sigma^{\alpha\beta} + \sigma^{33}_{,3} = -h\rho f^3, \tag{8.16b}$$

where $f^i \mathbf{a}_i$ is the gravitational acceleration. The constitutive relations for a Newtonian fluid with viscosity η are

$$\tau^{ij} = -pg^{ij} + 2\eta g^{ik} g^{jl} e_{kl}, \tag{8.17}$$

where p is the pressure and e_{ij} is the strain rate tensor given by (GZ, p. 381)

$$2e_{\alpha\beta} = \mu_\beta^\lambda (u_\lambda|_\alpha - b_{\lambda\alpha} u_3) + \mu_\alpha^\lambda (u_\lambda|_\beta - b_{\lambda\beta} u_3) \tag{8.18a}$$

$$2e_{\alpha 3} = u_{3,\alpha} + u_{\alpha,3} + b_\alpha^\lambda (u_\lambda - z u_{\lambda,3}), \tag{8.18b}$$

$$e_{33} = u_{3,3}. \tag{8.18c}$$

Finally, incompressibility of the fluid requires $g^{ij} e_{ij} = 0$, or equivalently (Exercise 8.4)

$$(hu_3)_{,3} + \left[a^{\alpha\beta} + z\left(b^{\alpha\beta} - 2Ha^{\alpha\beta}\right)\right] u_\alpha|_\beta = 0. \tag{8.19}$$

8.1.3 Shallow-Sheet Solution

The governing equations of the previous subsection are valid for any flow in the vicinity of a surface defined by its metric and curvature tensors and do not yet constitute a thin-sheet theory. Nevertheless, we can use those equations to develop a basic understanding of how viscous sheets respond to applied loads. In particular, our analysis will reveal the fundamental scalings for the two distinct types of deformation that loaded sheets can experience: membrane (i.e., stretching/shortening and in-plane shear) and inextensional (i.e., pure bending). Knowledge of these scalings is crucial for the derivation of thin-sheet theory to follow.

8.1 Theory of Thin Viscous Sheets

To determine the scalings, it is sufficient to consider a so-called shallow sheet whose midsurface departs from a reference plane by an amount much smaller than its principal radii of curvature. The metric tensor for such a sheet is nearly equal to that of a plane. Accordingly, the equations governing the flow in a shallow sheet can be simplified by neglecting the effects of curvature on the metric tensor while retaining them elsewhere in the equilibrium equations. Such an approximation is consistent because the former effects are proportional to the square of the curvature, whereas the latter are linear in the curvature. The solution presented next corresponds to a more accurate version of the Donnell–Mushtari–Vlasov theory of shallow elastic shells (Niordson, 1985, ch. 15).

Let x_α be orthogonal lines-of-curvature coordinates parallel to the directions of the sheet's principal curvatures $b_1^1 \equiv k_1$ and $b_2^2 \equiv k_2$. The components of the metric tensor of the shallow sheet are then $a_{11} = a_{22} = 1$, $a_{12} = a_{21} = 0$. If we suppose further that k_1 and k_2 are constant, then the coefficients of the continuity and momentum equations become independent of x_1 and x_2. Consider the flow driven by a normal stress $P \cos q_1 x_1 \cos q_2 x_2$ applied to the sheet's upper surface $z = T/2$. Then the velocity components $(u_1, u_2, u_3) \equiv (u, v, w)$ and the pressure p within the sheet have the forms

$$u = \frac{PL}{\eta} \sin q_1 x_1 \cos q_2 x_2 \sum_{j=0}^{J} u_j \hat{z}^j, \tag{8.20a}$$

$$v = \frac{PL}{\eta} \cos q_1 x_1 \sin q_2 x_2 \sum_{j=0}^{J} v_j \hat{z}^j, \tag{8.20b}$$

$$w = \frac{PL}{\eta} \cos q_1 x_1 \cos q_2 x_2 \sum_{j=0}^{J} w_j \hat{z}^j, \tag{8.20c}$$

$$p = P \cos q_1 x_1 \cos q_2 x_2 \sum_{j=0}^{J-1} p_j \hat{z}^j, \tag{8.20d}$$

where $2\pi L \equiv 2\pi \left(q_1^2 + q_2^2 \right)^{-1/2}$ is the wavelength of the applied load, $\hat{z} = z/T$ and u_j, v_j, w_j and p_j are dimensionless coefficients. A value $J = 4$ is sufficient to predict correctly all quantities of interest in the thin-sheet limit $\epsilon \to 0$, where

$$\epsilon = T/L \ll 1. \tag{8.21}$$

Substitution of (8.20) into the governing equations (8.16) through (8.19) and application of the boundary condition on the normal stress yields a set of nineteen linear algebraic equations for the coefficients u_j, v_j, w_j and p_j. In writing the solutions, it is convenient to define dimensionless curvatures $(\mathcal{K}_1, \mathcal{K}_2) = L(k_1, k_2)$ and dimensionless wavenumbers $(\mathcal{Q}_1, \mathcal{Q}_2) = L(q_1, q_2)$, where $\mathcal{Q}_1^2 + \mathcal{Q}_2^2 = 1$ in virtue of the definition of L. Also, let $\mathcal{G} \equiv \mathcal{K}_1 \mathcal{K}_2$ be the dimensionless Gaussian curvature, $\mathcal{H} \equiv (\mathcal{K}_1 + \mathcal{K}_2)/2$ be the dimensionless mean curvature and $\mathcal{K} \equiv (\mathcal{K}_1^2 + \mathcal{K}_2^2)^{1/2}$ be the dimensionless curvature modulus, respectively.

The solutions required for our purposes are

$$w_0 = \frac{3}{\epsilon \chi}, \qquad u_0 = \frac{3\mathcal{Q}_1 \left[2(\mathcal{K}_1 - \mathcal{K}_2) + 3\mathcal{I}\right]}{2\epsilon \chi}, \qquad (8.22a)$$

$$u_1 = -\frac{3\mathcal{Q}_1 \left[2\left(\mathcal{K}_1^2 - \mathcal{G} - 1\right) + 3\left(\mathcal{K}_1^2 \mathcal{Q}_2^2 + \mathcal{G}\mathcal{Q}_1^2\right)\right]}{2\chi}, \qquad (8.22b)$$

$$p_0 = \frac{6\mathcal{I} + \epsilon^2 \mathcal{H} \left(3\mathcal{K}^2 - \mathcal{G} - 3\right)}{2\epsilon \chi}, \qquad p_1 = -\frac{3\left[2 - 2\mathcal{I}\mathcal{H} + 3\mathcal{I}^2 - 2\mathcal{K}^2\right]}{\chi}, \qquad (8.22c)$$

where

$$\mathcal{I} = \mathcal{K}_1 \mathcal{Q}_2^2 + \mathcal{K}_2 \mathcal{Q}_1^2, \qquad (8.22d)$$

$$\chi = 9\mathcal{I}^2 + \epsilon^2 \left[\left(1 - \mathcal{K}^2\right)^2 + \mathcal{G}\left(1 - \mathcal{K}^2 + \mathcal{G}\right)\right]. \qquad (8.22e)$$

Only the leading-order terms in ϵ are retained in each of the numerators and denominators of (8.22a)–(8.22e). The solutions for v_0 and v_1 are obtained from those for u_0 and u_1 by interchanging the subscripts 1 and 2.

The two terms in the denominator χ correspond to the membrane and inextensional (bending-dominated) modes of deformation. In the membrane limit, the first term is dominant, and the sheet responds to a load by stretching and in-plane shear ($w_0 \sim \epsilon^{-1}$). The second term is dominant in the inextensional limit, where the sheet responds by bending ($w_0 \sim \epsilon^{-3}$). In general, a sheet responds to a load by a combination of both mechanisms. The relative importance of the bending response can be measured by a bending number B ($0 \leq B \leq 1$), defined as the ratio of the rate of energy dissipation due to bending to the total (membrane plus bending) dissipation rate. Direct calculation using the shallow-sheet solution yields

$$B = \frac{\epsilon^2 F(\mathcal{K}, \mathcal{G}, \phi)}{\mathcal{I}^2 + \epsilon^2 F(\mathcal{K}, \mathcal{G}, \phi)}, \qquad (8.23)$$

where F is a complicated function of $\mathcal{K}$, $\mathcal{G}$ and $\phi \equiv \tan^{-1}(\mathcal{Q}_2/\mathcal{Q}_1)$ that is of order unity in the limit $\mathcal{K} \to 0$. The angle ϕ ranges from 0 when the load varies only in the x_1 direction to $\pi/2$ when it varies only in the x_2 direction. Equation (8.23) shows that loaded viscous sheets deform primarily by stretching when $|\mathcal{I}| > \epsilon$, and primarily by bending when $|\mathcal{I}| < \epsilon$.

Figure 8.3 shows B as a function of ϕ and the reduced Gaussian curvature $r = \mathcal{G}/\mathcal{K}^2 = G/K^2$, for three values of $\mathcal{K}$ and $\epsilon = 0.01$. The parameter r spans the

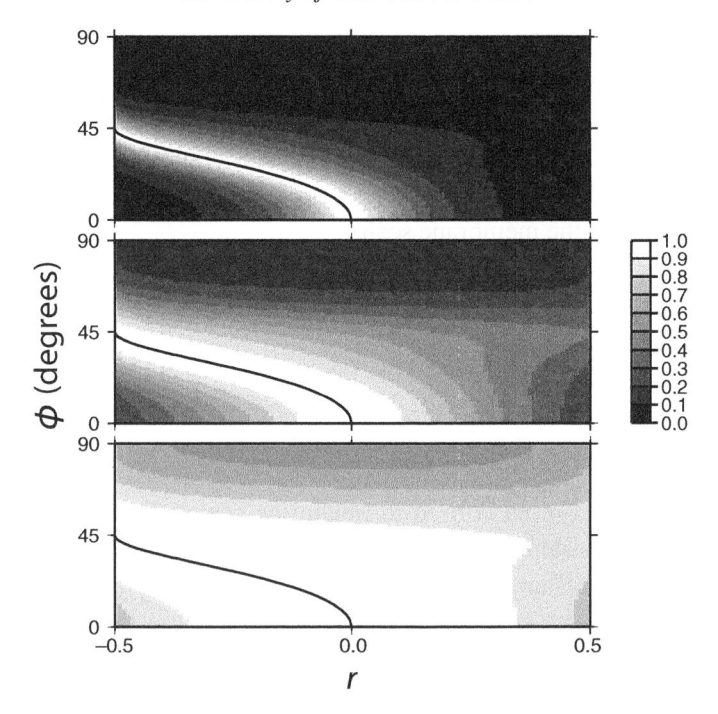

Figure 8.3 Bending number B defined by (8.23) for a harmonically loaded shallow sheet as a function of the reduced Gaussian curvature $r = G/K^2$ and $\phi = \tan^{-1}(Q_2/Q_1)$, for $\epsilon = 0.01$ and $\mathcal{K}/\epsilon = 0.3$ (bottom), $\mathcal{K}/\epsilon = 1.0$ (middle) and $\mathcal{K}/\epsilon = 3.0$ (top). The black line is the inextensional line $\mathcal{I} = 0$. See § 8.1.3 for discussion. Figure reproduced from Figure 2 of Ribe (2002).

entire range of Gaussian curvature, from $r = -1/2$ for a catenoidal sheet through $r = 0$ for a flat or cylindrical sheet to $r = 1/2$ for a spherical sheet. Figure 8.3 shows that $\mathcal{K} = \epsilon$ is the critical curvature modulus for which stretching and bending are of roughly equal importance. In general, $\mathcal{K} > \epsilon$ favors stretching and $\mathcal{K} < \epsilon$ favors bending, except in the neighborhood of an 'inextensional' line $\sin^2 \phi = k_2/(k_2-k_1)$ (black lines in Figure 8.3), where $\mathcal{I}$ vanishes. Along this line, the sheet responds to the applied load by bending no matter how large $\mathcal{K}$ may be. The existence of the inextensional line explains the fact, well known to structural engineers, that the behaviour of elastic shells with negative Gaussian curvature cannot in general be modelled using membrane theory alone. It is important to note that inextensionality depends both on the sheet's geometry and on the loading distribution. For example, the deformation of a cylindrical sheet ($r = 0$) is inextensional if the loading varies only in the azimuthal direction ($\phi = 0$), but not if it varies also in the axial direction.

The shallow-sheet solution obtained earlier reveals how the variables u, v, w and p scale as functions of ϵ. The solutions show, first, that the quadratic terms u_2, v_2 and p_2 are always small relative to the corresponding constant terms (u_0, v_0, p_0)

and linear terms (u_1, v_1, p_1). Accordingly, the scales for u, v and p are $\max(u_0, u_1)$, $\max(v_0, v_1)$ and $\max(p_0, p_1)$, respectively. Second, because $w_0 \gg w_1, w_2$, the scale for w is simply that for w_0. There are two distinct scaling regimes, according to whether the first or the second term in (8.22e) is dominant. Suppose first that $\mathcal{I} = O(1)$, so that the second (bending) term in (8.22e) is negligible. Assuming $\mathcal{K} = O(1)$, we obtain the membrane scaling

$$u, v, w \sim \frac{PL}{\eta}\epsilon^{-1}, \quad p \sim P\epsilon^{-1}. \tag{8.24}$$

If on the other hand $|\mathcal{I}| \ll \epsilon$, we find the inextensional scaling

$$u, v, w \sim \frac{PL}{\eta}\epsilon^{-3}, \quad p \sim P\epsilon^{-2}. \tag{8.25}$$

The assumption $\mathcal{K} = O(1)$ used to obtain the scales (8.24) and (8.25) is not strictly speaking consistent with the shallow-sheet approximation, which requires the dimensionless curvatures to be small. To test the validity of (8.24) and (8.25), Ribe (2002) determined exact analytical solutions for the flow in harmonically loaded cylindrical and spherical sheets of constant thickness. Those solutions show that the scales (8.24) and (8.25) are correct for sheets with arbitrary curvature up to $\mathcal{K} = O(1)$.

8.1.4 Global Force and Torque Balance

Armed with an understanding of how loaded viscous sheets deform, we now begin our march towards thin-sheet theory proper. The first step is to obtain equations for global force balance by integrating (8.16a) and (8.16b) across the sheet, yielding

$$n^{\alpha\beta}|_\alpha - b_\alpha^\beta q^\alpha + \mathcal{P}^\beta = 0, \tag{8.26a}$$

$$q^\alpha|_\alpha + b_{\alpha\beta} n^{\alpha\beta} + \mathcal{P}^3 = 0, \tag{8.26b}$$

where

$$n^{\alpha\beta} = \int_{-T/2}^{T/2} \sigma^{\alpha\beta}\, dz, \quad q^\alpha = \int_{-T/2}^{T/2} \sigma^{\alpha 3}\, dz \tag{8.27}$$

are resultants of the in-plane stresses and the shear stresses, respectively, and

$$\mathcal{P}^j = f^j \int_{-T/2}^{T/2} \rho h\, dz + \mathcal{F}_+^j + \mathcal{F}_-^j, \tag{8.28}$$

is the load vector per unit midsurface area. The load vector includes the gravitational body force (first term) and the tractions applied to the upper (+) and lower (−) surfaces of the sheet (second and third terms).

8.1 Theory of Thin Viscous Sheets

The equation for global torque balance is obtained by multiplying (8.16a) by z and then integrating, yielding

$$m^{\alpha\beta}|_\alpha - q^\beta + \mathcal{M}^\beta = 0, \tag{8.29}$$

where

$$m^{\alpha\beta} = \int_{-T/2}^{T/2} \sigma^{\alpha\beta} z\, dz, \tag{8.30}$$

is the bending moment tensor and

$$\mathcal{M}^\beta = f^\beta \int_{-T/2}^{T/2} \rho h z\, dz + \frac{T}{2}\left(\mathcal{F}_+^\beta - \mathcal{F}_-^\beta\right) \tag{8.31}$$

is the applied moment vector.

The global force balance equations (8.26) can now be simplified by using (8.29) to eliminate the shear stress resultant q^β and by introducing (symmetric) effective stress resultant and bending moment tensors (Budiansky and Sanders, 1967; Niordson, 1985)

$$N^{\alpha\beta} = n^{\alpha\beta} + b_\lambda^\beta m^{\lambda\alpha}, \quad M^{\alpha\beta} = \frac{1}{2}\left(m^{\alpha\beta} + m^{\beta\alpha}\right). \tag{8.32}$$

The resulting equations are

$$N^{\alpha\beta}|_\alpha - 2b_\lambda^\beta M^{\lambda\alpha}|_\alpha - b_\lambda^\beta|_\alpha M^{\lambda\alpha} + \mathcal{P}^\beta - b_\alpha^\beta \mathcal{M}^\alpha = 0, \tag{8.33a}$$

$$M^{\alpha\beta}|_{\alpha\beta} - b_{\alpha\lambda}b_\beta^\lambda M^{\alpha\beta} + b_{\alpha\beta}N^{\alpha\beta} + \mathcal{P}^3 + \mathcal{M}^\alpha|_\alpha = 0. \tag{8.33b}$$

In most applications of thin-sheet theory, the applied moment $\mathcal{M}^\alpha$ is small, and the last terms on the LHSs of (8.33a) and (8.33b) can be neglected.

8.1.5 Thin-Sheet Constitutive Relations

The next step is to determine the appropriate constitutive relations for thin viscous sheets. In the limit $\epsilon \to 0$, a viscous sheet behaves as a surface of effectively zero thickness with finite resistances to stretching and bending, measured, respectively, by the stress resultant and bending moment tensors $N^{\alpha\beta}$ and $M^{\alpha\beta}$. The thin-sheet constitutive relations specify how $N^{\alpha\beta}$ and $M^{\alpha\beta}$ depend on the two tensors that describe the rate of deformation of the midsurface: the strain rate tensor

$$\Delta_{\alpha\beta} = \frac{1}{2}\left(U_\alpha|_\beta + U_\beta|_\alpha\right) - b_{\alpha\beta}U_3 \tag{8.34}$$

and the rate of change of curvature tensor

$$\Omega_{\alpha\beta} = U_3|_{\alpha\beta} - b_\alpha^\lambda b_{\lambda\beta}U_3 + b_\alpha^\lambda U_\lambda|_\beta + b_\beta^\lambda U_\lambda|_\alpha + b_\beta^\lambda|_\alpha U_\lambda \tag{8.35}$$

where $U_i \equiv u_i(z = 0)$ is the fluid velocity at the midsurface. The tensors $\Delta_{\alpha\beta}$ and $\Omega_{\alpha\beta}$ need not be known in advance; they appear naturally in the course of the asymptotic expansion procedure discussed next.

The thin-sheet constitutive relations can be determined by expanding the primitive variables u_i and p in powers of the small parameter ϵ. Related expansion techniques have long been used in elastic thin-shell theory (e.g., Goldenveizer, 1963; Sanchez-Palencia, 1990). The powers of ϵ in the leading terms of these expansions are those revealed by the shallow-sheet scaling analysis of the previous section, where we saw that distinct scalings exist for membrane (stretching) and inextensional (bending) deformations. We therefore require separate asymptotic expansions for these two limits. Much unnecessary labor can be avoided by expanding u_i and p directly in double power series in ϵ and the dimensionless normal coordinate $\hat{z}$.

In view of the scalings (8.24), the appropriate expansions in the membrane limit are

$$u_i = \frac{PL}{\eta\epsilon} \sum_{m=0}\sum_{n=0} \epsilon^m \hat{z}^n u_i^{(mn)}, \quad p = \frac{P}{\epsilon} \sum_{m=0}\sum_{n=0} \epsilon^m \hat{z}^n p^{(mn)}, \tag{8.36}$$

where the coefficients $u_i^{(mn)}$ and $p^{(mn)}$ are dimensionless functions of the lateral coordinates x_α. The analogous expansions in the inextensional limit are

$$u_i = \frac{PL}{\eta\epsilon^3} \sum_{m=0}\sum_{n=0} \epsilon^m \hat{z}^n u_i^{(mn)}, \quad p = \frac{P}{\epsilon^2} \sum_{m=0}\sum_{n=0} \epsilon^m \hat{z}^n p^{(mn)}. \tag{8.37}$$

We now substitute each of the expansions (8.36) and (8.37) into the governing equations (8.16) through (8.19) and require terms proportional to the same powers of ϵ and $\hat{z}$ in each equation to vanish separately. For each of the two expansions, the result is a set of coupled linear algebraic equations for the coefficients $u_i^{(mn)}$ and $p^{(mn)}$ that can be solved sequentially. The details of the derivation are omitted, but can be found in Ribe (2002). The final result of both expansions is

$$N^{\alpha\beta} = 4\eta T \mathcal{A}^{\alpha\beta\lambda\rho} \Delta_{\lambda\rho} + \eta T^3 \mathcal{C}^{\alpha\beta\lambda\rho} \Omega_{\lambda\rho}, \tag{8.38a}$$

$$M^{\alpha\beta} = -\frac{\eta T^3}{3} \left(\mathcal{A}^{\alpha\beta\lambda\rho} \Omega_{\lambda\rho} + \mathcal{B}^{\alpha\beta\lambda\rho} \Delta_{\lambda\rho} \right), \tag{8.38b}$$

where

$$\mathcal{A}^{\alpha\beta\lambda\rho} = \frac{1}{4}\left(a^{\alpha\lambda}a^{\beta\rho} + a^{\alpha\rho}a^{\beta\lambda}\right) + \frac{1}{2}a^{\alpha\beta}a^{\lambda\rho}, \tag{8.39a}$$

$$\mathcal{B}^{\alpha\beta\lambda\rho} = \frac{1}{4}\left[8H\mathcal{A}^{\alpha\beta\lambda\rho} - 2a^{\alpha\beta}b^{\lambda\rho} - 4a^{\lambda\rho}b^{\alpha\beta} - 3\left(a^{\alpha\rho}b^{\beta\lambda} + a^{\beta\rho}b^{\alpha\lambda}\right)\right], \tag{8.39b}$$

$$\mathcal{C}^{\alpha\beta\lambda\rho} = \frac{1}{12}\left[8H\mathcal{A}^{\alpha\beta\lambda\rho} - 2a^{\alpha\beta}\left(3Ha^{\lambda\rho} + 2b^{\lambda\rho}\right)\right.$$
$$\left. - a^{\alpha\rho}b^{\beta\lambda} - a^{\beta\rho}b^{\alpha\lambda} - 2\left(a^{\alpha\lambda}b^{\beta\rho} + a^{\beta\lambda}b^{\alpha\rho}\right) - 5a^{\lambda\rho}b^{\alpha\beta}\right], \qquad (8.39c)$$

and $\Delta_{\lambda\rho}$ and $\Omega_{\lambda\rho}$ are defined by (8.34) and (8.35), respectively. A relatively simple form of thin-sheet theory without coupling between stretching and bending is obtained by setting $\mathcal{B}^{\alpha\beta\lambda\rho} = \mathcal{C}^{\alpha\beta\lambda\rho} = 0$, in which case the stress resultant tensor $N^{\alpha\beta}$ depends only on the strain rate tensor $\Delta_{\lambda\rho}$ while the bending moment tensor $M^{\alpha\beta}$ depends only on the rate of change of curvature tensor $\Omega_{\lambda\rho}$. The full coupled theory including $\mathcal{B}^{\alpha\beta\lambda\rho}$ and $\mathcal{C}^{\alpha\beta\lambda\rho}$ is more complicated, but the error of its predictions is reduced to $O(\epsilon^2)$ for all values of $\mathcal{I}$ (Ribe, 2002).

8.1.6 Evolution Equations

The last required element of thin viscous sheet theory is the set of kinematic equations that describe the temporal evolution of the shape of the midsurface and of the sheet's thickness. As mentioned previously, an advantage of the general-coordinate formalism is that these coordinates can be regarded as Lagrangian markers that follow the motion of material points on the midsurface. The evolution equations then involve partial derivatives with respect to time at fixed values of the midsurface coordinates θ_1 and θ_2. These equations are (Ribe, 2002)

$$\frac{\partial \mathbf{r}_0}{\partial t} = \mathbf{U}, \quad \frac{\partial a_{\alpha\beta}}{\partial t} = 2\Delta_{\alpha\beta}, \quad \frac{\partial b_{\alpha\beta}}{\partial t} = \Omega_{\alpha\beta}, \quad \frac{\partial T}{\partial t} = -T\Delta^{\alpha}_{\alpha}. \qquad (8.40\text{a--d})$$

Equation (8.40a) states the obvious fact that the position of a point on the midsurface changes at a rate equal to its velocity. Equation (8.40b) implies that the metric tensor at a material point changes due to in-plane stretching/compression and shear. Equation (8.40c) states that the curvature tensor changes at a rate equal to the rate of change of curvature tensor. Finally, equation (8.40d) states that the rate of change of the sheet's thickness is proportional to minus the local rate of expansion Δ^{α}_{α} of the midsurface. Equation 8.40d omits a small higher-order term on the RHS that is present in eqn. (7.16) of Ribe (2002).

8.2 Thin-Shell Theory in Lines-of-Curvature Coordinates

We now turn to the more restricted version of thin-shell theory in which the equations are written in terms of lines-of-curvature coordinates (θ_1, θ_2). The fundamental quantities that describe the shape of the midsurface are the principal curvatures K_α and the Lamé parameters $A_\alpha = |\partial_\alpha \mathbf{x}_0|$, where $\partial_\alpha = \partial/\partial_{\theta_\alpha}$. For notational convenience, let $B_\alpha = 1/A_\alpha$. All vector and tensor quantities defined on the midsurface are expressed relative to a local orthonormal basis comprising two surface-parallel unit vectors $\mathbf{d}_1, \mathbf{d}_2$ and a normal vector $\mathbf{d}_3$ defined by

$$\mathbf{d}_1 = B_1 \partial_1 \mathbf{x}_0, \quad \mathbf{d}_2 = B_2 \partial_2 \mathbf{x}_0, \quad \mathbf{d}_3 = \mathbf{d}_1 \times \mathbf{d}_2. \tag{8.41}$$

In the following, z is a coordinate normal to the midsurface $z = 0$. Moreover, the repeated subscript α is not summed unless explicitly indicated, and $\beta = 2$ when $\alpha = 1$ and *vice versa*.

The general static equilibrium equations for a shell are (Novozhilov, 1959, p. 39)

$$\partial_\alpha (A_\beta T_\alpha) + \partial_\beta (A_\alpha S) + S \partial_\beta A_\alpha - T_\beta \partial_\alpha A_\beta$$
$$- K_\alpha \left[\partial_\alpha (A_\beta M_\alpha) - M_\beta \partial_\alpha A_\beta + 2 \partial_\beta (A_\alpha H) + 2 K_\alpha^{-1} K_\beta H \partial_\beta A_\alpha \right]$$
$$= -A_1 A_2 P_\alpha \quad (\alpha = 1 \text{ or } 2), \tag{8.42a}$$

$$\sum_{\alpha=1}^{2} \left\{ B_1 B_2 \partial_\alpha \left[B_\alpha \left(\partial_\alpha (A_\beta M_\alpha) - M_\beta \partial_\alpha A_\beta + \partial_\beta (A_\alpha H) + H \partial_\beta A_\alpha \right) \right] + K_\alpha T_\alpha \right\} = -P_3, \tag{8.42b}$$

where

$$T_\alpha = \int_{-h/2}^{h/2} (1 - K_\beta z) \sigma_{\alpha\alpha} dz, \quad S = \int_{-h/2}^{h/2} (1 - K_1 K_2 z^2) \sigma_{12} dz, \tag{8.43a}$$

$$M_\alpha = \int_{-h/2}^{h/2} z(1 - K_\beta z) \sigma_{\alpha\alpha} dz, \quad H = \int_{-h/2}^{h/2} z[1 - (K_1 + K_2) z/2] \sigma_{12} dz, \tag{8.43b}$$

P_i is the total load vector (per unit midsurface area) and $\sigma_{\gamma\lambda}$ is the usual Cauchy stress tensor. Note that the notation of Novozhilov (1959) differs in some ways from that used in our derivation of thin-sheet theory in general coordinates. In particular, the thickness of the shell is now h rather than T, and the symbol H now denotes the moment defined in (8.43b) rather than the mean curvature. Also, Novozhilov's radii of curvature R_α have been replaced in the preceding equations by $-K_\alpha^{-1}$, where the negative sign ensures consistency with the definition of curvature in the general-coordinate derivation. With this convention, the curvature of a surface is positive if the normal vector points towards the center of curvature. Thus, e.g., the curvature of a spherical shell with an outward radial normal vector is negative.

The essential content of (8.42) is that a loaded shell can deform in two distinct ways: in-plane stretching and shear (membrane deformation), the intensity of which is measured by the effective stress resultants T_α and S, and by bending, which is measured by the effective bending moments M_α and H. The quantities T_α and S are analogous to the effective stress resultants $N^{\alpha\beta}$ in the general-coordinate derivation of thin viscous-sheet theory, and the quantities M_α and H are analogous to the effective bending moments $M^{\alpha\beta}$. In general both the membrane and bending modes are present, in a proportion that depends in a complicated way on the midsurface shape and on the distribution of the applied loads, as we already saw in Figure 8.3.

Equations (8.42) are valid for a shell of any material. To solve them, we need constitutive relations that link T_α, S, M_α and H to the displacement ζ_i (for an elastic shell) or the velocity U_i (for a viscous sheet) of the midsurface. For an elastic shell with Young's modulus E and Poisson's ratio ν, these are (Novozhilov, 1959, pp. 24, 48)

$$T_\alpha = \frac{Eh}{1-\nu^2}(\epsilon_\alpha + \nu\epsilon_\beta), \quad M_\alpha = \frac{Eh^3}{12(1-\nu^2)}(\kappa_\alpha + \nu\kappa_\beta), \qquad (8.44a)$$

$$S = \frac{Eh}{2(1+\nu)}\omega, \quad H = \frac{Eh^3}{12(1+\nu)}\tau, \qquad (8.44b)$$

where

$$\epsilon_\alpha = B_\alpha \partial_\alpha \zeta_\alpha + B_1 B_2 \zeta_\beta \partial_\beta A_\alpha - K_\alpha \zeta_3, \qquad (8.45a)$$

$$\omega = \sum_{\alpha=1}^{2} A_\beta B_\alpha \partial_\alpha (B_\beta \zeta_\beta), \qquad (8.45b)$$

$$\kappa_\alpha = -B_\alpha \partial_\alpha (B_\alpha \partial_\alpha \zeta_3 + K_\alpha \zeta_\alpha) - B_1 B_2 \partial_\beta A_\alpha (B_\beta \partial_\beta \zeta_3 + K_\beta \zeta_\beta), \qquad (8.45c)$$

$$\tau = -B_1 B_2 \partial_{12}^2 \zeta_3$$
$$+ \sum_{\alpha=1}^{2} \{B_1 B_2 B_\alpha (\partial_\beta A_\alpha) \partial_\alpha \zeta_3 - K_\alpha (B_\beta \partial_\beta \zeta_\alpha - B_1 B_2 \zeta_\alpha \partial_\beta A_\alpha)\} \qquad (8.45d)$$

are the six independent quantities that describe the deformation of the midsurface: the elongations ϵ_α and the changes of curvature κ_α in the two coordinate directions, the in-plane shear deformation ω and the torsional (twist) deformation τ. The analogous expressions for a viscous sheet are obtained by applying the transformation (5.4) to (8.44) and (8.45). The constitutive relations (8.44) include no coupling between stretching and bending: the stress resultants T_α and S do not depend on κ_α or τ, while the bending moments M_α and H do not depend on ϵ_α or ω. The general-coordinate analog of (8.44) is therefore (8.38) with $B^{\alpha\beta\lambda\rho} = C^{\alpha\beta\lambda\rho} = 0$.

8.3 Geodynamic Applications of Thin-Shell Theory

The equations (8.42) through (8.45) include all the special cases commonly considered in geodynamics. Let us consider several of these.

8.3.1 Flat Elastic Plate

The first is that of a flat ($K_1 = K_2 = 0$, $A_1 = A_2 = 1$) elastic plate with constant thickness h. The (uncoupled) equations governing the flexural and membrane modes are

$$\frac{Eh^3}{12(1-\nu^2)} \nabla_h^4 \zeta_3 = P_3, \tag{8.46a}$$

$$\frac{Eh}{2(1-\nu^2)} \left[(1-\nu)\nabla_h^2 \boldsymbol{\zeta} + (1+\nu)\nabla_h(\nabla_h \cdot \boldsymbol{\zeta})\right] = -P_1 \mathbf{e}_1 - P_2 \mathbf{e}_2 \tag{8.46b}$$

where $\boldsymbol{\zeta}$ is the 2-D (in-plane) displacement vector, ζ_3 is the normal displacement and $\mathbf{P}$ is the 3-D load vector. The quantity

$$\frac{Eh^3}{12(1-\nu^2)} \equiv D \tag{8.47}$$

is called the flexural rigidity.

The thin-plate equation (8.46a) has been used in countless geodynamic studies, too numerous to cite here. For purposes of illustration, I limit myself to two examples. First, (8.46a) has been widely used to model the flexure of the lithosphere due to topographic loading (e.g., Watts, 1978, 2001). In this case the normal load P_3 includes not only the topographic load itself, but also a term $\propto \zeta_3$ that represents the gravitational restoring force on the deflection. This application is the subject of Exercise 8.5 (the case of a flat elastic plate that is not thin has already been treated in § 5.2). The second example is the flexure of the lithosphere at subduction zones, where an outer topographic rise or 'flexural bulge' with an amplitude of several hundred m is often present seaward of the trench (Watts and Talwani, 1974). Theoretical studies of this problem (Le Pichon et al., 1973; Caldwell et al., 1976; Parsons and Molnar, 1976) typically start from an equation for a slightly curved 1-D plate subject to a constant in-plane compressional force per unit length $-T$, viz.,

$$D \frac{d^4 w}{dx^4} - T \frac{d^2 w}{dx^2} + (\rho_m - \rho_w) g w = 0. \tag{8.48}$$

Equation (8.48) is obtained from (8.42b), (8.44) and (8.45) by setting $A_1 = A_2 = 1$, $K_2 = \partial_2 = 0$, $K_1 = \partial_{11}^2 \zeta_3$, $P_3 = -(\rho_m - \rho_w) g \zeta_3$, $T_1 = T$, $\theta_1 = x$ and $\zeta_3 = w$ and neglecting the higher-order term $-B_\alpha \partial_\alpha (K_\alpha \zeta_\alpha)$ in (8.45c). However, for geodynamically reasonable values of T the second term in (8.48) is negligible for the problem at hand. The solution of the resulting equation that is bounded as $x \to \infty$ (i.e., seaward) and satisfies $w(0) = 0$ is (Caldwell et al., 1976)

$$w = A \sin \frac{x}{\alpha} \exp(-x/\alpha), \quad \alpha = \left[\frac{4D}{(\rho_m - \rho_w)g}\right]^{1/4}. \tag{8.49}$$

8.3 Geodynamic Applications of Thin-Shell Theory

The undetermined constant A can be related to the bending moment applied to the subducted part of the plate at some distance from the trench. The length α is called the 'flexural parameter'. Caldwell et al. (1976) showed that (8.49) fits well the topographic profiles seaward of several trenches in the western Pacific ocean. Turcotte et al. (1978) and McAdoo et al. (1978) subsequently generalized the analysis to an elastic/perfectly plastic plate.

Finally, the 2-D (x-y plane) generalization of (8.48) is

$$D\nabla_h^4 w - T_x \frac{\partial^2 w}{\partial x^2} - 2T_{xy}\frac{\partial^2 w}{\partial x \partial y} - T_y \frac{\partial^2 w}{\partial y^2} + (\rho_m - \rho_w)gw = p(x,y), \quad (8.50)$$

where T_x and T_y are the in-plane extensional forces (units N m^{-1}) in the x- and y-directions, respectively, T_{xy} is the in-plane shear force and p is the applied normal load. Equation (8.50) is written relative to arbitrary coordinates x and y that need not be lines-of-curvature coordinates. Wessel (1996) used (8.50) with $T_{xy} = 0$ to determine analytical Green functions for semi-infinite plates acted on by point and edge loads and applied them to seamount loads near fracture zones and variable edge loads at subduction zones.

8.3.2 Membrane Flow in a Flat Viscous Sheet

A well-known example of this case is the thin-sheet model for continental deformation of England and McKenzie (1983), whose eqn. (16) is just the viscous analog of (8.46b) with a power-law rheology of the form (4.1f) and expressions for P_1 and P_2 representing the lateral forces arising from variations in crustal thickness. Wdowinski et al. (1989) and Husson and Ricard (2004) applied the flat viscous-sheet equation to continental deformation above subduction zones. Ricard and Husson (2005) supplemented the flat viscous sheet equation by evolution equations for the moments of the vertical distributions of the compositional and thermal density anomalies (Lemery et al., 2000, see also § 7.3) and performed a linear stability analysis (LSA) to demonstrate the existence of propagating wave solutions ('tectonic waves'). Schmalholz (2011) used a 1-D thin-sheet model to study the gravity-driven necking instability (slab breakoff) of a falling vertical sheet with a power-law rheology. Duretz et al. (2012) extended the model of Schmalholz (2011) by adding a second 1-D equation describing the evolution of the temperature due to viscous dissipational heating.

8.3.3 Spherical Shell

Another important special case is that of a spherical shell with radius R and principal curvatures $K_1 = K_2 = -1/R$. If $\theta_1 = \theta$ (colatitude) and $\theta_2 = \phi$ (longitude), then $A_1 = R$ and $A_2 = R\sin\theta$. The equilibrium equations (8.42) reduce to

$$[(M_1 + RT_1) \sin \theta]_\theta - (M_2 + RT_2) \cos \theta + 2H_\phi = -P_1 R^2 \sin \theta, \quad (8.51a)$$

$$(M_2 + RT_2)_\phi + [(2H + RS) \sin \theta]_\theta + (2H + RS) \cos \theta = -P_2 R^2 \sin \theta, \quad (8.51b)$$

$$[M_{1\theta\theta} - M_1 + M_2 - R(T_1 + T_2)] \sin \theta + M_{2\phi\phi} \csc \theta$$
$$+ (2M_1 - M_2)_\theta \cos \theta + 2(H \cot \theta + H_\theta)_\phi = -P_3 R^2 \sin \theta, \quad (8.51c)$$

where subscripts θ and ϕ denote partial derivatives. The constitutive relations (8.44) remain valid, but with kinematic parameters that take the simplified forms

$$\epsilon_1 = R^{-1}(\zeta_{1\theta} + \zeta_3), \quad (8.52a)$$

$$\epsilon_2 = R^{-1}(\zeta_{2\phi} \csc \theta + \zeta_1 \cot \theta + \zeta_3), \quad (8.52b)$$

$$\omega = R^{-1}(\zeta_{2\theta} - \zeta_2 \cot \theta + \zeta_{1\phi} \csc \theta), \quad (8.52c)$$

$$\kappa_1 = R^{-2}(-\zeta_{3\theta\theta} + \zeta_{1\theta}), \quad (8.52d)$$

$$\kappa_2 = R^{-2}(-\zeta_{3\phi\phi} \csc^2 \theta + (\zeta_1 - \zeta_{3\theta}) \cot \theta + \zeta_{2\phi} \csc \theta), \quad (8.52e)$$

$$\tau = R^{-2}[(-\zeta_{3\theta\phi} + \zeta_{3\phi} \cot \theta + \zeta_{1\phi}) \csc \theta + \zeta_{2\theta} - \zeta_2 \cot \theta]. \quad (8.52f)$$

Many early geodynamic applications of spherical-shell theory were purely geometric and did not attempt to solve the equilibrium equations. A classic example is the 'indented ping-pong-ball' model of Frank (1968), which likens the shape of the lithosphere in subduction zones to that of the circular dent surrounding a hole in a deformed ping-pong ball. However, Bevis (1986) found that the geometries of many Wadati–Benioff zones (WBZs) are inconsistent with models (like Frank's) that assume perfectly inextensible shells. Scholz and Page (1970) and Bayly (1982) suggested that a subducting spherical shell should buckle along the strike of the trench to accommodate the reduction of the space available due to the spherical geometry as the slab penetrates deeper. Laravie (1975) proposed a geometric model that predicts different degrees of lateral strain in the slab depending on the dip of the WBZ and the radius of curvature of the arc. Schettino and Tassi (2012) reviewed the inadequacies of the ping-pong-ball model and proposed an alternative kinematic model that predicts the amount of lateral strain and the strain rate in the slab as a function of its geometry and subduction velocity. On the dynamic side, Turcotte (1974) used equations equivalent to those above in the membrane limit $M_1 = M_2 = H = 0$ to estimate the magnitude of the stresses generated in the lithosphere when it moves relative to Earth's equatorial bulge. He found that such stresses may be large enough to cause propagating fractures. Turcotte et al. (1981) studied the deflection of a spherical elastic shell produced by a topographic load using the approximate thin-shell equation (Kraus, 1967)

$$D(\mathcal{B}^6 + 4\mathcal{B}^4)\zeta_3 + EhR^2(\mathcal{B}^2 + 2)\zeta_3 = R^4(\mathcal{B}^2 + 1 - \nu)P_3, \quad (8.53)$$

where D is the flexural rigidity and the operator $\mathcal{B}^2$ is defined by (4.25). In (8.53), the term proportional to D corresponds to flexural support of the load, while the term proportional to EhR^2 corresponds to membrane support. Turcotte et al. (1981) concluded that membrane stresses play an important role in the support of topographic loads on the moon and on Mars, but not on Earth. Tanimoto (1997) and Tanimoto (1998) solved the complete (membrane plus flexural) equations for a normally loaded spherical elastic shell with negative buoyancy proportional to the normal displacement. He concluded that the displacement is well predicted by flat-sheet models due to the dominance of buoyancy forces, but that the state of stress is strongly influenced by the spherical geometry. Mahadevan et al. (2010) used scaling analysis and numerical solutions of approximate equations for shallow spherical caps to investigate the causes of the curvature and segmentation of subduction zones, with an emphasis on instabilities at the shell's edge. They concluded that doubly curved shells behave during subduction in a way fundamentally different than do flat plates.

8.3.4 Finite-Amplitude Deformation of a 2-D Viscous Sheet

A fourth limiting case is the finite-amplitude deformation of a 2-D viscous sheet (Ribe, 2001). By symmetry, $\partial_2 = u_2 = \kappa_2 = \epsilon_2 = \omega = \kappa_2 = \tau = S = H = 0$ for this case. If $\theta_1 \equiv s$ is the arclength along the midsurface, then $A_1 = A_2 = 1$. The general equilibrium equations (8.42) then reduce to

$$N' - KM' + P_1 + \Delta \rho g_1 h = 0, \tag{8.54a}$$

$$M'' + KN + P_3 + \Delta \rho g_3 h = 0, \tag{8.54b}$$

where primes denote differentiation with respect to s, $K = K_1$, $N = T_1$ and $M = M_1$. In (8.54) the load terms have been explicitly written as sums of the tractions exerted by the outer fluid (third term) and the buoyancy force (fourth term).

The relative simplicity of the 2-D thin-sheet equations (8.54) makes them amenable to an illuminating physical interpretation in terms of elementary force and torque balance. To this end, we write (8.54) in the alternative form

$$N' - KQ + P_1 + \Delta \rho g_1 h = 0, \tag{8.55a}$$

$$Q' + KN + P_3 + \Delta \rho g_3 h = 0, \tag{8.55b}$$

$$M' = Q. \tag{8.55c}$$

The quantities $N(s)$, $Q(s)$ and $M(s)$ are defined by

$$N = \int_{-h/2}^{h/2} \sigma_{ss} \mathrm{d}z, \quad Q = \int_{-h/2}^{h/2} \sigma_{sz} \mathrm{d}z, \quad M = \int_{-h/2}^{h/2} z \sigma_{ss} \mathrm{d}z. \tag{8.56}$$

166   Long-Wave Theories, 2

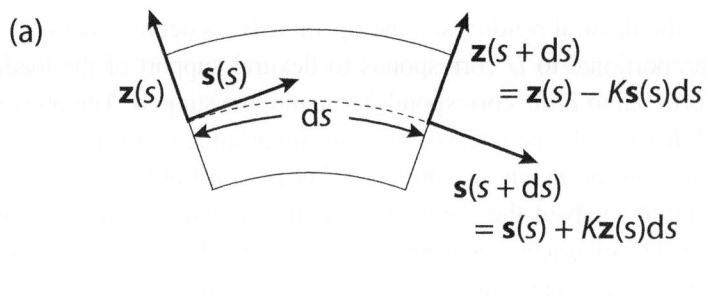

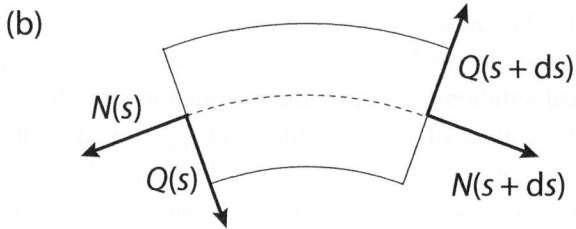

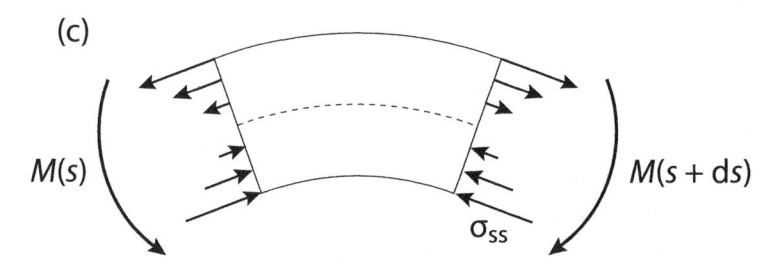

Figure 8.4 Geometry and static equilibrium of an element of a 2-D thin viscous sheet with arcwise extent ds. (a) Definition of the midsurface-parallel and midsurface-perpendicular unit vectors $\mathbf{s}(s)$ and $\mathbf{z}(s)$. (b) Resultants of the internal viscous forces acting on the element. (c) Bending moments acting on the element. The bending moment arises from the linear variation of the fiber stress σ_{ss} across the element when the deformation is dominated by bending.

The new quantity Q is the resultant of the shear force acting on cross sections of the sheet.

We now consider the static equilibrium of an element of the sheet with arcwise extent ds (Figure 8.4). Let $\mathbf{s}(s)$ and $\mathbf{z}(s)$ be unit vectors that are parallel to and perpendicular to the midsurface, respectively (Figure 8.4a). Referring to Figure 8.4b, we see that the internal viscous force acting on the element is

$$N(s+ds)\mathbf{s}(s+ds) - N(s)\mathbf{s}(s) + Q(s+ds)\mathbf{z}(s+ds) - Q(s)\mathbf{z}(s)$$
$$\equiv \mathbf{N}'ds \quad \text{where} \quad \mathbf{N} = N\mathbf{s} + Q\mathbf{z}. \tag{8.57}$$

Expanding the derivative $\mathbf{N}'$ and using the relations $\mathbf{s}' = K\mathbf{z}$ and $\mathbf{z}' = -K\mathbf{s}$, we obtain

$$\mathbf{N}' = (N' - KQ)\mathbf{s} + (Q' + KN)\mathbf{z}. \tag{8.58}$$

8.3 Geodynamic Applications of Thin-Shell Theory

Comparing this expression with (8.55a) and (8.55b), we see that the latter two equations can be written compactly as a single-vector force balance equation

$$\mathbf{N}' + \mathbf{P} + \Delta\rho g h = \mathbf{0} \qquad (8.59)$$

where $\mathbf{P} = P_1\mathbf{s} + P_3\mathbf{z}$ and $\mathbf{g} = g_1\mathbf{s} + g_3\mathbf{z}$.

To interpret (8.55c), we consider the torque acting on the element. If the deformation of the element is dominated by bending, the fiber stress σ_{ss} varies linearly across it, as shown in Figure 8.4c. The bending moment M defined by (8.56) is therefore nonzero and has the sense shown by the curved arrows in Figure 8.4c. The total counterclockwise torque about the midpoint of the element is

$$-M(s+ds) + M(s) + Q(s)\frac{ds}{2} + Q(s+ds)\frac{ds}{2}$$

$$\approx -M(s+ds) + M(s) + Q(s)ds$$

$$= (-M' + Q)ds. \qquad (8.60)$$

Comparing this result with (8.55c), we see that the latter equation is just the condition that the torque on the element must vanish.

The next step is to write the vector $\mathbf{N}$ in terms of the midsurface velocity $\mathbf{U}$, using the constitutive relations (8.44)–(8.45). Reducing these relations to 2-D form, setting $\nu = 1/2$ and $E = 3\eta$ and replacing ζ_i by the midsurface velocity U_i, we obtain

$$N = 4\eta h(U_1' - KU_3) \equiv 4\eta h \left(\mathbf{U}' \cdot \mathbf{s}\right), \qquad (8.61a)$$

$$M = -\frac{\eta h^3}{3}(U_3' + KU_1)' \equiv -\frac{\eta h^3}{3}\left(\mathbf{U}' \cdot \mathbf{z}\right)', \qquad (8.61b)$$

$$\mathbf{N} = 4\eta h \left(\mathbf{U}' \cdot \mathbf{s}\right)\mathbf{s} - \frac{\eta}{3}\left[h^3 \left(\mathbf{U}' \cdot \mathbf{z}\right)'\right]'\mathbf{z}. \qquad (8.62)$$

More accurate forms of the constitutive laws (8.61a) and (8.61b) including the coupling between stretching and bending were derived by Ribe (2001). Finally, the evolution of the midsurface position $\mathbf{X}(s,t)$ and the sheet thickness $h(s,t)$ are described by the kinematic equations

$$\frac{D\mathbf{X}}{Dt} = \mathbf{U}, \quad \frac{Dh}{Dt} = -h\mathbf{U}' \cdot \mathbf{s}, \qquad (8.63)$$

where D/Dt is a convective derivative that follows the motion of material points on the (stretching) midsurface (Buckmaster et al., 1975). The convective derivative is necessary because the arclength s is not a material coordinate (compare with (8.40)).

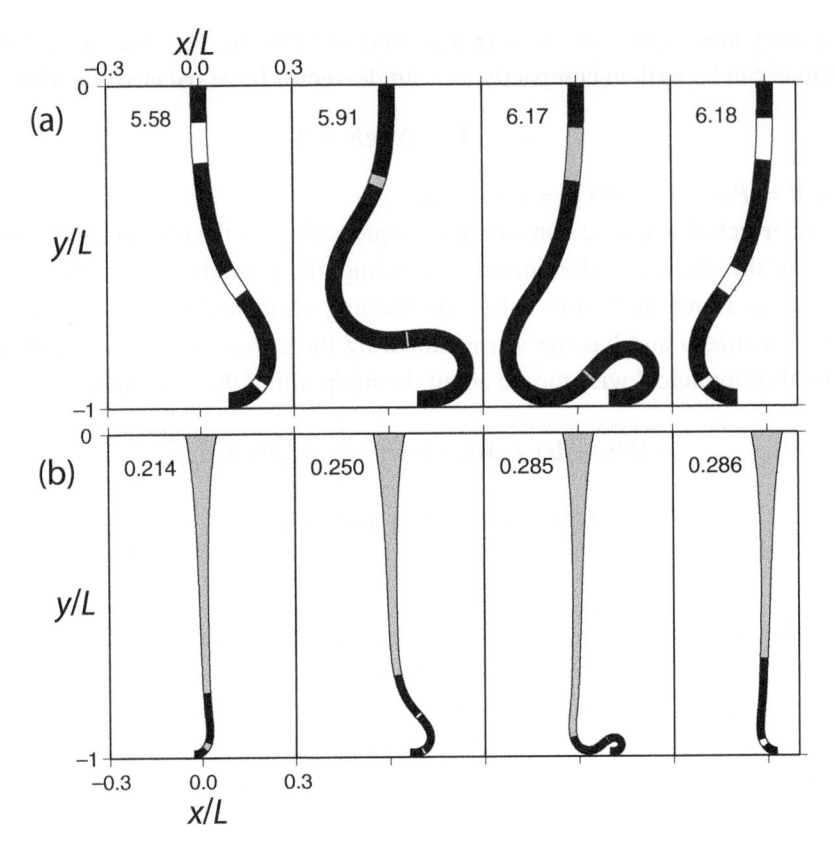

Figure 8.5 Modes of folding of a 2-D sheet of viscous fluid falling without inertia onto a rigid surface (§ 8.3.4). (a) Viscous mode; (b) gravitational mode. One half-period of the folding is shown in both cases. Greyshades indicate the mode of deformation that accounts for at least 50% of the total rate of viscous dissipation: stretching (grey), compression (white) and bending (black). The total fall height is L, and the numbers are dimensionless times tU_1/L, where U_1 is the longitudinal velocity of the sheet at the contact point. The third panel in each part shows the formation of a new contact point, and the fourth shows the sheet just after the portion downstream from the new contact has been removed. Reproduced from Figures 2 and 3 of Ribe (2003).

Ribe (2003) used the preceding equations to determine scaling laws for the periodic buckling instability of a viscous sheet falling onto a horizontal rigid surface from a height L. If inertia is negligible, folding can occur in either of two modes – viscous or gravitational – depending on the relative importance of gravitational and viscous forces. Figure 8.5 shows these two modes over times equal to one-half the folding period. In the viscous mode, the buoyancy of the fluid is negligible relative to viscous forces, and the sheet deforms predominantly by bending. In the gravitational mode, the upper part of the sheet is stretched downward by gravity, and gravity also balances the viscous resistance to bending in the lowermost part of the sheet where the folding occurs.

Ribe et al. (2007) suggested that periodic folding instabilities may be the cause of the apparent widening of subducted lithosphere imaged beneath some subduction zones, a conclusion supported by 2-D numerical simulations that show periodic folding instabilities in a mantle with a viscosity increase at 660 km depth (Lee and King, 2011; Čížková and Bina, 2013; Dannberg et al., 2017). Buffett (2006) used the 2-D thin-sheet equations to derive an expression for the bending-induced horizontal force that resists the slab pull on a subducting plate. Buffett and Becker (2012) used concepts from thin-sheet theory together with a laboratory-based rheological model to argue that subducting lithosphere behaves as a plastic material having an effective power-law rheology (4.1f) with $n \approx 14$.

As a concluding remark, I note that Medvedev and Podladchikov (1999) have derived a higher-order extended thin-sheet approximation (ETSA) that includes layer-parallel shear (lubrication flow) in addition to deformation by stretching and bending. The ETSA allows the viscosity to vary both along and across the sheet, but is limited to nearly flat sheets.

8.4 Immersed Viscous Sheets

Another situation of geodynamic relevance for which thin-sheet theory proves useful is a highly viscous sheet immersed in a second fluid with a lower density and viscosity. This two-fluid configuration is a simple model for the behaviour of subducting oceanic lithosphere in an ambient mantle. The buoyancy force that drives the motion is here balanced by a combination of the viscous tractions exerted by the ambient fluid (which are formally included in the load terms P_i in (8.42)) and the internal viscous force that resists the deformation of the sheet by stretching and bending.

8.4.1 Thin-Sheet Scaling Analysis of Free Subduction

Ribe (2010) and Li and Ribe (2012) used the boundary-element method (BEM) in two and three dimensions, respectively, to study the dynamics of a subducting viscous sheet having initially the geometry of Figure 1.2c. A simple scaling analysis based on thin viscous-sheet theory reveals the dimensionless parameter that determines whether the sheet's dynamics is controlled by its own viscosity or by the viscosity of the ambient fluid. Let l_b be the 'bending length', i.e., the length of that portion of the sheet in which deformation is dominated by bending (as opposed to stretching/shortening). For subducting lithosphere on Earth, l_b is the sum of the slab length and the length of the flexural bulge seaward of the trench

(Ribe, 2010). Also, let V be the vertical velocity of the slab's leading end at the edge $y = w/2$ of the sheet. The traction (normal plus shear) applied to the bending portion of the sheet by the external fluid $\sim \eta_1 V/l_b$, which when integrated over the area of the bending portion ($\sim l_b w$) gives a total force $F_{\text{ext}} \sim \eta_1 V w$. Now according to the thin-sheet theory described in § 8.3.4, the internal traction that resists bending is $Q' \sim \eta_2 h^3 V/l_b^4$, which corresponds to a total (integrated) force $F_{\text{int}} \sim \eta_2 h^3 V w/l_b^3$. Finally, because the negative buoyancy of the horizontal part of the sheet is compensated by normal stresses in the lubrication layer, the effective buoyancy force $F_b \sim h l w g \Delta \rho$ is due entirely to the slab of length l. In the limit of negligible bending resistance, the balance $F_b \sim F_{\text{ext}}$ implies

$$V \sim \frac{h l g \Delta \rho}{\eta_1} \equiv V_{\text{Stokes}}, \qquad (8.64)$$

which is of the order of the Stokes sinking speed of a tabular body of thickness h and lateral dimension l. Now the ratio $F_{\text{int}}/F_{\text{ext}}$ of the resisting viscous forces is

$$\frac{F_{\text{int}}}{F_{\text{ext}}} = \frac{\eta_2}{\eta_1} \left(\frac{h}{l_b}\right)^3 = St, \qquad (8.65)$$

where St is the sheet's 'flexural stiffness'. Figure 8.6 shows a plot of V/V_{Stokes} vs. St for 95 instantaneous 3-D BEM solutions, where the ratio w/l of the sheet width to the slab length is varied from 1.0 to 8.0. The slab dip is fixed at 60° to ensure geometric similarity of the sheet's midsurface. Also shown for comparison are the predictions of a two-dimensional BEM model (Ribe, 2010). In both 2-D

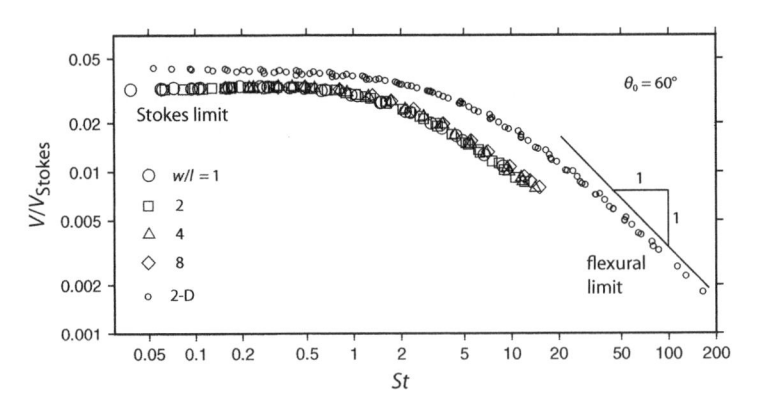

Figure 8.6 Sinking speed V of a subducting viscous sheet normalized by the Stokes velocity V_{Stokes}, as a function of the sheet's flexural stiffness St for several values of the normalized sheet width w/l and $\theta_0 = 60°$. V_{Stokes} and St are defined by (8.64) and (8.65). The top and bottom curves are the BEM predictions in 2-D and 3-D, respectively. In 3-D, V is defined as the sinking speed at the lateral edge of the sheet. See § 8.4.1 for discussion. Redrawn from Figure 5 of Li and Ribe (2012).

and 3-D, all the numerical points collapse onto a single master curve that exhibits two distinct limits. In the Stokes limit (slope 0), the sheet's internal resistance to bending is negligible, and the slab's negative buoyancy is balanced by the tractions exerted by the ambient fluid. In the flexural limit (slope $= -1$), by contrast, the external tractions are negligible, and the sheet's negative buoyancy is balanced by the internal resistance to bending. Thus the sinking speed is controlled by the external viscosity η_1 in the Stokes limit and by the internal viscosity η_2 in the flexural limit.

8.4.2 A Hybrid Boundary-Integral/Thin-Sheet Model for an Immersed Sheet

In the foregoing immersed-sheet problem, thin-sheet theory was used *a posteriori* to interpret solutions obtained using the BEM. However, it is possible to develop a model in which thin-sheet theory is built in from the start (Xu and Ribe, 2016). Figure 8.7a shows a sketch of the model, which is in two dimensions for simplicity. A thin sheet of length L and thickness h has viscosity η_2 and density $\rho_1 + \Delta\rho$ and

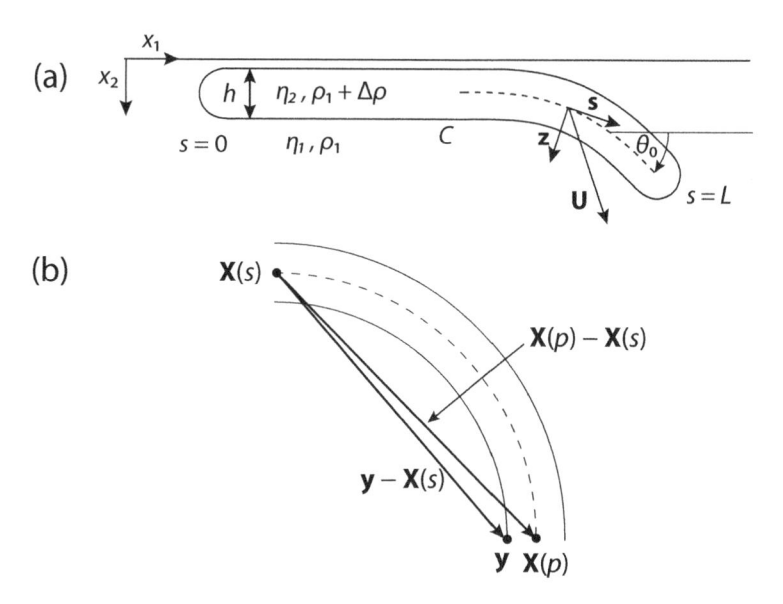

Figure 8.7 Geometry of the 2-D boundary-integral/thin-sheet model (§ 8.4.2). (a) Definition sketch. A thin sheet of length L and thickness h has viscosity η_2 and density $\rho_1+\Delta\rho$ and is immersed in a second fluid with viscosity η_1 and density ρ_1. The unit vectors **s** and **z** are parallel to and perpendicular to the sheet's midsurface (dashed line), respectively. The arclength along the sheet's midsurface is s, and **U**(s) is the midsurface velocity. The interface between the two fluids is C. (b) For a thin sheet, the vector $\mathbf{y} - \mathbf{X}(s) \approx \mathbf{X}(p) - \mathbf{X}(s)$, where $\mathbf{X}(p)$ is the position on the midsurface that is closest to **y**.

is immersed in a second fluid with viscosity η_1 and density ρ_1. The fluid velocity on the sheet's midsurface is **U**.

The starting point of the derivation is the general boundary-integral representation (4.101) for the flows in fluids 1 and 2. Now add these two equations and apply the velocity matching condition $\mathbf{u}_1 = \mathbf{u}_2$ on C to obtain

$$\chi_1(\mathbf{x})\mathbf{u}_1(\mathbf{x}) + \chi_2(\mathbf{x})\mathbf{u}_2(\mathbf{x}) = \int_C \left(\frac{\mathbf{f}_2}{\eta_2} - \frac{\mathbf{f}_1}{\eta_1} \right) \cdot \mathbf{J}(\mathbf{y} - \mathbf{x}) d\ell, \qquad (8.66)$$

where $d\ell(\mathbf{y})$ has been shortened to $d\ell$. Now apply the stress matching condition $\mathbf{f}_2(\mathbf{y}) = \mathbf{f}_1(\mathbf{y}) + \Delta\rho(\mathbf{g} \cdot \mathbf{y})\mathbf{n}$, whereupon (8.66) becomes

$$\chi_1(\mathbf{x})\mathbf{u}_1(\mathbf{x}) + \chi_2(\mathbf{x})\mathbf{u}_2(\mathbf{x})$$

$$= \frac{1-\gamma}{\eta_2} \int_C \mathbf{f}_1 \cdot \mathbf{J}(\mathbf{y} - \mathbf{x}) d\ell + \frac{\Delta\rho}{\eta_2} \int_C (\mathbf{g} \cdot \mathbf{y})\mathbf{n} \cdot \mathbf{J}(\mathbf{y} - \mathbf{x}) d\ell. \qquad (8.67)$$

Evaluate (8.67) on the midsurface $\mathbf{x} = \mathbf{X}(s)$ to obtain

$$\mathbf{U}(s) = \frac{1-\gamma}{\eta_2} \int_C \mathbf{f}_1 \cdot \mathbf{J}(\mathbf{y} - \mathbf{X}(s)) d\ell + \frac{\Delta\rho}{\eta_2} \int_C (\mathbf{g} \cdot \mathbf{y})\mathbf{n} \cdot \mathbf{J}(\mathbf{y} - \mathbf{X}(s)) d\ell, \qquad (8.68)$$

where $\mathbf{U}(s) = \mathbf{U}(\mathbf{X}(s))$. To simplify the first integral on the RHS of (8.68), note that for a thin sheet $\mathbf{y} - \mathbf{X}(s) \approx \mathbf{X}(p) - \mathbf{X}(s)$, where p is the arclength of the point on the midsurface that is closest to $\mathbf{y}$ (Figure 8.7b). This substitution is valid except in a region of arcwise extent $\sim h$ around the point $\mathbf{X}(s)$. Now write the integral as the sum of contributions from the upper (C_+) and lower (C_-) surfaces of the sheet, ignoring the small contributions from the rounded ends:

$$\int_C \mathbf{f}_1 \cdot \mathbf{J}(\mathbf{y} - \mathbf{X}(s)) d\ell \approx \int_{C_+} \mathbf{f}_+ \cdot \mathbf{J}(\mathbf{y} - \mathbf{X}(s)) d\ell + \int_{C_-} \mathbf{f}_- \cdot \mathbf{J}(\mathbf{y} - \mathbf{X}(s)) d\ell$$

$$\approx \int_0^L \left[\mathbf{f}_+(p) + \mathbf{f}_-(p) \right] \cdot \mathbf{J}(\mathbf{X}(p) - \mathbf{X}(s)) dp$$

$$= -\int_0^L \left[\mathbf{N}' + h\Delta\rho\mathbf{g} \right] \cdot \mathbf{J}(\mathbf{X}(p) - \mathbf{X}(s)) dp \qquad (8.69)$$

where the thin-sheet force balance (8.59) has been used in the last step. The second integral on the RHS of (8.68) can be simplified by approximating it as a line integral over a distribution of Stokeslets of strength $h\mathbf{g}$ per unit length of the midsurface:

$$\int_C (\mathbf{g} \cdot \mathbf{y})\mathbf{n} \cdot \mathbf{J}(\mathbf{y} - \mathbf{X}) d\ell \approx \int_0^L h\mathbf{g} \cdot \mathbf{J}(\mathbf{X}(p) - \mathbf{X}(s)) dp. \qquad (8.70)$$

Substituting (8.69) and (8.70) into (8.68), we obtain the final boundary-integral/thin-sheet equation

$$\mathbf{U}(s) = \frac{1}{\eta_2} \int_0^L \left[\gamma \mathbf{g} h(p) \delta \rho + (\gamma - 1) \mathbf{N}'(p) \right] \cdot \mathbf{J}(\mathbf{X}(p) - \mathbf{X}(s)) dp. \quad (8.71)$$

Now recall that the explicit expression for $\mathbf{N}$ is (8.62), which involves arcwise derivatives of $\mathbf{U}$. Thus (8.71) is a Fredholm integral equation of the second kind for $\mathbf{U}(s)$. Finally, the evolution of the sheet's geometry is described by the kinematic equations (8.63). The equations (8.71), (8.62) and (8.63) were solved numerically by Xu and Ribe (2016), who showed that the model agrees with the full BEM approach in predicting curves of V/V_{Stokes} vs. St like those of Figure 8.6.

8.4.3 Semi-Analytical Subduction Models

This is an appropriate place to mention two semi-analytical models of subduction that offer simpler (if more limited) alternatives to the full boundary-element models discussed earlier. Dvorkin et al. (1993) proposed a 3-D model of the flow in the mantle wedge above a slab of finite length and width subducting at a shallow angle with a given velocity. Using the lubrication approximation (§ 7.1), they reduced the problem to a 2-D elliptic equation for the pressure in the horizontal coordinates (x, y). The model predicts a reduced (relative to an infinitely wide slab) hydrodynamic lifting torque (§ 4.3.3) due to inflow of material from the sides, but the slab's geometry and motion are imposed rather than predicted. Royden and Husson (2006) proposed a 3-D model of a thin flexible (elastic or viscous) negatively buoyant slab subject to (approximately determined) stresses due to poloidal flow above and below it and toroidal flow around it. The model allows the time evolution of a subduction system to be followed numerically and predicts self-consistent (i.e., nonimposed) trench motion and slab geometry.

Exercises

8.1 Verify that the base vectors defined by (8.10) satisfy $\mathbf{g}_\alpha \cdot \mathbf{g}^\beta = \delta_\alpha^\beta$.

8.2 Consider general coordinates $(\theta_1, \theta_2, \theta_3)$ that coincide with the standard cylindrical coordinates (r, θ, z). Determine the expressions for all the nonzero Christoffel symbols of the second kind. Use these results to write out explicitly the partial derivative $\mathbf{v}_{,2}$ given by (8.14b). Finally, write out $\mathbf{v}_{,2}$ in terms of the physical velocity components $u = v_1$, $v = v_2/r$, $w = v_3$ and the unit vectors $\mathbf{e}_r$, $\mathbf{e}_\theta$, $\mathbf{e}_z$. Check that your result is correct by expanding the derivative $\partial_\theta (u \mathbf{e}_r + v \mathbf{e}_\theta + w \mathbf{e}_z)$. (Note: the physical components of **v** are those relative to unit vectors in the directions $\mathbf{g}^i$. The physical component v and the covariant component v_2 are not the same because $\mathbf{g}^2$ is not a unit vector.)

8.3 Show that the covariant derivative $g_{ij}|_r$ of the metric tensor g_{ij} is zero. You will need the following definitions:

$$A_{ij}|_r = A_{ij,r} - \Gamma^m_{ir} A_{mj} - \Gamma^m_{jr} A_{im},$$

$$\Gamma_{ijs} = \frac{1}{2}(g_{is,j} + g_{js,i} - g_{ij,s}),$$

$$\Gamma^r_{ij} = g^{rs}\Gamma_{ijs},$$

$$g^{ms} g_{mj} = \delta^s_j,$$

where Γ_{ijs} and Γ^r_{ij} are the Christoffel symbols of the first and second kind, respectively.

8.4 [SM] Show that the equation $g^{ij} e_{ij} = 0$ is equivalent to the continuity equation (8.19). Hint: don't hesitate to write out explicitly the sums over subscript/superscript pairs.

8.5 Using the 1-D version of the thin-plate equation (8.46a), determine the gravitational admittance $Z(k)$ for the configuration shown in Figure 5.1a. Comment on the difference between this thin-plate admittance and the thick-plate admittance (5.17).

8.6 [SM] Consider a spherical elastic shell of radius R and thickness $h \ll R$, loaded by a normal stress $P_3 = P Y^0_l(\theta)$ applied to its outer surface, where $Y^0_l(\theta)$ is an axisymmetric spherical harmonic of degree l and order 0. Using the shell equations in lines-of-curvature coordinates, determine the displacement $\zeta_1 \mathbf{e}_\theta + \zeta_3 \mathbf{e}_r$ of the shell's midsurface, assuming that the material of the shell is incompressible ($\nu = 1/2$). Then calculate the ratio of the bending energy to the total (membrane plus bending) energy of the deformed shell. Make a plot of this ratio as a function of spherical harmonic degree $l \in [2, 20]$, assuming a typical terrestrial value of $h/R = 100/6370 \approx 0.0157$. Note: the bending energy U_b and the membrane energy U_m per unit midsurface area are

$$U_b = \frac{Eh^3 \left[(\kappa_1 + \kappa_2)^2 - 2(1-\nu)(\kappa_1\kappa_2 - \tau^2)\right]}{24(1-\nu^2)}, \tag{8.72}$$

$$U_m = \frac{Eh \left[(\epsilon_1 + \epsilon_2)^2 - 2(1-\nu)(\epsilon_1\epsilon_2 - \omega^2/4)\right]}{2(1-\nu^2)}. \tag{8.73}$$

9
Theory of Two-Phase Flow

Media comprising two rheologically distinct phases in contact are widespread in nature: examples include partially molten rocks and glacial ice, 'mushy zones' in magma chambers, soils, sediments and slurries. The physics of such media is considerably more complicated than that of single-phase media for two reasons. First, in most natural two-phase media the rheologically strong phase is not rigid, as in the classical theory of porous media, but rather deformable. Second, the two phases interact with each other both mechanically and energetically, so that mass, momentum and energy are not conserved separately for each phase.

Important contributions to the theory of geological two-phase flow have been made by numerous authors; a partial list would include Drew and Segel (1971), Sleep (1974), McKenzie (1984) and Fowler (1985b, 1990). More recently, Bercovici et al. (2001a) have proposed a generalized two-phase flow theory that allows for surface energy on the interface between the phases and for the generation of damage by viscous deformational work. The following development is based on Bercovici et al. (2001a) (henceforth BRS01), with corrections and improvements proposed by Bercovici and Ricard (2003) (henceforth BR03).

Several simplifying assumptions underlie the theory of BRS01: (1) inertia is negligible, so that the two-phase medium deforms by creeping flow; (2) each phase behaves as a highly viscous Newtonian fluid; (3) each (incompressible) phase has a constant density and viscosity and (4) the two-phase mixture remains isotropic on average, i.e., pores and grains are not collectively elongated in a preferred direction. In keeping with many previous studies, I refer to the two phases as the 'matrix' (denoted by a subscript m) and the 'fluid' (subscript f). However, following BRS01 the theory will be developed in such a way that the equations remain valid if the roles of the matrix and the fluid are interchanged (material invariance). I also use the notation

$$\bar{q} = \phi q_f + (1-\phi) q_m, \quad \Delta q = q_m - q_f, \tag{9.1}$$

where q is any property defined in both the matrix and the fluid and ϕ is the volume fraction of the fluid phase (porosity).

9.1 Geometrical Properties of Two-Phase Media

As a preliminary, it is important to understand the meaning of averaging in a two-phase medium. An elemental volume δV over which averaging is performed must be at the same time large enough to contain many grains and pores and small enough so that average physical properties may be regarded as continuous functions of position. If, for example, the grains and pores have typical dimensions ≈ 1 mm, then a reasonable value for δV is $\approx 10 - 100$ cm^3. To formalize the averaging process, define a distribution function $\theta(\mathbf{x})$ such that $\theta = 1$ inside the fluid pores and 0 inside the matrix. The porosity of the volume δV is then

$$\phi = \frac{1}{\delta V} \int_{\delta V} \theta \, dV. \tag{9.2}$$

Similarly, we define average velocities $\mathbf{v}_f$ of the fluid and $\mathbf{v}_m$ of the matrix as

$$\phi \mathbf{v}_f = \frac{1}{\delta V} \int_{\delta V} \hat{\mathbf{v}}_f \theta \, dV, \quad (1-\phi)\mathbf{v}_m = \frac{1}{\delta V} \int_{\delta V} \hat{\mathbf{v}}_m (1-\theta) \, dV, \tag{9.3}$$

where $\hat{\mathbf{v}}_f$ and $\hat{\mathbf{v}}_m$ are the true fluid and matrix velocities at the scale of the pores. The velocity $\mathbf{v}_f$ is often called the interstitial velocity, while the product $\phi \mathbf{v}_f$ is called the Darcy velocity.

An important property of a two-phase mixture is the interfacial area density α, defined as the average interfacial area per unit volume. The interface location and orientation are given by $\nabla \theta$, which is a Dirac δ-function centered on the interface times the unit normal to the interface. The interface area within a volume δV is then

$$\delta A_{\text{int}} = \int_{\delta V} |\nabla \theta| \, dV. \tag{9.4}$$

However, in a mixture formalism the location and orientation of the interface are unknown, and so we assume the existence of an average (and isotropic) interfacial area per unit volume, $\alpha = \delta A_{\text{int}}/\delta V$. The quantity α necessarily depends on the porosity because it vanishes if the medium comprises only a single phase, i.e., when $\phi = 0$ or 1. BRS01 assume the form

$$\alpha = \alpha_0 \phi^a (1-\phi)^b \tag{9.5}$$

where α_0, a and b are constants that depend on the material properties of the phases. The constant $\alpha_0 \propto d^{-1}$, where d is the characteristic pore or grain size. Hier-Majumder et al. (2006) proposed a different model for α for partially molten rocks, taking into account the influence of the melt geometry on grain boundaries.

The next quantity to be defined is the average interface curvature, to which the surface tension between the phases is proportional. A simple conceptual example shows that the sum of the principal curvatures of the interface is just $d\alpha/d\phi$. Consider a volume δV containing N spherical pores each with radius r. Then $\phi = 4\pi r^3 N/(3\delta V)$ and $\alpha = 4\pi r^2 N/\delta V$, whence $d\alpha/d\phi = (d\alpha/dr)/(d\phi/dr) = 2/r$, which is the sum of the two principal curvatures. Note that the curvature is defined to be positive when the interface is concave to the fluid phase. Using (9.5), we find

$$\frac{d\alpha}{d\phi} = \alpha_0 a \phi^{a-1}(1-\phi)^{b-1}\left(1 - \frac{\phi}{\phi_c}\right) \tag{9.6}$$

where $\phi_c = a/(a+b)$ is the critical porosity at which the curvature changes sign. Because the average curvature becomes infinite in the limits $\phi \to 0$ and $\phi \to 1$, we require $a < 1$ and $b < 1$. However, a and b can be very different from one another for real systems. For silicate systems with an interconnected melt phase, for example, the curvature becomes negative at low porosity, implying $a \ll b$.

9.2 Conservation Laws

9.2.1 Conservation of Mass

We now turn to the derivation of the conservation laws for a two-phase medium, starting with conservation of mass. For a fixed volume V bounded by a surface S with outward unit normal $\mathbf{n}$, the rate of change of fluid mass within the volume is

$$\frac{d}{dt}\int_V \rho_f \phi \, dV = -\int_S \rho_f \phi \mathbf{v}_f \cdot \mathbf{n} \, dS + \int_V \Gamma \, dV, \tag{9.7}$$

where Γ is the rate per unit volume at which matrix is being transformed into fluid (e.g., by melting). Transforming the first integral on the right-hand side (RHS) of (9.7) into a volume integral using the divergence theorem and taking d/dt inside the integral on the left-hand side (LHS), we obtain

$$\int_V \frac{\partial}{\partial t}(\rho_f \phi) \, dV = -\int_V \nabla \cdot (\rho_f \phi \mathbf{v}_f) \, dV + \int_V \Gamma \, dV. \tag{9.8}$$

The preceding equation must hold for any volume V, and so

$$\frac{\partial \phi}{\partial t} + \nabla \cdot (\phi \mathbf{v}_f) = \frac{\Gamma}{\rho_f}, \tag{9.9}$$

where we have used the fact that ρ_f is a constant. A similar argument applied to the matrix yields

$$\frac{\partial(1-\phi)}{\partial t} + \nabla \cdot [(1-\phi)\mathbf{v}_m] = -\frac{\Gamma}{\rho_m}. \tag{9.10}$$

Note that (9.9) and (9.10) exhibit material invariance: each can be transformed into the other via the transformations $f \leftrightarrow m$, $\phi \leftrightarrow 1 - \phi$ and $\Gamma \leftrightarrow -\Gamma$.

9.2.2 Conservation of Momentum

Now that we have seen how conservation laws may be derived from integral balances, the remaining laws will be written directly in differential form. Conservation of momentum for the fluid phase requires

$$0 = -\nabla(\phi P_f) + \nabla \cdot (\phi \tau_f) - \rho_f \phi g \mathbf{e}_z + \mathbf{h}_f, \qquad (9.11)$$

where P_f is the pressure averaged over the fluid volume, τ_f is the viscous stress tensor averaged over the volume and $\mathbf{h}_f$ is the interaction force exerted by the matrix on the fluid across their common interface. The corresponding conservation law for the matrix is

$$0 = -\nabla[(1-\phi)P_m] + \nabla \cdot [(1-\phi)\tau_m] - \rho_m(1-\phi)g\mathbf{e}_z + \mathbf{h}_m, \qquad (9.12)$$

which is obtained from (9.11) via the transformation $f \to m$, $\phi \to 1 - \phi$. The viscous stress tensors are

$$\tau_j = \eta_j \left(\nabla \mathbf{v}_j + [\nabla \mathbf{v}_j]^T - \frac{2}{3} \nabla \cdot \mathbf{v}_j \mathbf{I} \right), \qquad (9.13)$$

where $j = f$ or m, η_f and η_m are the true viscosities of the fluid and the matrix, respectively, a superscript T denotes the tensor transpose and $\mathbf{I}$ is the identity tensor.

One of the more subtle points of two-phase flow theory is how to define the interaction forces $\mathbf{h}_f$ and $\mathbf{h}_m$ (Drew and Segel, 1971; McKenzie, 1984). To approach this problem, we begin with the equation of conservation of momentum for the entire mixture, which is

$$0 = \nabla \overline{P} + \nabla \cdot \overline{\tau} - \overline{\rho} g \mathbf{e}_z + \frac{1}{\delta V} \int_{C_i} \tilde{\sigma} \mathbf{t} \, d\ell, \qquad (9.14)$$

where the last term represents the force of surface tension on the mixture. In that term, $\tilde{\sigma}$ is the true surface tension coefficient, C_i is the curve that traces the intersection between the interface and the surface of the control volume, $d\ell$ is a line element along C_i and $\mathbf{t}$ is a unit vector that is tangent to the interface and normal to $d\ell$ (Figure 9.1). BRS01 (Appendix A1) show that the last term in (9.14) can be rewritten as $(1/\delta V) \int_{\delta A} \sigma \alpha \mathbf{n} \, dA$, where δA is the surface of the control volume, $\mathbf{n}$ is the unit normal to that surface (Figure 9.1) and σ is a reduced (by a factor of order unity) surface tension coefficient. With the help of the divergence theorem, (9.14) can be rewritten as

$$0 = \nabla \overline{P} + \nabla \cdot \overline{\tau} - \overline{\rho} g \mathbf{e}_z + \nabla(\sigma \alpha). \qquad (9.15)$$

9.2 Conservation Laws

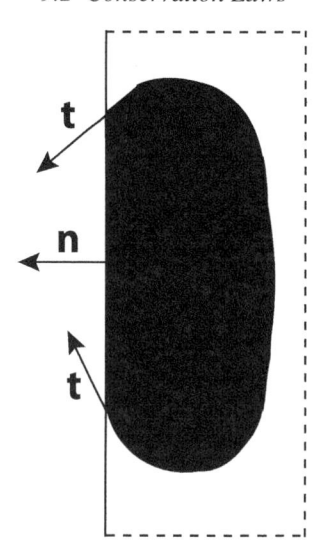

Figure 9.1 Side view of a cross-section of a portion of a control volume containing matrix (black or white) and fluid (white or black). The vertical black line is a boundary of the control volume, and the dashed lines indicate that the portion shown is connected to the rest of the volume. The unit tangent vector **t** and unit normal vector **n** introduced in the text following (9.14) are shown. Redrawn from Figure 2 of Bercovici et al. (2001a) by permission of John Wiley and Sons.

The surface tension term in (9.15) can be expanded as

$$\nabla(\sigma\alpha) = \sigma\frac{d\alpha}{d\phi}\nabla\phi + \alpha\nabla\sigma. \tag{9.16}$$

The first term on the RHS of (9.16) represents the surface tension force due to interface curvature, while the second term is the Marangoni force associated with gradients in σ due to temperature variations or the presence of surfactants.

Now the sum of (9.11) and (9.12) must equal (9.15), which requires

$$\mathbf{h}_f + \mathbf{h}_m = \nabla(\sigma\alpha). \tag{9.17}$$

Without surface tension, therefore, the interaction forces of the two phases are equal and opposite, i.e., $\mathbf{h}_f = -\mathbf{h}_m$. We therefore write

$$\mathbf{h}_f = \boldsymbol{\chi} + \omega\nabla(\sigma\alpha), \tag{9.18a}$$

$$\mathbf{h}_m = -\boldsymbol{\chi} + (1-\omega)\nabla(\sigma\alpha) \tag{9.18b}$$

where $\boldsymbol{\chi}$ is the part of the interaction force that is equal and opposite in the two phases and ω is a surface energy partitioning function. BR03 show that an appropriate form for ω is

$$\omega = \frac{\phi \eta_f}{\phi \eta_f + (1-\phi)\eta_m}. \tag{9.19}$$

Referring to (9.18), we see that the expression (9.19) implies that the surface energy is attached more to the mechanically stronger (matrix) phase than to the weaker (fluid) phase, since $\omega \ll 1$ when $\eta_m \gg \eta_f$.

The force χ must take into account the viscous interaction due to relative motion between the phases and the pressure acting at the interface. The simplest expression for the viscous contribution that satisfies Galilean invariance is $C\Delta\mathbf{v}$, where C is an interfacial drag coefficient that will be determined later. The pressure contribution must be such as to allow for a state of no motion in which the two pressures are constant everywhere. BRS01 posit the form

$$\chi = C\Delta\mathbf{v} + [\lambda P_f + (1-\lambda)P_m]\nabla\phi, \tag{9.20}$$

where λ is an unknown weighting function. To determine λ we substitute (9.18) and (9.20) into (9.11) and (9.12) and take the limit of no motion $\Delta\mathbf{v} = \tau_m = \tau_f = g = \nabla P_f = \nabla P_m = 0$, which yields

$$(1-\lambda)\Delta P \nabla\phi + \omega\sigma \nabla\alpha = 0, \tag{9.21a}$$

$$\lambda \Delta P \nabla\phi + (1-\omega)\sigma \nabla\alpha = 0, \tag{9.21b}$$

where $\Delta P = P_m - P_f$. Now the two equations (9.21) can both be true only if $\lambda = 1 - \omega$, in which case both reduce to the Laplace static equilibrium surface tension condition

$$\Delta P + \sigma \frac{d\alpha}{d\phi} = 0. \tag{9.22}$$

The momentum conservation equations (9.11) and (9.12) now take the forms

$$0 = -\phi[\nabla P_f + \rho_f g \mathbf{e}_z] + \nabla \cdot [\phi \tau_f]$$
$$+ C\Delta\mathbf{v} + \omega[\Delta P \nabla\phi + \nabla(\sigma\alpha)], \tag{9.23a}$$

$$0 = -(1-\phi)[\nabla P_m + \rho_m g \mathbf{e}_z] + \nabla \cdot [(1-\phi)\tau_m]$$
$$- C\Delta\mathbf{v} + (1-\omega)[\Delta P \nabla\phi + \nabla(\sigma\alpha)]. \tag{9.23b}$$

The next step is to determine a materially invariant form of the interfacial drag coefficient C. This can be done by considering the balance of forces at the fluid–matrix interface. This balance can be written as

$$\eta_m \frac{\mathbf{v}_m - \mathbf{v}_i}{\delta_m^2} = \eta_f \frac{\mathbf{v}_i - \mathbf{v}_f}{\delta_f^2}, \tag{9.24}$$

9.2 Conservation Laws

where $\mathbf{v}_i$ is the interface velocity and δ_j is the typical size of an element of phase j ($= f$ or m). Now solve (9.24) for $\mathbf{v}_i$, use the result to evaluate the viscous force scale (either side of (9.24)) and equate the result to the interfacial force $C\Delta\mathbf{v}$ to obtain

$$C(\mathbf{v}_m - \mathbf{v}_f) = \frac{\eta_m \eta_f (\mathbf{v}_m - \mathbf{v}_f)}{\eta_f \delta_m^2 + \eta_m \delta_f^2}. \qquad (9.25)$$

Now assume that δ_f and δ_m are functions only of the porosity. Then to preserve material symmetry, both must be related to the same function δ, i.e.,

$$\delta_f = \delta(\phi), \quad \delta_m = \delta(1-\phi). \qquad (9.26)$$

Now we require that in the limit $\eta_f \ll \eta_m$ we recover Darcy's law, in which $C = \eta_f \phi^2/k(\phi)$, where $k(\phi)$ is the permeability. Thus we have

$$\lim_{\eta_f/\eta_m \to 0} C = \frac{\eta_f \phi^2}{k(\phi)} = \frac{\eta_f}{\delta^2(\phi)}, \qquad (9.27)$$

which implies $\delta(\phi) = \sqrt{k(\phi)}/\phi$. Combining this result with (9.25) and (9.26) we obtain

$$C = \frac{\eta_m \eta_f \phi^2 (1-\phi)^2}{\eta_f k(1-\phi)\phi^2 + \eta_m k(\phi)(1-\phi)^2}. \qquad (9.28)$$

9.2.3 Conservation of Energy

We next turn to the equations for conservation of energy. Here the novel feature is that internal energy resides not just in the fluid and the matrix, but also on the interface. Let ε_f and ε_m be the internal energy per unit mass of the fluid and matrix, respectively, and let ξ_i be the energy per unit interface area. Then the rate of change of energy (internal plus interfacial) is

$$\frac{\partial}{\partial t}\left[\phi\rho_f \varepsilon_f + (1-\phi)\rho_m \varepsilon_m + \alpha \xi_i\right] = Q - \nabla \cdot \mathbf{q}$$
$$- \nabla \cdot [\phi\rho_f \varepsilon_f \mathbf{v}_f + (1-\phi)\rho_m \varepsilon_m \mathbf{v}_m + \alpha \xi_i \mathbf{v}_\omega]$$
$$+ \nabla \cdot [-\phi P_f \mathbf{v}_f - (1-\phi)P_m \mathbf{v}_m + \phi \mathbf{v}_f \cdot \boldsymbol{\tau}_f + (1-\phi)\mathbf{v}_m \cdot \boldsymbol{\tau}_m + \sigma \alpha \mathbf{v}_\omega]$$
$$- \phi \mathbf{v}_f \cdot (\rho_f g \mathbf{e}_z) - (1-\phi)\mathbf{v}_m \cdot (\rho_m g \mathbf{e}_z). \qquad (9.29)$$

The first term (Q) on the RHS of (9.29) is the rate of heat production per unit volume, and $\mathbf{q}$ is the diffusive energy flux vector (e.g., the heat flux vector). The third term represents the rate of energy transport across the surface of the volume, where the interfacial energy is transported at the effective interfacial velocity

$$\mathbf{v}_\omega = \omega \mathbf{v}_f + (1-\omega)\mathbf{v}_m. \qquad (9.30)$$

The fourth term represents the rate at which work is done by surface forces acting on the surface of the volume, and the last two terms represent the rate at which work is done by the body force on the interior of the volume. In keeping with the character of slow viscous flow, changes of kinetic energy are neglected in (9.29).

To simplify (9.29), we first recall that both phases are assumed to be incompressible, so that their internal energies ε_f and ε_m are functions of temperature alone. We can therefore write $d\varepsilon_f = c_f dT$ and $d\varepsilon_m = c_m dT$, where the temperature T is assumed the same in both phases and c_f and c_m are the heat capacities. Moreover, the interfacial energy ξ_i is related to the reduced surface tension σ by (BRS01, Appendix A2)

$$\xi_i = \sigma - T\frac{d\sigma}{dT}, \tag{9.31}$$

where $-d\sigma/dT > 0$ is the entropy per unit area on the interface. We now simplify (9.29) in the usual way by expanding the derivatives and using the continuity equations (9.9) and (9.10) and the momentum equations (9.23). After rather lengthy algebra (Exercise 9.1), the result is (Šrámek et al., 2007)

$$\phi \rho_f c_f \frac{D_f T}{Dt} + (1-\phi)\rho_m c_m \frac{D_m T}{Dt} - T\frac{D_\omega}{Dt}\left(\alpha \frac{d\sigma}{dT}\right) - T\alpha \frac{d\sigma}{dT}\nabla \cdot \mathbf{v}_\omega$$

$$= Q - \nabla \cdot \mathbf{q} + \Psi - \left(\Delta P + \sigma \frac{d\alpha}{d\phi}\right)\frac{D_\omega \phi}{Dt} + \Gamma\left(\Delta \varepsilon + \frac{P_m}{\rho_m} - \frac{P_f}{\rho_f}\right), \tag{9.32}$$

where $\Delta \varepsilon = \varepsilon_m - \varepsilon_f$,

$$\frac{D_j}{Dt} = \frac{\partial}{\partial t} + \mathbf{v}_j \cdot \nabla \quad (j = f, m, \text{ or } \omega), \tag{9.33}$$

and

$$\Psi = C\Delta \mathbf{v} \cdot \Delta \mathbf{v} + \phi \nabla \mathbf{v}_f : \boldsymbol{\tau}_f + (1-\phi)\nabla \mathbf{v}_m : \boldsymbol{\tau}_m \tag{9.34}$$

is the rate of viscous dissipation of energy per unit volume.

Equation (9.32) is arranged so that terms related to temporal entropy variations are on the LHS, while the terms on the RHS are related to entropy sources and fluxes. In addition to the usual terms representing heat production (Q), heat flux ($\nabla \cdot \mathbf{q}$) and viscous dissipation (Ψ), the RHS of (9.32) contains two additional terms. The fourth term is proportional to the quantity $\Delta P + \sigma d\alpha/d\phi$, which is zero in the motionless equilibrium state when Laplace's condition (9.22) on the pressure drop across a curved interface holds. The quantity $\Delta P + \sigma d\alpha/d\phi$ is thus the nonequilibrium pressure difference, and the fourth term on the RHS of (9.32) represents the rate at which this pressure difference does work by changing the porosity. The fifth term on the RHS is a source term proportional to the rate at which one phase transforms into the other and can be rewritten as $\Gamma \Delta \mathcal{H} \equiv \Gamma(\mathcal{H}_m - \mathcal{H}_f)$, where $\mathcal{H}_j = \varepsilon_j + P_j/\rho_j$ are the specific enthalpies of the two phases.

The last equations required to close the system are expressions for the pressure difference ΔP and the phase transformation rate Γ. Using a combination of micromechanical models and nonequilibrium thermodynamics, Bercovici and Ricard (2003) and Šrámek et al. (2007) derived the phenomenological laws

$$\Delta P = -\sigma \frac{d\alpha}{d\phi} - K_0 \frac{\eta_f + \eta_m}{\phi(1-\phi)} \left(\frac{D_\omega \phi}{Dt} - \frac{\rho_i}{\rho_f \rho_m} \Gamma \right), \quad (9.35)$$

$$\Gamma = L_{22} \left(\Delta\varepsilon - T\Delta s - P_i \frac{\Delta\rho}{\rho_f \rho_m} - \sigma \frac{d\alpha}{d\phi} \frac{\rho_i}{\rho_f \rho_m} \right), \quad (9.36)$$

where K_0 is a dimensionless constant of order unity, $\rho_i = (1-\omega)\rho_f + \omega\rho_m$, L_{22} is an undetermined phenomenological coefficient, $\Delta s \equiv s_m - s_f$ is the difference of the specific entropies of the two phases and $P_i = (1-\omega)P_f + \omega P_m$. In practice, the nonequilibrium relation (9.36) is replaced by a simpler equilibrium formulation in which the rate of the phase change (e.g., melting) is controlled by the energy equation (Šrámek et al., 2007).

To summarize, the two-phase equations we have derived involve the eleven variables ϕ, $\mathbf{v}_f$, $\mathbf{v}_m$, P_f, P_m, T and Γ. These are constrained by eleven equations: two continuity equations (9.9) and (9.10), six momentum equations (9.23), the energy equation (9.32) and the phenomenological laws (9.35) and (9.36), respectively. The system of equations is therefore properly closed.

To conclude, I note several important extensions of the two-phase equations we have derived. The first, due to BRS01, is to two-phase systems undergoing damage. The basic idea is that the deformational work done on the interface by deviatoric stresses ($= \Psi$) comprises two parts: a fraction $1 - f$ (say) that is dissipated as heat, and a fraction f that goes towards (reversibly) deforming the interface, and which is stored as surface energy rather than dissipated. BRS01 refer to the partitioning of a fraction f of Ψ toward the production of surface energy as 'damage'. Further discussion of this matter is beyond our scope here; the interested reader is referred to BRS01, Bercovici et al. (2001b) and BR03. A second extension is to a matrix with a more realistic nonlinear viscoelastoplastic rheology (Yarushina and Podladchikov, 2015). Yet a third extension is to systems involving multicomponent thermodynamics and disequilibrium melting, as discussed later in our treatment of the reactive infiltration instability.

9.3 The Geodynamic Limit $\eta_f \ll \eta_m$

Among the most important examples of two-phase systems in geodynamics are porous mantle rocks through which low-viscosity fluids (e.g., melt or metasomatic fluids) percolate. In this section, the equations derived earlier are specialized to such cases, for which $\eta_f \ll \eta_m$ and $\omega \to 0$. For simplicity, we shall consider only the mechanical equations (conservation of mass and momentum), leaving considerations of heat transport aside. For still further simplicity, we shall assume that there is no mass transfer between the phases, i.e., $\Gamma = 0$.

The equations (9.9) and (9.10) for conservation of mass now take the simplified forms

$$\frac{\partial \phi}{\partial t} + \nabla \cdot [\phi \mathbf{v}_f] = 0, \qquad \frac{\partial (1-\phi)}{\partial t} + \nabla \cdot [(1-\phi)\mathbf{v}_m] = 0. \qquad (9.37\text{a,b})$$

The equations (9.23) for conservation of momentum become

$$0 = -\phi[\nabla P_f + \rho_f g \mathbf{e}_z] + C\Delta \mathbf{v}, \qquad (9.38\text{a})$$

$$0 = -(1-\phi)[\nabla P_m + \rho_m g \mathbf{e}_z] + \nabla \cdot [(1-\phi)\boldsymbol{\tau}_m]$$
$$- C\Delta \mathbf{v} + \Delta P \nabla \phi + \nabla(\sigma \alpha), \qquad (9.38\text{b})$$

where

$$C = \frac{\eta_f \phi^2}{k(\phi)}. \qquad (9.39)$$

The last remaining equation is (9.35), which for $\Gamma = 0$ and $\eta_m \gg \eta_f$ takes the form

$$\Delta P = -\sigma \frac{d\alpha}{d\phi} - \frac{K_0 \eta_m}{\phi(1-\phi)} \frac{D_m \phi}{Dt}, \qquad (9.40)$$

where we have used the fact that $D/D_\omega t = D/D_m t$ when $\omega = 0$. But now (9.37b) implies $D_m \phi / Dt = (1-\phi)\nabla \cdot \mathbf{v}_m$, whereupon (9.40) becomes

$$\Delta P = -\sigma \frac{d\alpha}{d\phi} - \frac{K_0 \eta_m}{\phi} \nabla \cdot \mathbf{v}_m. \qquad (9.41)$$

In the absence of surface tension ($\sigma = 0$), (9.41) becomes

$$\Delta P = -\eta_b \nabla \cdot \mathbf{v}_m, \qquad (9.42)$$

where

$$\eta_b = \frac{K_0 \eta_m}{\phi} \qquad (9.43)$$

is an effective bulk viscosity. Equation (9.42) is equivalent to an analogous equation in the theory of McKenzie (1984) if ΔP is replaced by minus the difference of the average normal stresses in the matrix and the fluid. Thus the limit $\eta_f/\eta_m \to 0$ of the mechanical equations of BRS01 and BR03 is formally identical to the mechanical equations of McKenzie (1984). However, McKenzie (1984) assumed that the bulk viscosity is constant, whereas η_b must in fact tend to infinity as $\phi \to 0$ to prevent compaction beyond $\phi = 0$ to negative porosities (Schmeling, 2000).

9.4 One-Dimensional Model Problems

Much insight into the behavior of two-phase systems can be gained by considering simple one-dimensional model problems in which all the variables depend only on a single depth coordinate z and the time t. For this case, the mass conservation equations (9.37) reduce to

$$\dot{\phi} + [\phi w_f]' = 0, \qquad -\dot{\phi} + [(1-\phi)w_m]' = 0, \qquad (9.44\text{a,b})$$

where primes denote $\partial/\partial z$ and dots denote $\partial/\partial t$. The two equations (9.44) can be added and integrated once with respect to z to yield

$$\phi w_f + (1-\phi)w_m = W(t), \qquad (9.45)$$

where $w_j = \mathbf{v}_j \cdot \mathbf{e}_z$ and $W(t)$ is an arbitrary function of time. Equations (9.38) for conservation of momentum take the forms

$$0 = -\phi(P'_f + \rho_f g) + \frac{\eta_f \phi^2}{k(\phi)}(w_m - w_f), \qquad (9.46\text{a})$$

$$0 = -(1-\phi)(P'_m + \rho_m g) - \frac{\eta_f \phi^2}{k(\phi)}(w_m - w_f)$$

$$+ \left(\Delta P + \sigma \frac{d\alpha}{d\phi}\right)\phi' + \frac{4}{3}\eta_m[(1-\phi)w'_m]', \qquad (9.46\text{b})$$

where the reduced surface tension σ has been assumed constant. Finally, (9.41) becomes

$$\Delta P + \sigma \frac{d\alpha}{d\phi} = -\frac{K_0 \eta_m}{\phi} w'_m. \qquad (9.47)$$

Alternate forms of the 1-D two-phase flow equations have been derived by Connolly and Podladchikov (1998) for a viscoelastic matrix and Connolly and Podladchikov (2017) for a matrix with a nonlinear (power-law) rheology.

9.4.1 Gravitational Compaction of a Half-Space

As a first model problem, consider a two-phase half-space with initially constant porosity ϕ_0 resting on an impermeable surface $z = 0$ (Richter and McKenzie, 1984). For simplicity, assume that the surface tension $\sigma = 0$. Our task is to determine the buoyancy-driven compaction rate $\dot{\phi}(z,0)$ throughout the half-space at the initial instant $t = 0$.

Because $\phi(z,0)$ is constant, equations (9.45), (9.46) and (9.47) are four coupled equations for the variables w_f, w_m, P_f and P_m. The impermeability conditions

$w_f(0,t) = w_m(0,t) = 0$ imply that $W(t) = 0$ in (9.45). Reducing the four equations to a single ordinary differential equation for w_m, we obtain

$$\eta_m \left(\frac{4}{3} + \frac{K_0}{\phi_0} \right) w_m'' - \frac{\eta_f}{(1-\phi_0)k_0} w_m = g\Delta\rho, \qquad (9.48)$$

where $k_0 = k(\phi_0)$. The solution that is finite as $z \to \infty$ is

$$w_m = \frac{(1-\phi_0)k_0 g \Delta\rho}{\eta_f} \left[\exp(-z/\delta_c) - 1 \right] \qquad (9.49)$$

where

$$\delta_c = \left[\frac{(1-\phi_0)k_0 \eta_m (4/3 + K_0/\phi_0)}{\eta_f} \right]^{1/2} \qquad (9.50)$$

is the 'compaction length' (McKenzie, 1984). Ricard (2015) estimates δ_c to lie in the range 30 cm (for dry siliceous melt) to 8 km (for basaltic melt). The initial compaction rate $\dot{\phi}(z,0)$ is determined from (9.44b) and is

$$\dot{\phi}(z,0) = -\frac{\phi_0}{t_c} \exp(-z/\delta_c) \qquad (9.51)$$

where

$$t_c = \frac{\phi_0 \eta_f \delta_c}{(1-\phi_0)^2 k_0 g \Delta\rho} \qquad (9.52)$$

is a characteristic compaction time. Compaction is thus confined initially to a BL of thickness δ_c adjoining the impermeable surface. The thickness of the compaction BL increases with time, but its structure for $t > 0$ must be determined numerically. Above the BL, the two-phase medium is in a state of uniform fluidization in which the driving buoyancy force is balanced by the Darcy resistance of the matrix, with velocities

$$w_m = -\frac{\phi_0}{(1-\phi_0)} w_f = -\frac{(1-\phi_0)k_0 g \Delta\rho}{\eta_f}. \qquad (9.53)$$

9.4.2 Solitary Waves

Richter and McKenzie (1984) and Scott and Stevenson (1984) showed that the one-dimensional two-phase equations admit a solitary wave solution in which a localized high-porosity pulse propagates upward without change of shape over a low-porosity background. To determine this solution, we begin by using (9.45) with $W(t) = 0$, (9.46a) and (9.47) to eliminate P_f, P_m and w_f from (9.46b), again assuming $\sigma = 0$. The result is

9.4 One-Dimensional Model Problems

$$0 = -\frac{\eta_f}{k(\phi)} w_m - (1-\phi)g\Delta\rho + \left[\eta_m\left(\frac{4}{3} + \frac{K_0}{\phi}\right)(1-\phi)w'_m\right]', \qquad (9.54)$$

which together with (9.44b) constitutes a pair of coupled equations for w_m and ϕ. The permeability is given by the Kozeny-Carman relation

$$k(\phi) = \beta d^2 \phi^n \qquad (9.55)$$

where d is the grain size, β is a constant of order unity and n is a power-law exponent (typically $= 3$).

Next, we assume that $\phi \ll 1$, which allows us to set $1-\phi \approx 1$ and $4/3 + K_0/\phi \approx K_0/\phi$. Our pair of coupled equations then becomes

$$\dot{\phi} = w'_m, \qquad (9.56a)$$

$$0 = -\frac{\eta_f}{k(\phi)} w_m - g\Delta\rho + \eta_m K_0 \left(\frac{w'_m}{\phi}\right)'. \qquad (9.56b)$$

Now define dimensionless (starred) variables

$$\phi^* = \frac{\phi}{\phi_0}, \quad w^*_m = \frac{w_m \eta_f}{k_0 g \Delta\rho} \equiv w_m/w_c,$$

$$z^* = z \left(\frac{\phi_0 \eta_f}{k_0 K_0 \eta_m}\right)^{1/2} \equiv z/\delta_c, \quad t^* = \frac{t w_c}{\phi_0 \delta_c}. \qquad (9.57)$$

Substituting (9.57) into (9.56b) and dropping the stars, we obtain

$$\left(\frac{w'_m}{\phi}\right)' - \frac{w_m}{\phi^n} = 1. \qquad (9.58)$$

The form of (9.56a) remains unchanged after nondimensionalization. Elimination of w_m between (9.56a) and (9.58) yields

$$\dot{\phi} = \left\{\phi^n \left[\left(\frac{\dot{\phi}}{\phi}\right)' - 1\right]\right\}'. \qquad (9.59)$$

We now seek a propagating wave solution of the form

$$\phi = F(\zeta) \quad \text{where} \quad \zeta = z - ct \qquad (9.60)$$

and c is the (dimensionless) wavespeed. Substituting (9.60) into (9.59) we find

$$cF' = \left\{F^n \left[c\left(\frac{F'}{F}\right)' + 1\right]\right\}', \qquad (9.61)$$

which must be solved subject to $F(\pm\infty) - 1 = F'(\pm\infty) = F''(\pm\infty) = 0$. Integrating (9.61) once subject to these conditions and rearranging, we obtain

$$\frac{c(F-1)+1-F^n}{cF^n} = \left(\frac{F'}{F}\right)' \equiv \frac{F}{2}\frac{d}{dF}\left[F^{-2}(F')^2\right]. \tag{9.62}$$

A second integration subject to the boundary conditions yields

$$\frac{1-n+c[n(1-F)-1]}{n(n-1)cF^n} - \frac{1-n-c}{n(n-1)c} - \frac{\ln F}{c} = \frac{(F')^2}{2F^2}. \tag{9.63}$$

Now the maximum amplitude $A_{\max}$ of the wave occurs where $F' = 0$. The dispersion relation is obtained by setting $F' = 0$ and $F = A_{\max}$ in (9.63) and solving the resulting equation for c, yielding

$$c = \frac{(n-1)[A_{\max}^n(n \ln A_{\max} - 1) + 1]}{n(1-A_{\max}) + A_{\max}^n - 1}. \tag{9.64}$$

Equation (9.64) agrees with equation (8) of Scott and Stevenson (1984). Moreover, for $n = 2$ (9.64) is identical to the dispersion relation (7.90) for a solitary wave in a viscous conduit.

The conduit equation (7.82) and the two-phase flow equation (9.59) are special cases of a more general 'magma equation' (Scott and Stevenson, 1984)

$$\phi_t + (\phi^n)_z - (\phi^n(\phi^{-m}\phi_t)_z)_z = 0, \tag{9.65}$$

where m is the exponent in the relation $\eta_{\text{bulk}} \propto \phi^{-m}$ between the bulk viscosity of the matrix and the porosity. The conduit equation corresponds to $n = 2$, $m = 1$ while the most common form of the two-phase flow equation has $n = 3$, $m = 1$.

The mathematical properties of the magma equation have been carefully studied by geodynamicists and by applied mathematicians interested in nonlinear waves. Here I cite just a few highlights of the extensive relevant literature. Whitehead and Helfrich (1986) showed that in the limit of small amplitude the magma equation reduces to the Korteweg-de Vries equation

$$\frac{\partial y}{\partial t} + y\frac{\partial y}{\partial x} + \frac{\partial^3 y}{\partial x^3} = 0, \tag{9.66}$$

where y, t and x are the scaled amplitude, time and spatial coordinate, respectively. Barcilon and Richter (1986) and Harris (1996) showed that the magma equation has exactly two conservation laws, which means that magma solitary waves are not true solitons, i.e. that they are not exactly conserved through collisions. Spiegelman (1993) studied the properties of linear and nonlinear waves in two-phase media without and with melting and showed that the governing equations in the zero compaction-length limit admit shock wave solutions. Simpson and Weinstein (2008) proved that magma solitary waves are asymptotically stable to 1-D perturbations. A recent survey of the literature on the magma and conduit equations can be found in Maiden and Hoefer (2016).

In conclusion, I note that Yarushina et al. (2015) have proposed a theory for solitary waves in a viscoplastic matrix with dilatant brittle failure, focussing on the asymmetry between compaction and decompaction.

9.5 Solutions in Two and Three Dimensions

The equations of two-phase flow are difficult to solve analytically in 2-D and 3-D, and numerical methods are generally required. Nevertheless, some useful solutions exist for the special case in which the porosity (and by extension the permeability) is uniform over the flow field, and no melting is occurring ($\Gamma = 0$). Here I discuss two of these solutions that are direct generalizations of the classical single-phase solutions for corner flow and for Stokes flow around a sphere (§ 4.3).

Spiegelman and McKenzie (1987) studied two-phase corner flow models for mid-ocean ridges (Figure 4.3b) and subduction zones (Figure 4.3a). Movement of the melt is the sum of advection by the matrix, vertical flow driven by the melt's buoyancy and flow in response to the 'piezometric' pressure gradient $\nabla \mathcal{P} \equiv \eta_m \nabla^2 \mathbf{v}_m$ associated with matrix shear. Since $\mathcal{P} \to -\infty$ as $r \to 0$, melt is sucked towards the corner. The flow domain is divided into a broad 'melt extraction zone' from which melt is focussed towards the corner and a region outside this zone where buoyancy drives melt upward to the base of the overlying plate. Spiegelman and McKenzie (1987) proposed that this explains how melt can be drawn from a broad area to feed narrowly focussed volcanism at ridges and island arcs.

Rudge (2014) derived analytical solutions for compacting two-phase flow around a rigid impermeable sphere. He considered the case of a sphere embedded in a general linear flow comprising uniform flow, rotation and straining (see Exercise 4.10). The key dimensionless parameter is the ratio of the sphere radius a to the compaction length δ_c. For the classic Stokes case of a sphere in a uniform flow and $a/\delta_c \gg 1$, the local compaction rate $\nabla \cdot \mathbf{v}_m$ is nonzero only in thin boundary layers fore and aft of the sphere. If, however, $a/\delta_c \ll 1$, then compaction is broadly distributed. Rudge (2014) derived expressions for the drag, torque and stresslet on the sphere for arbitrary values of a/δ_c and the bulk viscosity/shear viscosity ratio.

9.6 Instabilities of Two-Phase Flow

While simple one-dimensional solutions of the equations of two-phase flow are useful, many of them turn out to be unstable to small perturbations. This section briefly reviews three of these instabilities.

The first case of interest is the 1-D solitary waves discussed earlier. Scott and Stevenson (1986) and Barcilon and Lovera (1989) showed analytically that they are unstable to 2-D perturbations. Wiggins and Spiegelman (1993) subsequently showed numerically that both 1-D and 2-D waves are unstable to 3-D perturbations. This suggests that melt/matrix solitary waves should be quasi-spherical 3-D structures.

A second example of a two-phase instability is a purely mechanical instability that leads to the formation of melt-rich bands separated by melt-poor domains. Its mechanism was first identified by Stevenson (1989), who demonstrated that the essential factor is a matrix shear viscosity η that decreases with increasing porosity ϕ. To explain the mechanism of the instability, Stevenson (1989) imagined a two-phase medium deforming in uniform pure shear at a rate $\dot\epsilon_0$, with the maximum rates of extension and shortening in the x- and z-directions, respectively. Because the shear stress σ_{xz} associated with the basic state is zero, force balance requires that the stress component $\sigma_{xx} \equiv -P_l + 2\eta\dot\epsilon_0$ be independent of x, where P_l is the fluid pressure. Now suppose that there is some small spatial fluctuation in porosity along the x-direction. Since η is lower where ϕ is larger, the constancy of σ_{xx} requires P_l to be lower where ϕ is larger and vice versa. But melt will flow into low-pressure regions, i.e. it will accumulate where it is already concentrated, leading to segregation of melt into bands parallel to the direction (x) of the maximum compressive stress. This simple analysis is borne out by a more detailed linear stability analysis (LSA) of the two-phase governing equations (Stevenson, 1989). However, that analysis fails to predict a preferred wavelength for the instability because the fastest growth occurs in the limit of vanishingly small wavelength.

Study of the instability of Stevenson (1989) received a new impulse from the experimental discovery by Holtzman et al. (2003) that simple shear deformation of partially molten rocks causes melt to segregate into melt-rich bands oriented at $\approx 20°$ to the shear plane ($\approx 25°$ to the direction of maximum compressive stress). This discovery inspired numerous LSAs aimed at predicting the orientation and preferred wavelength of the bands (Spiegelman, 2003; Katz et al., 2006; Takei and Holtzman, 2009; Takei and Hier-Majumder, 2009; Butler, 2009, 2010, 2012; Takei and Katz, 2013; Rudge and Bercovici, 2015; Takei and Katz, 2015; Taylor-West and Katz, 2015; Bercovici and Rudge, 2016; Gebhardt and Butler, 2016). It turns out that a number of different physical factors can explain the existence of low-angle bands, including a shear-thinning (Katz et al., 2006) or anisotropic (Takei and Holtzman, 2009) matrix viscosity, anisotropic permeability (Taylor-West and Katz, 2015) and two-phase damage rheology (Rudge and Bercovici, 2015). Prediction of a preferred wavelength was achieved by Takei and Hier-Majumder (2009) and Bercovici and Rudge (2016), whose models included the effect of surface tension.

The third instability, the so-called reactive infiltration instability (RII), is not purely mechanical but instead involves a feedback between porosity and chemistry. This instability was first discussed in the context of nondeformable porous crustal rocks dissolved by aqueous solutions flowing through them (e.g. Ortoleva et al., 1987). Its mechanism is simple: if the porosity at a point in the rock is larger than elsewhere, the local Darcy flux of unsaturated solution will be greater there, causing faster dissolution that further increases the porosity. This positive feedback leads to the growth of high-porosity channels in the rock.

The theory of the RII was extended to the mantle melting context by Aharonov et al. (1995). In their model, a steady flux of melt moves upward through a compactible matrix whose solubility in the melt increases linearly with height. The rate of dissolution is governed by first-order kinetics, i.e., it is proportional to the difference between the concentration of the soluble phase (pyroxene) in the melt and the local equilibrium concentration. LSA of this model (Aharonov et al., 1995) shows that two types of instabilities can occur: travelling dissolution waves with complex growth rates; and an absolute instability consisting of stationary channels. Spiegelman et al. (2001) extended the model of Aharonov et al. (1995) to include matrix shear and chemical diffusion in the melt. Hesse et al. (2011) performed a LSA of a model with matrix upwelling and local chemical equilibrium between the melt and matrix. They determined a regime diagram for compaction–dissolution waves versus stationary channels as a function of the ratio of matrix to melt velocities, the solubility change per compaction length and the depletion of the incoming upwelling mantle. Finally, Rudge et al. (2011) and Keller and Katz (2016) were motivated by the RII to derive extended two-phase flow theories that include multicomponent thermodynamics and disequilibrium melting.

Exercises

9.1 Starting from the primitive form (9.29) of the energy equation, derive (9.32).

9.2 A two-phase layer of thickness h with initially constant porosity ϕ_0 fills a container with an impermeable base $z = 0$. The upper surface $z = h$ of the layer is in contact with a porous piston or screen that allows passage of the fluid phase while preventing passage of the matrix phase. The two phases have the same density, so no buoyancy-driven compaction occurs. At time $t = 0$, the screen is set in motion with a (downward) velocity $-W_0$. Determine the force per unit area acting on the screen at the initial instant $t = 0$, assuming that the fluid pressure is zero at the screen.

10

Hydrodynamic Stability and Thermal Convection

Not every correct solution of the governing equations of fluid mechanics can exist in nature or in the laboratory. To be observable, the solution must also be stable, i.e., able to maintain itself against the small disturbances or perturbations that are ubiquitous in any physical environment. Whether this is the case can be determined by linear stability analysis (LSA), wherein one solves the linearized equations that govern infinitesimal perturbations of a solution of interest (the basic state) to determine the conditions under which these perturbations grow (instability) or decay (stability). Typically, one expands the perturbations in normal modes that satisfy the equations and boundary conditions and then determines which modes have a rate of exponential growth with a positive real part. However, LSA describes only the initial growth of perturbations and is no longer valid when their amplitude is sufficiently large that nonlinear interactions between the modes become important. Various nonlinear stability methods have been developed to describe the dynamics of this stage.

In this chapter I illustrate these methods for two geodynamically important instabilities: the Rayleigh–Taylor instability (RTI) of a buoyant layer and Rayleigh–Bénard convection (RBC) between isothermal surfaces.

10.1 Rayleigh–Taylor Instability

We have already encountered the RTI in § 3.4, where it served as an example of how intermediate asympotic limits of a function can be identified and interpreted. I now outline the LSA that leads to the expression (3.20) for the growth rate of infinitesimal perturbations.

The model geometry is shown in Figure 3.3. The flow in both fluids satisfies the Stokes equations $\nabla p = \eta \nabla^2 \mathbf{u}$, which must be solved subject to the appropriate boundary and matching conditions. Let the velocities in the two layers be

10.1 Rayleigh–Taylor Instability

$\mathbf{u}_n \equiv (u_n, v_n, w_n)$. The vanishing of the normal velocity and the shear stress at $z = -h_0$ requires

$$w_0(-h_0) = \partial_z u_0(-h_0) = \partial_z v_0(-h_0) = 0, \tag{10.1}$$

where the arguments x_1, x_2 and t of the variables have been suppressed. The velocity must vanish at $z = \infty$, which requires

$$u_1(\infty) = v_1(\infty) = w_1(\infty) = 0. \tag{10.2}$$

Finally, continuity of the velocity and traction at the interface requires

$$[\mathbf{u} \cdot \mathbf{t}] = [\mathbf{u} \cdot \mathbf{n}] = [\mathbf{t} \cdot \sigma \cdot \mathbf{n}] = [\mathbf{n} \cdot \sigma \cdot \mathbf{n}] = 0, \tag{10.3}$$

where $[\ldots]$ denotes the jump in the enclosed quantity from fluid 1 to fluid 0 across the interface $z = \zeta$, σ is the stress tensor,

$$\mathbf{n} = (1 + |\nabla_h \zeta|^2)^{-1/2}(\mathbf{e}_z - \nabla_h \zeta) \tag{10.4}$$

is the unit vector normal to the interface, and $\mathbf{t}$ is any unit vector tangent to the interface. The final relation required is the kinematic condition

$$\partial_t \zeta + \mathbf{u}(\zeta) \cdot \nabla_h \zeta = w(\zeta), \tag{10.5}$$

which expresses the fact that the interface $z = \zeta$ is a material surface.

Although the Stokes equations themselves are linear, the conditions (10.3) and (10.5) are inherently nonlinear if the deformation ζ of the interface is finite. Therefore to carry out a LSA we must linearize (10.3) and (10.5) by taking ζ to be infinitesimal. Let us illustrate the procedure for the most complicated of our five conditions, the normal stress matching condition $[\mathbf{n} \cdot \sigma \cdot \mathbf{n}] = 0$. We begin by assuming that the slope of the interface is very small ($|\nabla_h \zeta| \ll 1$), so that $\mathbf{n} \approx \mathbf{e}_z$. The normal stress matching condition then becomes

$$\sigma_{zz}^{(0)}(\zeta) - \sigma_{zz}^{(1)}(\zeta) = 0, \tag{10.6}$$

where the superscripts (0) and (1) are the fluid indices and the argument ζ indicates that the normal stress is evaluated at $z = \zeta$. Using the definition of the normal stress, we may write (10.6) as

$$-p_0(\zeta) + 2\eta_0 \partial_z w_0(\zeta) + p_1(\zeta) - 2\eta_1 \partial_z w_1(\zeta) = 0. \tag{10.7}$$

Now write the pressures p_n as sums of a hydrostatic part $\rho_n g z$ and a modified pressure $\hat{p}_n$ that is directly related to the motion of the fluid:

$$p_n(\zeta) = \hat{p}_n(\zeta) + \rho_n g \zeta. \tag{10.8}$$

Substituting (10.8) into (10.7) and defining a modified normal stress $\hat{\sigma}_{zz}^{(n)} = -\hat{p}_n + 2\eta_n \partial_z w_n$, we obtain

$$\hat{\sigma}_{zz}^{(0)}(\zeta) - \hat{\sigma}_{zz}^{(1)}(\zeta) = -g\Delta\rho\zeta. \tag{10.9}$$

Because the hydrostatic pressure gradients in the two fluids are different, the modified normal stress jumps across the interface by an amount $-g\Delta\rho\zeta$ precisely because the total normal stress is continuous there. To linearize (10.9), we perform a two-term Taylor series expansion of the left-hand side (LHS) about $z=0$ to obtain

$$\hat{\sigma}_{zz}^{(0)}(0) + \zeta\partial_z\hat{\sigma}_{zz}^{(0)}(0) - \hat{\sigma}_{zz}^{(1)}(0) - \zeta\partial_z\hat{\sigma}_{zz}^{(1)}(0) = -g\Delta\rho\zeta. \tag{10.10}$$

However, because the flow is driven entirely by the buoyancy associated with the disturbance of the interface, $\hat{\sigma}_{zz}^{(n)}(0)$ and $\partial_z\hat{\sigma}_{zz}^{(n)}(0) \sim O(\zeta)$. The second and fourth terms on the LHS of (10.10) are therefore $O(\zeta^2)$ and can be neglected. The matching condition then becomes

$$[\hat{\sigma}_{zz}]_0 = -g\Delta\rho\zeta, \tag{10.11}$$

where $[\ldots]_0$ indicates a jump across the undeformed position $z=0$ of the interface. Similar reasoning applied to the other matching conditions in (10.3) and the kinematic condition (10.5) yields the linearized forms

$$[u]_0 = [v]_0 = [w]_0 = [\sigma_{xz}]_0 = [\sigma_{yz}]_0 = 0, \tag{10.12}$$

$$\partial_t\zeta = w(0). \tag{10.13}$$

To lowest order in ζ, the flow is purely poloidal (Ribe, 1998) and can be described by two poloidal scalars $\mathcal{P}_n$ (one for each of the layers $n=0$ and 1) that satisfy

$$\nabla^4\mathcal{P}_n = 0 \tag{10.14}$$

(see § 4.2.2). Let the exponentially growing deformation of the interface be $\zeta = \zeta_0 f(x,y)e^{st}$, where s is the growth rate and f is a planform function satisfying $\nabla_h^2 f = -k^2 f$, where $k \equiv |\mathbf{k}|$ is the magnitude of the horizontal wavevector $\mathbf{k}$. Then the solutions for $\mathcal{P}_n$ must have the form

$$\mathcal{P}_n = P_n(z)f(x,y)e^{st}. \tag{10.15}$$

Substitution of (10.15) into (10.14), (10.1), (10.2), (10.11) and (10.12) yields the two-point BVP

$$(D^2 - k^2)^2 P_n = 0, \tag{10.16a}$$

$$P_1(-h) = D^2 P_1(-h) = P_2(\infty) = DP_2(\infty) = 0, \tag{10.16b}$$

$$[P] = [DP] = [\eta(D^2 + k^2)P] = [\eta D(D^2 - 3k^2)P] + g\Delta\rho\zeta_0 = 0, \tag{10.16c}$$

10.1 Rayleigh–Taylor Instability

where $D = d/dz$. The general solution of (10.16a) is

$$P_n = (A_n + B_n z)e^{-kz} + (C_n + D_n z)e^{kz}. \tag{10.17}$$

Substitution of (10.17) into (10.16b) and (10.16c) yields eight coupled linear equations, which can be solved for the constants $A_n - D_n$. The growth rate is then determined from the transformed kinematic condition $\zeta_0 s = k^2 P_0(0)$, yielding (3.20).

The RTI has been invoked as a model for a variety of mantle processes. The simplest versions of the model involve two Newtonian fluids. Motivated by the problem of the initiation of mantle plumes, Whitehead and Luther (1975) performed a LSA of a buoyant fluid layer bounded below by a free-slip surface and above by an infinite half-space of fluid with a different viscosity. Canright and Morris (1993) extended the model of Whitehead and Luther (1975) to the case of two layers of finite depth bounded by free-slip surfaces, using a LSA to determine the growth rate and performing a detailed scaling analysis (see § 3.4). Ribe (1998) used a weakly nonlinear analysis of the same configuration to study planform selection and the direction of superexponential growth (spouting). Conrad and Molnar (1997) performed a LSA of a dense layer between a rigid surface and a fluid half-space. LSAs of a system comprising crustal and lithospheric mantle layers over a passive asthenosphere without and with horizontal shortening have been performed by Neil and Houseman (1999) and Molnar and Houseman (2004, 2013, 2015) and applied to intraplate orogeny. Turning to a different geometry, Lister and Kerr (1989) studied the instability of a rising horizontal cylinder of buoyant fluid, motivated in part by suggestions that the RTI in this geometry might explain the characteristic spacing of island-arc volcanoes (Marsh and Carmichael, 1974) and of volcanic centers along mid-ocean ridges (Whitehead et al., 1984).

Important generalizations of the RTI are to more complicated rheologies including shear thinning (power-law), viscoelastic and anisotropic fluids. Motivated by the problem of subduction initiation, Canright and Morris (1993) studied the nonlinear RTI of a power-law layer bounded above by a free-slip surface and below by an effectively inviscid half-space. They found that the longtime evolution of the layer thickness h is

$$h \propto h_0 \left(\frac{t^* - t}{\tau_0} \right)^{-1/n}, \quad t^* \propto \tau_0 \equiv \left(\frac{B}{h_0 g \Delta \rho} \right)^n, \tag{10.18}$$

where n is the power-law index, B is the rheological stiffness (see Equation 4.1f) and the omitted factors of proportionality depend on the form of the initial perturbation. The finite-time singularity at $t = t^*$ corresponds to catastrophic detachment of the thickened portion of the layer (see also § 7.3 for the Newtonian case $n = 1$). Houseman and Molnar (1997) used a heuristic argument to obtain an analogous relation for a power-law layer bounded by a rigid surface, a boundary condition appropriate for the delamination of the lowermost lithosphere. Their expression $h \propto (t^* - t)^{1/(1-n)}$ differs from (10.18) because a rigid boundary condition implies a flow dominated by vertical shear, rather than by plug flow as in the problem of Canright and Morris (1993). Conrad and Molnar (1997) performed a LSA of a power-law layer adjoining a rigid boundary in the presence of a small background strain. Molnar et al. (1998) generalized the analysis of Houseman and Molnar (1997) to allow B to decrease exponentially with depth, and Harig et al. (2008) performed a LSA of a similar model with a free-slip upper surface. Kaus and Becker (2007) used a semi-analytical method to study the influence of elasticity on the RTI. Lev and Hager (2008) used the propagator matrix method to investigate how RTI of the lithosphere is influenced by transversely anisotropic viscosity.

A final important generalization of the RTI model is to systems whose rheology and density vary strongly with depth, which are useful models for the delamination of the lowermost lithosphere. Bassi and Bonnin (1988) proposed a semi-analytical method for the RTI of an extending lithosphere comprising three or four power-law layers within each of which the stiffness B is either constant or varies exponentially with depth. Conrad and Molnar (1997) used a similar method to analyze the RTI for situations with exponential variation of viscosity and/or linear variation of density. Conrad and Molnar (1999) generalized the analysis of Houseman and Molnar (1997) to the case of a lithosphere with arbitrary depth variations of B and of the temperature T. They predict that the amplitude ζ of the instability grows at a rate

$$\frac{1}{\zeta}\frac{d\zeta}{dt} \propto \left[\frac{\rho_m g \alpha (T_m - T_0)}{B_m}\right]^n h_0 F_n \zeta^{n-1}, \tag{10.19a}$$

$$F_n = \int_0^\infty \left[\frac{1 - \hat{T}(\hat{z})}{\hat{B}(\hat{T}(\hat{z}))}\right]^n d\hat{z}, \tag{10.19b}$$

$$\hat{z} = \frac{z + h_0}{h_0}, \quad \hat{T} = \frac{T - T_0}{T_m - T_0}, \quad \hat{B}(\hat{T}) = \frac{B(T)}{B_m}, \tag{10.19c}$$

where g is the gravitational acceleration, α is the thermal expansivity, T_0 is the (cold) temperature at the upper surface $\hat{z} = 0$, h_0 is the characteristic thickness of the boundary layer, T_m is the (hot) temperature below the boundary layer, $\rho_m = \rho(T_m)$ and $B_m = B(T_m)$. The quantity F_n is the depth integral of the nth power of the ratio of the dimensionless buoyancy ($= 1 - \hat{T}$) to the dimensionless stiffness $\hat{B}$. This definition reflects the fact that negatively buoyant material that is very stiff is prevented from participating in the instability, for which reason Conrad and Molnar (1999) called F_n the 'available buoyancy'.

To end this discussion of strongly depth-dependent structures, I note that Mondal and Korenaga (2018) have introduced an efficient propagator-matrix technique for the RTI of multilayered structures with strong vertical (Newtonian) viscosity variations.

In closing, it is perhaps of interest to note that the RTI problem solved at the beginning of this section includes as a special case a simple model for the glacial isostatic adjustment (GIA) of Earth's surface after removal of an ice load. In particular, we consider the level $z = 0$ in Figure 3.3 to be the undeformed position of Earth's surface and suppose that the fluid above it has zero viscosity and density ($\rho_1 = \eta_1 = 0$). The density difference across the interface $z = \zeta$ is then $\Delta\rho = -\rho_0 < 0$, so that the amplitude of the surface deformation decays rather than grows. The decay rate is given in the first row ($\gamma \equiv \eta_1/\eta_0 \ll \epsilon^3$) and the sixth column of Table 3.1 and is

$$s = -\frac{\rho_0 g}{2k\eta_0} \equiv -\frac{g}{2k\nu_0}, \tag{10.20}$$

where ν_0 is the kinematic viscosity of fluid 0. We now use this relation to estimate the viscosity of Earth's mantle, using data from Haskell (1935) for the uplift of Fennoscandia. From Haskell's Table I, the uplift rate between 5000 B.C. and 2000 B.C. is $-s \approx 7.2 \times 10^{-12}$ s^{-1}. From his figure 3, the half-wavelength of the surface deformation is $\pi/k \approx 2000$ km, whence $\nu_0 \equiv -g/2ks \approx 4.3 \times 10^{17}$ m^2 s^{-1}. This is to be compared with Haskell's own estimate $\nu_0 \approx 2.9 \times 10^{17}$ m^2 s^{-1} obtained assuming a more realistic axisymmetric geometry.

10.2 Rayleigh–Bénard Convection

RBC is the paradigmatic case of a pattern-forming instability. The classic Rayleigh–Bénard configuration (Figure 10.1) is a fluid layer confined between horizontal planes $z = \pm d/2$ held at fixed temperatures $\mp \Delta T/2$. The planes $z = \pm d/2$ may be either rigid or free-slip.

Most studies of RBC make use of the Boussinesq approximation, whereby one neglects temperature- and pressure-induced density variations everywhere in the governing equations except in the body-force term in the momentum equation. Accordingly, the equation of continuity reduces to the incompressibility condition $\nabla \cdot \mathbf{u} = 0$. In the body-force term, the density is assumed to depend linearly on temperature as $\rho = \rho_0(1 - \alpha T)$, where ρ_0 is the reference density at $T = 0$ and α is the coefficient of thermal expansion such that $\alpha \Delta T \ll 1$. For simplicity one typically assumes that the fluid's viscosity η and thermal diffusivity κ are constants, although this is not a requirement of the Boussinesq approximation *per se*

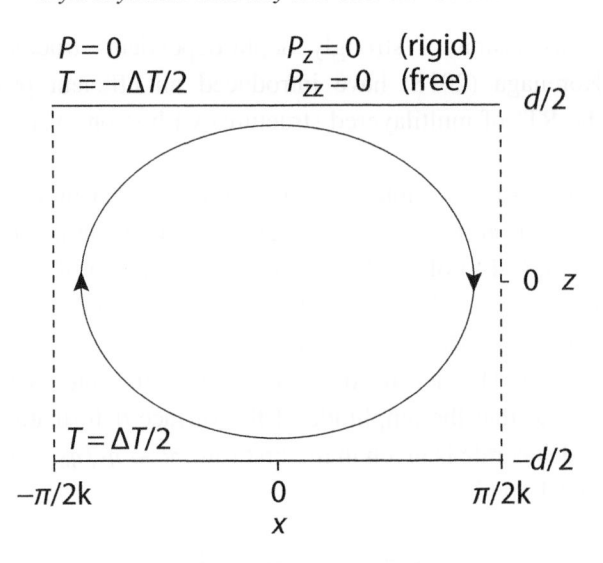

```
   P = 0              P_z = 0   (rigid)
   T = −ΔT/2          P_zz = 0  (free)
                                              d/2

                                              0  z


   T = ΔT/2
                                             −d/2
  −π/2k              0            π/2k
                     x
```

Figure 10.1 Geometry of Rayleigh–Bénard convection (§ 10.2). Fluid with constant kinematic viscosity ν, thermal diffusivity κ and thermal expansivity α is confined between horizontal planes $z = \pm d/2$ held at temperatures $\mp \Delta T/2$. The characteristic horizontal wavenumber of the convection pattern is k. For the special case of 2-D rolls, the roll axis is parallel to the y-direction. The boundary conditions on the poloidal scalar $\mathcal{P}$ for free-slip and rigid boundaries are indicated above the top boundary, and subscripts denote partial differentiation.

(see § 11.1). In the geodynamically relevant limit of negligible acceleration and inertia, the equations governing the convection are

$$\nabla \cdot \mathbf{u} = 0, \tag{10.21a}$$

$$0 = -\nabla p - \rho_0(1 - \alpha T)g\mathbf{e}_z + \eta \nabla^2 \mathbf{u}, \tag{10.21b}$$

$$\partial_t T + \mathbf{u} \cdot \nabla T = \kappa \nabla^2 T, \tag{10.21c}$$

where all symbols have their usual meanings.

The basic state is a motionless ($\mathbf{u} = \mathbf{0}$) layer with a linear (conductive) temperature profile $T = -z\Delta T/d$. Because the viscosity is constant, buoyancy forces generate a purely poloidal flow

$$u_i = \mathcal{L}_i \mathcal{P}, \quad \mathcal{L}_i = \delta_{i3}\nabla^2 - \partial_i\partial_3, \tag{10.22}$$

where $\mathcal{P}$ is the poloidal scalar defined by (4.20). The dimensionless equations satisfied by $\mathcal{P}$ and the perturbation θ of the conductive temperature profile are

$$\nabla^4 \mathcal{P} = -Ra\,\theta, \tag{10.23a}$$

$$\theta_t + \mathcal{L}_j \mathcal{P} \partial_j \theta - \nabla_h^2 \mathcal{P} = \nabla^2 \theta, \tag{10.23b}$$

10.2 Rayleigh–Bénard Convection

where

$$Ra = \frac{g\alpha d^3 \Delta T}{\nu \kappa} \tag{10.24}$$

is the Rayleigh number. In (10.23) and for the remainder of this section and § 10.3, all variables have been nondimensionalized using d, d^2/κ, κ/d and ΔT as scales for length, time, velocity and temperature, respectively, Moreover, the notations (x_1, x_2, x_3) and (x, y, z) are equivalent. The operator ∇_h^2 that appears in the primitive equation (4.22a) for $\mathcal{P}$ is omitted in (10.23a) because it corresponds to a simple multiplication by $-k^2$ when $\mathcal{P}$ is a periodic function of (x, y) with wavenumber k, as in the solutions to be discussed later. The third term in (10.23b) represents the advection of the conductive temperature gradient ($= -1$) by the vertical velocity $u_3 \equiv \nabla_h^2 \mathcal{P}$. The required boundary conditions on $z = \pm 1/2$ are $\theta = 0$ together with the conditions (4.23) on $\mathcal{P}$.

10.2.1 Linear Stability Analysis

The initial/boundary value problem describing the evolution of small perturbations to the basic state is obtained by linearizing (10.23) about $(\mathcal{P}, \theta) = (0, 0)$, reducing the resulting equations to a single equation for $\mathcal{P}$ by cross-differentiation and recasting the boundary conditions $\theta(\pm 1/2) = 0$ in terms of $\mathcal{P}$ with the help of (10.23a). For the (analytically simpler) case of free-slip boundaries, the result is

$$\left[\nabla^4(\nabla^2 - \partial_t) - Ra\nabla_h^2\right]\mathcal{P} = 0, \tag{10.25a}$$

$$\mathcal{P} = \partial_{33}^2 \mathcal{P} = \partial_{3333}^4 \mathcal{P} = 0 \text{ at } z = \pm 1/2. \tag{10.25b}$$

Equation (10.25) admits normal mode solutions of the form

$$\mathcal{P} = P(z) f(x, y) e^{st} \tag{10.26}$$

where s is the growth rate and $f(x, y)$ is the planform function satisfying $\nabla_h^2 f = -k^2 f$. Substituting (10.26) into (10.25), we obtain

$$\left[(D^2 - k^2)^2 (D^2 - k^2 - s) + Ra\, k^2\right] P = 0, \tag{10.27a}$$

$$P = D^2 P = D^4 P = 0 \text{ at } z = \pm 1/2, \tag{10.27b}$$

where $D = d/dz$. Equation (10.27) constitutes an eigenvalue problem that has a solution only for particular values of the growthrate s. The solution is

$$P = \sin n\pi \left(z + \frac{1}{2}\right), \quad s = \frac{Ra\, k^2}{(n^2\pi^2 + k^2)^2} - n^2\pi^2 - k^2, \tag{10.28a,b}$$

where the index n defines the vertical wavelength of the mode. The growth rate s becomes positive when Ra exceeds a value $Ra_0(k)$ that corresponds to marginal stability. This occurs first for the mode $n = 1$, for which

$$Ra_0 = \frac{(\pi^2 + k^2)^3}{k^2}. \tag{10.29}$$

The most unstable wavenumber k_c and the corresponding critical Rayleigh number Ra_c are found by minimizing $Ra_0(k)$, yielding

$$Ra_c = \frac{27\pi^4}{4} \approx 657.5, \quad k_c = \frac{\pi}{\sqrt{2}} \approx 2.22. \tag{10.30}$$

The corresponding results for convection between rigid surfaces must be obtained numerically (Chandrasekhar, 1981, pp. 36–42) and are $Ra_c \approx 1707.8$, $k_c \approx 3.117$. The marginally stable Rayleigh number $Ra_0(k)$ is shown for both free-slip and rigid surfaces in Figure 10.3a. For future reference, note that (10.28b) with $n = 1$ can be written as

$$s = (\pi^2 + k^2) \frac{Ra - Ra_0}{Ra_0}. \tag{10.31}$$

10.3 Order-Parameter Equations for Finite-Amplitude Rayleigh–Bénard Convection

Linear stability analysis predicts the initial growth rate of a normal mode with wavenumber **k**. Typically, however, many different modes have the same or nearly the same growth rate: examples include convection rolls with the same wavenumber but different orientations, and rolls with the same orientation but slightly different wavenumbers within a narrow band around the most unstable wavenumber k_c. The question therefore arises: among a set of modes with equal (or nearly equal) growth rates, which mode or combination of modes is actually realized for a given (supercritical) Rayleigh number? The answer depends on a combination of two factors: the nonlinear coupling between different modes due to the nonlinear terms in the governing equations, and external biases imposed by the initial and/or boundary conditions. Linear stability analysis is powerless to help us here, and more complicated nonlinear theories are required.

An especially powerful method of this type is to reduce the full 3-D equations governing convection to 2-D equations for one or more order parameters that describe the degree of order or patterning in the system. Newell et al. (1993) identify four distinct classes of such equations, depending on whether the degree of supercriticality $\epsilon \equiv [(Ra - Ra_c)/Ra_c]^{1/2}$ is small or of order unity, and whether the horizontal spectrum of the allowable modes is discrete or (quasi-) continuous

10.3 Order-Parameter Equations

Table 10.1. *Order-parameter equations for Rayleigh–Bénard convection*

Spectrum	$\epsilon \ll 1$	$\epsilon = O(1)$
Discrete	Amplitude equations	Finite-amplitude convection rolls
Continuous	Envelope equations	Slowly modulated patterns

(Table 10.1). The spectrum is discrete for convection in an infinite layer with a periodic planform characterized by a single fundamental wavenumber k, and also for a layer of finite extent $L = O(d)$ when only a few of the allowable wavenumbers $2n\pi/L$ are unstable for a given Rayleigh number. The wavenumbers of the unstable modes become ever more closely spaced as L increases, until in the limit $L/d \gg 1$ they can be regarded as forming a quasi-continuous spectrum.

Historically, the first case to be studied (Malkus and Veronis, 1958) was that of a single mode with $\epsilon \ll 1$. Here the order parameter is the amplitude $A(t)$ of the dominant mode, whose temporal evolution is described by a nonlinear Landau equation (Landau, 1944). Subsequently, these weakly nonlinear results were extended to $\epsilon = O(1)$ by Busse (1967a), who used a Galerkin method to obtain numerical solutions for convection rolls and to examine their stability. The more complicated case of a continuous spectrum was first studied by Newell and Whitehead (1969) and Segel (1969), who derived the evolution equation governing the slowly varying (in time and space) amplitude envelope $A(\epsilon x, \epsilon^{1/2} y, \epsilon^2 t)$ of weakly nonlinear ($\epsilon \ll 1$) convection with modes contained in a narrow band surrounding the critical wavevector $\mathbf{k}_c = (k_c, 0)$ for straight parallel rolls. Finally, Newell et al. (1990) extended these results to $\epsilon = O(1)$ by deriving the phase diffusion equation that governs the slowly varying phase Φ of a convection pattern that locally has the form of straight parallel rolls.

I now discuss each of these four cases in turn.

10.3.1 Amplitude Equation for Convection Rolls

Consider convection between free-slip boundaries in the form of straight rolls with axes parallel to $\mathbf{e}_y$. The equations and boundary conditions satisfied by $\mathcal{P}$ and θ are (10.23), where $\mathcal{P}$ and θ are independent of y and $\partial_2 \equiv 0$.

The basic idea of weakly nonlinear analysis is to expand the dependent variables in powers of a small parameter ϵ that measures the degree of supercriticality. By substituting these expansions into the governing equations and gathering together the terms proportional to different powers of ϵ, one reduces the original nonlinear problem to a sequence of linear (but inhomogeneous) problems that can be solved sequentially. Thus we write

$$P = \sum_{n=1}^{\infty} \epsilon^n P_n, \quad \theta = \sum_{n=1}^{\infty} \epsilon^n \theta_n, \quad Ra = Ra_0 + \sum_{n=1}^{\infty} \epsilon^n Ra_n, \tag{10.32}$$

where the last expansion can be regarded as an implicit definition of the supercriticality $\epsilon(Ra)$. Now because the boundary conditions are the same on both surfaces, the hot and cold portions of the flow are mirror images of each other. The problem is therefore invariant under the transformation $\{P \to -P, \theta \to -\theta\}$ or (equivalently) $\epsilon \to -\epsilon$. Application of the latter transformation to the expanded form of (10.23a) shows that $Ra_n = 0$ for all odd n. To lowest order, therefore,

$$Ra - Ra_0 \approx \epsilon^2 Ra_2, \tag{10.33}$$

where $Ra_2 = O(1)$ is to be determined. Now comparison of (10.33) with (10.31) shows that $s \propto \epsilon^2$. We therefore introduce a slow time $T \equiv \epsilon^2 t$ that is of order unity during the initial stage of exponential growth, whence

$$\partial_t = \epsilon^2 \partial_T. \tag{10.34}$$

Substituting (10.32), (10.33) and (10.34) into (10.23) and collecting terms proportional to ϵ, we obtain

$$\nabla^4 P_1 + Ra_0 \theta_1 = 0, \quad \nabla^2 \theta_1 + \partial_{11}^2 P_1 = 0, \tag{10.35}$$

subject to the boundary conditions $P_1 = \partial_{33}^2 P_1 = \theta_1 = 0$ at $z = \pm 1/2$. This is just the linear stability problem, for which the solution is

$$P_1 = -\frac{C}{k^2} \cos \pi z \cos kx, \quad \theta_1 = \frac{C}{\pi^2 + k^2} \cos \pi z \cos kx, \quad Ra_0 = \frac{(\pi^2 + k^2)^3}{k^2}, \tag{10.36}$$

where the amplitude $C(T)$ of the vertical velocity $w_1 \equiv \partial_{11}^2 P_1$ remains to be determined. Next, we collect terms proportional to ϵ^2 to obtain

$$\nabla^4 P_2 + Ra_0 \theta_2 = -Ra_1 \theta_1 \equiv 0, \tag{10.37a}$$

$$\nabla^2 \theta_2 + \partial_{11}^2 P_2 = \mathcal{L}_j P_1 \partial_j \theta_1 \equiv -\frac{C^2 \pi}{2(\pi^2 + k^2)} \sin 2\pi z. \tag{10.37b}$$

Now because the inhomogeneous term in (10.37b) depends only on z, we must seek solutions of (10.37) of the forms

$$P_2 = \tilde{P}_2(x, z), \quad \theta_2 = \tilde{\theta}_2(x, z) + \overline{\theta}_2(z), \tag{10.38}$$

where $\tilde{P}_2$ and $\tilde{\theta}_2$ are the fluctuating (periodic in x) parts of the solution, and $\overline{\theta}_2$ is the mean (x-independent) part. The mean part of P_2 is set to zero because the poloidal scalar is arbitrary to within an additive function of z. Now the equations satisfied by $\tilde{P}_2$ and $\tilde{\theta}_2$ are just the homogeneous forms of (10.37), which are identical to

10.3 Order-Parameter Equations

the order ϵ equations (10.35). We may therefore set $\tilde{\mathcal{P}}_2 = \tilde{\theta}_2 = 0$ with no loss of generality. The solution for $\overline{\theta}_2$ is then obtained by integrating (10.37b) subject to the boundary conditions $\overline{\theta}_2(\pm 1/2) = 0$. We thereby find

$$\mathcal{P}_2 = 0, \quad \theta_2 \equiv \overline{\theta}_2 = \frac{C^2}{8\pi(\pi^2 + k^2)} \sin 2\pi z. \tag{10.39}$$

Physically, $\overline{\theta}_2(z)$ describes the average heating (cooling) of the upper (lower) half of the layer that is induced by the convection. Despite appearances, (10.39) is consistent with (10.37a) because $\mathcal{P}_2$ is arbitrary to within an additive function of z.

The parameter Ra_2 is still not determined, so we must proceed to order ϵ^3, for which the equations are

$$\nabla^4 \mathcal{P}_3 + Ra_0 \theta_3 = -Ra_2 \theta_1, \tag{10.40a}$$

$$\nabla^2 \theta_3 + \partial^2_{11} \mathcal{P}_3 = \partial_T \theta_1 + \mathcal{L}_j \mathcal{P}_1 \partial_j \theta_2. \tag{10.40b}$$

We now evaluate the right-hand sides (RHSs) of (10.40), set $\mathcal{P}_3 = \tilde{\mathcal{P}}_3(z, T) \cos kx$ and $\theta_3 = \tilde{\theta}_3(z, T) \cos kx$ and reduce the resulting equations to a single equation for $\tilde{\mathcal{P}}_3$, obtaining

$$\left[(D^2 - k^2)^3 + k^2 Ra_0 \right] \tilde{\mathcal{P}}_3 = -\left[\frac{Ra_0 (C^3 \cos 2\pi z + 4\dot{C})}{4(\pi^2 + k^2)} - Ra_2 C \right] \cos \pi z, \tag{10.41}$$

where $D = d/dz$ and a superposed dot denotes d/dT. Now the homogeneous form of (10.41) is identical to the eigenvalue problem at order ϵ and thus has a solution $\propto \cos \pi z$ that satisfies the boundary conditions. Adding an inhomogeneous term that is not orthogonal to $\cos \pi z$ will thus generate unbounded resonant solutions. The only way to avoid this is to require that the inhomogeneous term be orthogonal to all solutions of the homogeneous adjoint problem. This is known as the solvability condition. In the general case, this condition is found by multiplying the inhomogeneous equation by the solution of the homogeneous adjoint problem and integrating over the domain of the independent variable. For our (self-adjoint) problem, the homogeneous adjoint solution is $\propto \cos \pi z$, and the procedure just described yields

$$\int_{-1/2}^{1/2} \cos \pi z \left[(D^2 - k^2)^3 + k^2 Ra_0 \right] \tilde{\mathcal{P}}_3 \, dz = -\frac{[Ra_0(C^3 + 8\dot{C}) - 8(\pi^2 + k^2)Ra_2 C]}{16(\pi^2 + k^2)}. \tag{10.42}$$

Now the LHS of (10.42) is identically zero, as one can easily show by integrating repeatedly by parts and applying the boundary conditions on $\tilde{\mathcal{P}}_3$. The RHS of (10.42) must therefore vanish. Now by (10.32) and (10.36), the amplitude of the vertical velocity is $\epsilon C \equiv W$. Because only the product of ϵ and C is physically meaningful, the definition of ϵ itself – or, equivalently, of Ra_2 – is arbitrary. Making the most convenient choice $Ra_2 = Ra_0$, we find that the vanishing of the RHS of (10.42) yields the following evolution equation for the physical amplitude W:

$$\frac{dW}{dT} = (\pi^2 + k^2)W - \frac{Ra_0}{8(Ra - Ra_0)}W^3. \tag{10.43}$$

Initially, W grows exponentially with a growth rate $W^{-1}\dot{W} = \pi^2 + k^2$. At long times, however, $\dot{W} \to 0$ and the amplitude approaches a steady value

$$W(T \to \infty) = \left[\frac{8(\pi^2 + k^2)(Ra - Ra_0)}{Ra_0}\right]^{1/2}. \tag{10.44}$$

10.3.2 Finite-Amplitude Convection Rolls and Their Stability

The extension of the preceding results to strongly nonlinear ($\epsilon = O(1)$) rolls between rigid surfaces is due to Busse (1967a), whose development is followed here with some changes of notation. The first step is to determine steady roll solutions of (10.23) using a Galerkin method (Fletcher, 1984) whereby the temperature perturbation θ and the poloidal scalar $\mathcal{P}$ are expanded in orthogonal functions that satisfy the boundary conditions, and the coefficients are then chosen to satisfy the governing equations approximately.

One begins by expanding θ into a complete set of Fourier modes that satisfy the boundary conditions $\theta(x, \pm 1/2) = 0$:

$$\theta = \sum_{m=-\infty}^{\infty} \sum_{n=1}^{\infty} c_{mn} e^{imkx} f_n(z), \quad f_n(z) = \sin n\pi \left(z + \frac{1}{2}\right), \tag{10.45}$$

where $2\pi/k$ is the wavelength of the convection rolls and $c_{mn} = \overline{c_{-mn}}$, where the overbar denotes complex conjugation. Now substitute (10.45) into (10.23a) and solve the resulting equation to obtain

$$\mathcal{P} = -Ra \sum_{m,n} c_{mn} e^{imkx} Q_n(mk, z), \tag{10.46a}$$

$$Q_n(r, z) = \frac{f_n(z) + n\pi h_n(r)}{(r^2 + n^2\pi^2)^2}, \tag{10.46b}$$

$$h_n(r) = \begin{cases} (2SC + r)^{-1}(2zC \sinh rz - S \cosh rz) & \text{(n odd)} \\ (2SC - r)^{-1}(C \sinh rz - 2zS \cosh rz) & \text{(n even)} \end{cases} \tag{10.46c}$$

$$S = \sinh \frac{r}{2}, \quad C = \cosh \frac{r}{2}. \tag{10.46d}$$

10.3 Order-Parameter Equations

To determine the coefficients c_{mn}, we substitute (10.46) into (10.23b), multiply the result by $f_q(z)e^{-ipkx}$ and integrate over the fluid layer. By using for p and q all integers in the ranges of m and n, respectively, we obtain an infinite set of nonlinear algebraic equations for $c_{mn}(Ra)$ that can be truncated and solved numerically (Busse, 1967a).

Another advantage of the Galerkin method is that a stability analysis of the solution can easily be performed. The linearized equations governing infinitesimal perturbations $(\tilde{\mathcal{P}}, \tilde{\theta})$ to the steady roll solution $(\mathcal{P}_0, \theta_0)$ are

$$\nabla^4 \tilde{\mathcal{P}} = -Ra\tilde{\theta}, \tag{10.47a}$$

$$\tilde{\theta}_t + \mathcal{L}_j \mathcal{P}_0 \partial_j \tilde{\theta} + \partial_j \theta_0 \mathcal{L}_j \tilde{\mathcal{P}} - \nabla_h^2 \tilde{\mathcal{P}} = \nabla^2 \tilde{\theta}, \tag{10.47b}$$

subject to $\tilde{\theta} = \tilde{\mathcal{P}} = \partial_3 \tilde{\mathcal{P}} = 0$ at $z = \pm 1/2$. Now (10.47) are linear PDEs with coefficients that are periodic in the x-direction. The general solution can be written as the product of an exponential function of t, periodic functions of x and y and a function of x and z having the same periodicity as the stationary solution (Busse, 1967a). We thus write

$$\tilde{\theta} = \left\{ \sum_{m,n} \tilde{c}_{mn} e^{imkx} f_n(z) \right\} e^{i(ax+by)+st} \tag{10.48}$$

and note further that (10.47a) and the boundary conditions on $\tilde{\mathcal{P}}$ are satisfied exactly if

$$\tilde{\mathcal{P}} = -Ra \left\{ \sum_{m,n} \tilde{c}_{mn} e^{imkx} Q_n(\sqrt{(mk+a)^2 + b^2}, z) \right\} e^{i(ax+by)+st}. \tag{10.49}$$

Substituting (10.48) and (10.49) into (10.47b), multiplying by $f_q(z)e^{-i(pkx+ax+by)-st}$ and averaging over the fluid layer, we obtain an infinite system of linear equations for the coefficients $\tilde{c}_{mn}$. The system is then truncated, and the largest eigenvalue $s(Ra, k, a, b)$ is determined numerically, assuming that s is real in view of the fact that $\Im(s) = 0$ at the onset of convection (Busse, 1967a).

The results show that convection rolls are stable ($s < 0$) in an elongate region of the (Ra, k) space often called the Busse balloon (Figure 10.2). Above $Ra \approx 22600$, convection rolls of any wavenumber are unstable. The lower-left edge of the balloon (denoted by Z in Figure 10.2) represents the onset of a zigzag instability in the form of rolls oblique to the original ones. Around the rest of the balloon (denoted by C), the stability of rolls is limited by the cross-roll instability, which grows in the form of perpendicularly aligned rolls. The results of Busse (1967a) were extended to the case of free-slip boundaries by Straus (1972), Busse (1984) and Bolton and Busse (1985).

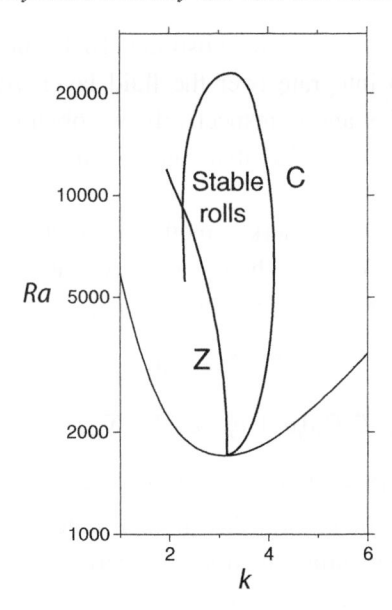

Figure 10.2 Region of stability in the Rayleigh number/wavenumber plane of 2-D convection rolls between rigid isothermal surfaces (Busse balloon). The portions of the boundary of the balloon labelled Z and C correspond to the onset of the zigzag and cross-roll instabilities, respectively. See § 10.3.2 for discussion. Figure redrawn from Figure 5 of Busse (1967a) by permission of John Wiley and Sons.

10.3.3 Envelope Equation for Modulated Convection Rolls

The analyses in § 10.3.1 and § 10.3.2 assume convection in the form of straight parallel rolls with a single dominant wavenumber k. In the laboratory, however, rolls often exhibit a more irregular pattern in which both the magnitude and direction of the wavevector $\mathbf{k} = (k_1, k_2)$ vary slowly as functions of time and position. To describe this behavior, Newell and Whitehead (1969) and Segel (1969) derived an envelope equation for modulated convection rolls whose wavevectors form a continuous spectrum within a narrow band centered on the wavevector $(k_c, 0)$ for straight parallel rolls. The envelope equation is derived via a multiscale expansion that accounts in a self-consistent way for both the fast and slow variations of the flow field in time and space. Richter (1973b) applied a similar method to convection modulated by long-wavelength variations in the boundary temperatures.

The first step is to determine the scales over which the slow (i.e., long-wavelength) spatial variations occur. Because the marginal stability curve $Ra_0(k)$ is a parabola in the vicinity of its minimum $(k, Ra_0) = (k_c, Ra_c)$, the wavenumbers $|\mathbf{k}|$ that become unstable when Ra exceeds Ra_c by an amount $\epsilon^2 Ra_c$ comprise a continuous band of width $\sim \epsilon$ centered on k_c (Figure 10.3a). However, the orientation of the rolls, measured by the ratio k_2/k_1, may also vary. Now the

10.3 Order-Parameter Equations

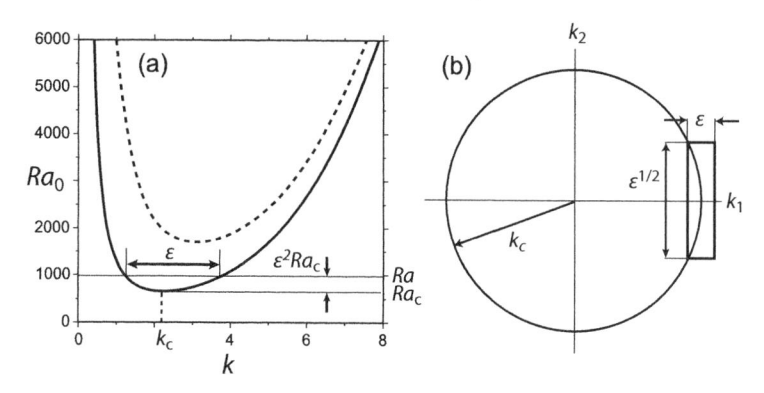

Figure 10.3 (a) Marginally stable Rayleigh number Ra_0 as a function of wavenumber k for RBC between free-slip (solid line) and rigid (dashed line) surfaces. The critical Rayleigh number Ra_c and wavenumber k_c are indicated for the free-slip case. For a slightly supercritical Rayleigh number $Ra = (1 + \epsilon^2)Ra_c$, a band of wavenumbers of width $\sim \epsilon$ is unstable. See § 10.2.1 for discussion. (b) The most unstable wavevector $\mathbf{k} = (k_1, k_2)$ for RBC lies on a circle of radius k_c. Changing $|\mathbf{k}|$ by an amount $\sim \epsilon$ in the vicinity of the wavevector $(k_c, 0)$ for rolls with axes parallel to the x_2-direction corresponds to changing k_1 and k_2 by amounts $\sim \epsilon$ and $\sim \epsilon^{1/2}$, respectively (rectangle). See § 10.3.3 for discussion. Part (b) redrawn from Figure 2 of Newell and Whitehead (1969) by permission of Cambridge University Press.

most unstable wavevector lies on a circle of radius k_c (Figure 10.3b). Therefore if we change $|\mathbf{k}|$ by an amount $\sim \epsilon$ in the vicinity of the straight-roll wavevector $(k_c, 0)$, the (maximum) corresponding changes of k_1 and k_2 are $\sim \epsilon$ and $\sim \epsilon^{1/2}$, respectively. The appropriate slow variables are therefore

$$X = \epsilon x, \quad Y = \epsilon^{1/2} y, \quad T = \epsilon^2 t. \tag{10.50}$$

where the expression for T derives from the argument preceding (10.34).

The essence of the multiscale procedure is to treat the flow fields (here, $\mathcal{P}$ and θ) as functions of both the fast variables (x, z) and the slow variables (X, Y, T). Accordingly, the asympotic expansions analogous to (10.32) are

$$\mathcal{P} = \sum_{n=1}^{\infty} \epsilon^n \mathcal{P}_n(X, Y, T, x, z) \quad \theta = \sum_{n=1}^{\infty} \epsilon^n \theta_n(X, Y, T, x, z). \tag{10.51}$$

Because the solution for steady straight rolls is independent of y and t, $\mathcal{P}_n$ and θ_n depend on these variables only through the variables Y and T that measure the slow modulation of the pattern. By the chain rule, derivatives of the expansions (10.51) with respect to the fast variables transform as

$$\partial_t \to \epsilon^2 \partial_T, \quad \partial_x \to \partial_x + \epsilon \partial_X, \quad \partial_y \to \epsilon^{1/2} \partial_Y, \quad \partial_z \to \partial_z. \tag{10.52}$$

We now substitute (10.51) into the governing equations (10.23) and collect terms proportional to like powers of ϵ, just as in § 10.3.1. The solutions at order ϵ analogous to (10.36) are

$$\mathcal{P}_1 = -\frac{1}{2k_c^2}\left[Ce^{ik_c x} + \overline{C}e^{-ik_c x}\right]\cos\pi z, \tag{10.53a}$$

$$\theta_1 = \frac{1}{2(\pi^2 + k_c^2)}\left[Ce^{ik_c x} + \overline{C}e^{-ik_c x}\right]\cos\pi z, \tag{10.53b}$$

where $C = C(X, Y, T)$ is the slowly varying envelope of the roll solution and overbars denotes complex conjugation. The solution at order ϵ^2 analogous to (10.39) is

$$\mathcal{P}_2 = 0, \quad \theta_2 = \frac{C\overline{C}}{8\pi(\pi^2 + k_c^2)}\sin 2\pi z. \tag{10.54}$$

The temperature θ_2 also contains free modes proportional to $\partial_X(C, \overline{C})$ and $\partial_{YY}(C, \overline{C})$, but these do not change the solvability condition at order ϵ^3 and can therefore be neglected. Evaluating this solvability condition with $k_c = \pi/\sqrt{2}$ (Exercise 10.3), we obtain the Newell–Whitehead–Segel equation for the envelope $W \equiv \epsilon C$ of the vertical velocity field:

$$\frac{\partial W}{\partial T} - \frac{4}{3}\left(\frac{\partial}{\partial X} - \frac{i}{\sqrt{2\pi}}\frac{\partial^2}{\partial Y^2}\right)^2 W = \frac{3\pi^2}{2}W - \frac{Ra_c}{8(Ra - Ra_c)}|W|^2 W. \tag{10.55}$$

Equation (10.55) differs from (10.43) by the addition of a diffusion-like term on the LHS, which represents the interaction of neighboring rolls via the buoyant torques they apply to each other. Equations generalizing (10.55) to N interacting wavepackets are given by Newell and Whitehead (1969).

10.3.4 Phase Diffusion Equation for Thermal Convection

The final case is that of large-amplitude ($\epsilon = O(1)$) convection in the form of modulated rolls, which can be described by a phase diffusion equation. The following derivation is based on Newell et al. (1990), henceforth NPS90.

The basic assumption is that the flow field consists locally of straight parallel rolls whose wavevector varies slowly over the fluid layer. This wavevector can be written as $\mathbf{k} \equiv \nabla\phi$, where ϕ is the phase of the roll pattern. The phase is the crucial dynamic variable far from onset because the amplitude is an algebraic function of the wavenumber and the Rayleigh number and no longer an independent parameter. The amplitude is therefore slaved to the phase gradients, and the phase diffusion equation alone suffices to describe the dynamics.

The problem involves two very different length scales, the layer depth d and the tank width $L \gg d$, and may therefore be solved using a multiscale expansion. Because the convection is strongly nonlinear, however, the relevant small parameter is no longer the degree of supercriticality (as in § 10.3.3), but rather the inverse aspect ratio $d/L \equiv \Gamma^{-1}$. Now the lateral scale of the variation of $\mathbf{k}$ is the tank width $L \equiv \Gamma d$, and its characteristic timescale is the lateral thermal diffusion time $L^2/\kappa \equiv \Gamma^2 d^2/\kappa$. Accordingly, the slow variables that characterize the modulation of the basic roll pattern are

$$X = \Gamma^{-1}x, \quad Y = \Gamma^{-1}y, \quad T = \Gamma^{-2}t. \tag{10.56}$$

The starting point of the analysis is the Galerkin representation (10.45)–(10.46) for large-amplitude steady rolls with axes parallel to the y-direction and wavevector $\mathbf{k}_0 = (k_0, 0)$. Let A be some measure of the amplitude of this solution, which can be calculated via the Galerkin procedure as a function of Ra and k_0. Next, we seek modulated roll solutions for the dependent variables $\mathbf{v} = (\mathbf{u}, \theta, p)$, where $\mathbf{u}$ is the velocity, θ is the temperature relative to the linear conductive profile and p is the pressure. The solutions are assumed to have the form

$$\mathbf{v} = \mathbf{F}(\phi = \Gamma \Phi(X, Y, T), z; A(X, Y, T)) \tag{10.57}$$

where $\mathbf{F}$ is an unknown function of the arguments indicated. In (10.57), Φ is a slow phase variable defined such that

$$\mathbf{k} = \nabla_x \phi = \nabla_X \Phi, \tag{10.58}$$

where ∇_x and ∇_X are the horizontal gradient operators with respect to the variables (x, y) and (X, Y), respectively. Note that the wavevector $\mathbf{k}$ defined by (10.58) is not required to be a small perturbation of the parallel-roll wavevector $\mathbf{k}_0$. Because derivatives act on functions of ϕ, z, X, Y and T, the chain rule implies the transformations

$$\partial_z \to \partial_z, \quad \nabla_x \to \mathbf{k}\partial_\phi + \Gamma^{-1}\nabla_X, \quad \partial_t \to \Gamma^{-1}(\partial_T \Phi)\partial_\phi + \Gamma^{-2}\partial_T,$$

$$\nabla^2 \to k^2 \partial_{\phi\phi}^2 + \partial_{zz}^2 + \Gamma^{-1} D \partial_\phi + \Gamma^{-2} \nabla_X^2, \tag{10.59}$$

where $k = |\mathbf{k}|$ and $D = 2\mathbf{k} \cdot \nabla_X + \nabla_X \cdot \mathbf{k}$.

Because of the slow modulation, (10.57) is no longer an exact solution of the Boussinesq equations. We therefore seek solutions in the form of an asympotic expansion in powers of Γ^{-1},

$$\mathbf{v} = \mathbf{v}_0 + \Gamma^{-1} \mathbf{v}_1 + \Gamma^{-2} \mathbf{v}_2 + \cdots \tag{10.60}$$

where $\mathbf{v}_0$ is the Galerkin solution for steady parallel rolls. Substituting (10.60) into the Boussinesq equations, using (10.59) and collecting terms proportional to Γ^{-1}, we obtain a BVP for $\mathbf{v}_1$, which has a solution only if a solvability condition is satisfied. That condition yields the phase diffusion equation

$$\frac{\partial \Phi}{\partial T} + \frac{1}{\tau(k)} \nabla_X \cdot [B(k)\mathbf{k}] = 0 \tag{10.61}$$

where the functions $\tau(k)$ and $B(k)$ are determined numerically. The diffusive character of (10.61) becomes clear when one recalls that $\mathbf{k} = \nabla_X \Phi$. In (10.61), an additional mean-drift advection term that is nonzero only for finite values of the Prandtl number has been neglected (see the first of equations (1.1) in NPS90). NPS90 show that a LSA of the straight parallel roll solution using (10.61) reproduces closely the portions of the boundary of the Busse balloon (§ 10.3.2) that correspond to long-wave instabilities.

10.4 Convection at High Rayleigh Number

Even though convection in the form of steady 2-D rolls is unstable at high Rayleigh numbers (§ 10.3.2), much can be learned by studying the structure of such rolls in the limit $Ra \to \infty$. Because temperature variations are now confined to thin thermal boundary layer (TBLs) around the edges of the cells, scaling analysis and BL theory are particularly effective tools for this purpose.

10.4.1 Scaling Analysis

The essential scalings for cellular convection at high Rayleigh number can be determined by a scaling analysis of the governing equations. The following derivation generalizes the analysis of McKenzie et al. (1974) to include both free-slip and rigid surface boundary conditions.

The scaling analysis envisions convection in the form of two-dimensional rolls, for which the upwelling and downwelling plumes are infinitely long sheets. We assume (in this subsection only) that the aspect ratio of the rolls does not differ much from unity. The geometry considered is shown in Figure 10.4. The analysis proceeds by determining six equations relating six unknown quantities: the thickness δ_p of the thermal plumes, the thickness δ_h of the horizontal TBLs, the maximum vertical velocity v_p and the vorticity ω in the plumes, the horizontal velocity u_h at the edge of the horizontal BLs and the heat flux q across the layer (per unit length along the roll axes). The heat flux carried by an upwelling plume is

$$q \sim \rho c_p v_p \Delta T \delta_p, \tag{10.62}$$

where c_p is the specific heat at constant pressure. Because the convection is steady, the flux (10.62) must equal that lost by conduction through the top horizontal BL, implying

10.4 Convection at High Rayleigh Number

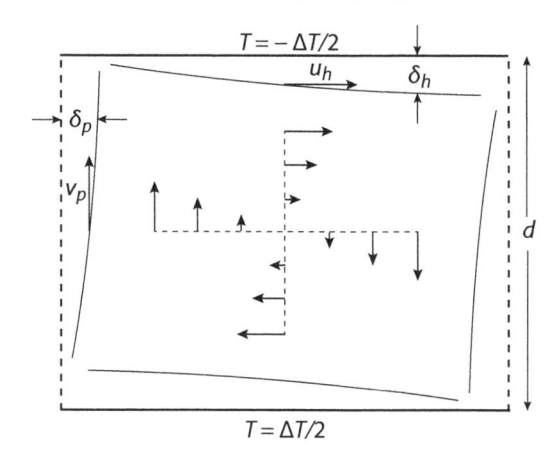

Figure 10.4 Geometry of a convection roll as considered in the scaling analysis of § 10.4.1. The upper (+) and lower (−) surfaces are held at temperatures $\mp\Delta T/2$ and may be either rigid or free-slip. The typical thicknesses of the plumes and the horizontal TBLs are δ_p and δ_h, respectively. The typical vertical and horizontal velocities at the edges of the plumes and the horizontal TBLs are v_p and u_h, respectively.

$$q \sim k_c d \Delta T / \delta_h, \tag{10.63}$$

where k_c is the thermal conductivity. The thickness δ_h of the horizontal BLs is controlled by the balance $uT_x \sim \kappa T_{yy}$ of advection and diffusion, which implies

$$u_h \Delta T / d \sim \kappa \Delta T / \delta_h^2. \tag{10.64}$$

Now u_h depends on whether the horizontal surfaces are free-slip or rigid. As in our earlier scaling analysis of heat transfer from a moving sphere (§ 2.3), the velocity parallel to the boundary is constant to lowest order across a TBL at a free-slip surface ($n = 0$ say) and varies linearly across a TBL at a rigid surface ($n = 1$). Therefore

$$u_h \sim v_p (\delta_h/d)^n. \tag{10.65}$$

The force balance in the plumes is scaled using the vorticity equation $\nu \nabla^2 \omega = g\alpha T_x$, which implies

$$\omega \sim g\alpha \Delta T \delta_p / \nu. \tag{10.66}$$

Finally, the vorticity ω in the plumes is of the same order as the rotation rate of the isothermal core, or

$$\omega \sim v_p/d. \tag{10.67}$$

Table 10.2. *Scaling laws for vigorous Rayleigh–Bénard convection*

Boundaries	$v_p d/\kappa$	δ_p/d	$Nu \equiv d/\delta_h$
Free-slip	$Ra^{2/3}$	$Ra^{-1/3}$	$Ra^{1/3}$
Rigid	$Ra^{3/5}$	$Ra^{-2/5}$	$Ra^{1/5}$

The simultaneous solution of (10.62)–(10.67) yields scaling laws for each of the six unknown quantities as a function of Ra. Table 10.2 shows the results for v_p, δ_p and the Nusselt number $Nu \equiv d/\delta_h$. The free-slip surface heat transfer law $Nu \sim Ra^{1/3}$ corresponds to a dimensional heat flux $q \equiv k_c \Delta T Nu/d$ that is independent of the layer depth d, whereas $q \propto d^{-2/5}$ for rigid surfaces. A revealing check of the results is to note that $v_p d/\kappa$ (column 2 of Table 10.2) is just the effective Péclet number Pe for the flow. The scaling laws for Nu (column 4) then imply $Nu \sim Pe^{1/(n+2)}$, where $n = 0$ for free-slip boundaries and 1 for rigid boundaries. This agrees with our previous expression (2.26) for the heat transfer from a hot sphere moving in a viscous fluid.

While the scaling laws just determined are illuminating, it must be emphasized again that convection in the form of steady rolls is not stable at high Rayleigh numbers. As discussed further in § 10.4.5, high-Rayleigh number convection is characterized by the quasi-periodic generation of plumes in the upper and lower BLs followed by local drainage of those layers. The net result of this process for both rigid and free-slip boundaries is a heat transfer $Nu \propto Ra^{1/3}$, i.e., a dimensional heat flux that is independent of the layer depth. The law $Nu \propto Ra^{1/5}$ that we derived for rigid boundaries is therefore unlikely to be observed in practice. The validity of the law $Nu \propto Ra^{1/3}$ for rigid boundaries has been further confirmed by boundary-layer analysis coupled with direct numerical simulation (Shishkina et al., 2017).

An alternative approach to the heat flux in thermally turbulent RBC at infinite Prandtl number is to use a variational method to determine a rigorous upper bound on $Nu(Ra)$. Different methods give different results, but all agree that the upper bound is $\sim Ra^{1/3}$ times a logarithmic correction (e.g., $Nu \lesssim Ra^{1/3}(\ln \ln Ra)^{1/3}$; Otto and Seis, 2011). The details of these approaches are highly technical and beyond our scope here; a good survey of the recent literature is given in Nobili and Otto (2017).

10.4.2 Flow in the Isothermal Core

Given the fundamental scales of Table 10.2, we can now carry out a more detailed analysis based on BL theory. This approach, pioneered by Turcotte (1967) and Turcotte and Oxburgh (1967) and extended by Roberts (1979), Olson and Corcos

10.4 Convection at High Rayleigh Number

(1980), and Jimenez and Zufiria (1987) (henceforth JZ87), is applicable to convection in a layer bounded by either free-slip or rigid surfaces. Here we consider only the former case, following JZ87 with some changes of notation. More recent asymptotic analyses of high Rayleigh-number RBC with both free-slip and rigid horizontal boundaries (Fowler, 2011; Vynnycky and Masuda, 2013) find excellent agreement with JZ87's results for the free-slip case.

The dimensional equations governing the flow are

$$\nu \nabla^4 \psi = -g\alpha T_x, \tag{10.68a}$$

$$u\theta_x + v\theta_y = \kappa \nabla^2 T, \tag{10.68b}$$

where T is the temperature, $\mathbf{u} = u\mathbf{e}_x + v\mathbf{e}_y \equiv -\psi_y\mathbf{e}_x + \psi_x\mathbf{e}_y$ is the velocity and ψ is the streamfunction. In the limit $Ra \to \infty$, each cell comprises a nearly isothermal core surrounded by thin vertical plumes and horizontal TBLs. Because temperature gradients are negligible in the core, the principal agency driving the flow within it is the shear stresses applied to its vertical boundaries by the plumes. Our first task is therefore to derive a boundary condition on the core flow that represents this agency.

Consider for definiteness the upwelling plume at the left boundary $x = -\beta d/2 \equiv x_-$ of the cell (Figure 4.4), and let its halfwidth be δ_p. The (upward) buoyancy force acting on the right half of this plume must be balanced by a downward shear stress at its edge $x = x_- + \delta_p$ (the shear stress on $x = x_-$ is zero by symmetry). This requires

$$\nu v_x|_{x=x_-+\delta_p} = -g\alpha \int_{x_-}^{x_-+\delta_p} (T - T_c)\mathrm{d}x, \tag{10.69}$$

where T_c ($\equiv 0$) is the temperature in the core. Multiplying (10.69) by v_p, we obtain

$$v_p v_x|_{x=x_-+\delta_p} = -\frac{\alpha g}{\nu \rho c_p}\left[\rho c_p \int_{x_-}^{x_-+\delta_p} v_p(T - T_c)\mathrm{d}x\right], \tag{10.70}$$

where we have used the fact that $v_p \equiv \psi_x$ is constant to first order across the plumes to take it inside the integral on the RHS. Now the quantity [...] in (10.69) is just the heat flux carried by the right half of the plume, which is $k_c \Delta T Nu/2$ where Nu is the Nusselt number. Moreover, because the plume is thin, $v_x \equiv \psi_{xx}$ can be evaluated at $x = x_-$ rather than at $x = x_- + \delta_p$. Equation (10.70) then becomes

$$\psi_x \psi_{xx}|_{x=x_-} = -\frac{1}{2}\frac{g\kappa\alpha\Delta T}{\nu}Nu, \tag{10.71}$$

which is a nonlinear and inhomogeneous boundary condition on the core flow. The corresponding condition at $x = +\beta d/2$ is obtained from (10.71) by reversing the sign of the RHS.

We now invoke the results of the scaling analysis (Table 10.2) to nondimensionalize the equations and boundary conditions for the core flow. We first write the scaling law for Nu as

$$Nu = C(\beta)Ra^{1/3} \tag{10.72}$$

where $C(\beta)$ (to be determined) measures the dependence of the heat transfer on the aspect ratio. Rewriting the equations and boundary conditions in terms of the dimensionless variables $(x', z') = (x, z)/d$ and $\psi' = \psi/\kappa\sqrt{C(\beta)}Ra^{2/3}$ and then dropping the primes, we obtain

$$\nabla^4 \psi = 0, \tag{10.73a}$$

$$\psi(x, \pm 1/2) = \psi_{yy}(x, \pm 1/2) = 0, \tag{10.73b}$$

$$\psi(\pm\beta/2, y) = \psi_x(\pm\beta/2, y)\psi_{xx}(\pm\beta/2, y) \mp 1/2 = 0. \tag{10.73c}$$

The problem (10.73) can be solved either numerically (JZ87) or using a superposition method (§ 4.4.1).

10.4.3 Thermal Boundary Layers and Heat Transfer

The next step is to determine the temperature in the plumes and the horizontal TBLs using BL theory. The temperature in all these layers is governed by the transformed BL equation (6.7). Turcotte and Oxburgh (1967) solved this equation assuming self-similarity, obtaining solutions of the form (6.9). However, Roberts (1979) pointed out that this is not correct because the fluid travelling around the margins of the cell sees a periodic boundary condition, which is alternatingly isothermal (along the horizontal boundaries) and insulating (in the plumes). Now because (6.7) is parabolic, its solution can be written in convolution-integral form in terms of an arbitrary boundary temperature $T_b(\tau)$ and an initial temperature profile $T(\psi, 0)$ at $\tau = 0$, where $\tau = \int U(s)ds$ is the time-like variable introduced in § 6.2.1 and the velocity $U(s)$ parallel to the boundary is determined (to within the unknown scale factor $C(\beta)$) from the core flow streamfunction ψ that satisfies (10.73). However, in the limit $\tau \to \infty$ corresponding to an infinite number of transits around the cell, the convolution integral describing the evolution of the initial profile vanishes, and the solution is (JZ87)

$$T(\psi, \tau) = -\frac{\psi}{2(\pi\kappa)^{1/2}} \int_0^\infty \frac{T_b(\tau - t)}{t^{3/2}} \exp\left(-\frac{\psi^2}{4\kappa t}\right) dt, \tag{10.74}$$

where all variables are dimensional. Now T_b is known only on the top and bottom surfaces. Along the (insulating) plume centerlines, it can be found by setting to zero the temperature gradient $\partial_\psi T|_{\psi=0}$ calculated from (10.74) and solving numerically the resulting integral equation. The temperature everywhere in the plumes and horizontal BLs can then be determined from (10.74). The final step is to determine the unknown scale factor $C(\beta)$ by matching the heat flow advected vertically by the plumes to the conductive heat flux across the upper boundary (JZ87, eqn. (23)).

The solution of JZ87 shows that the scale factor $C(\beta)$ (and hence the Nusselt number $Nu = CRa^{1/3}$) is maximum for $\beta \approx 0.75$, which may therefore be the preferred cell aspect ratio (Malkus and Veronis, 1958). A particularly interesting result from a geodynamic perspective is that the temperature profile in the TBL beneath the cold upper surface exhibits a hot asthenosphere, i.e., a range of depths where the temperature exceeds that of the cell's interior.

10.4.4 Structure of the Flow Near the Corners

The solutions described earlier break down near the corners of the cell, where the vorticity of the core flow solution becomes infinite and the BL approximation is no longer valid. The (dimensional) characteristic radius r_c of this corner region can be determined using a scaling argument. Consider for definiteness the corner above the upwelling plume, and let (r, ϕ) be polar coordinates with origin at the corner such that $\phi = 0$ on the upper surface and $\phi = \pi/2$ on the plume centerline. Close to the corner, the streamfunction has the self-similar form $\psi(r, \phi) = r^\lambda F(\phi)$ for some λ, where F is given by (4.47). Near the corner, the dimensional forms of the boundary conditions (10.73b) and (10.73c) are

$$\psi(r,0) = \psi(r, \pi/2) = \psi_{\phi\phi}(r, 0) = 0, \tag{10.75a}$$

$$r^{-3}\psi_\phi(r, \pi/2)\psi_{\phi\phi}(r, \pi/2) = \frac{1}{2}\frac{g\kappa\alpha\Delta T}{\nu}Nu. \tag{10.75b}$$

Equation (10.75b) immediately implies $\lambda = 3/2$, and the streamfunction that satisfies all the conditions (10.75) is

$$\psi = \frac{\kappa}{2}\left(\frac{r}{d}\right)^{3/2} C(\beta)^{1/2} Ra^{2/3} \left(\sin\frac{3}{2}\phi - \sin\frac{1}{2}\phi\right). \tag{10.76}$$

Note that the vorticity $\omega \sim -(\kappa/d^2)Ra^{2/3}(r/d)^{-1/2}$ becomes infinite at the corner. Now the balance between viscous forces and buoyancy in the corner region requires $\nu\nabla^4\psi \sim -g\alpha T_x$, or

$$\psi \sim \kappa Ra(r_c/d)^3. \tag{10.77}$$

Equating (10.77) with the scale $\psi \sim \kappa Ra^{2/3}(r_c/d)^{3/2}$ implied by (10.76), we find (Roberts, 1979)

$$r_c \sim Ra^{-2/9}d. \tag{10.78}$$

A corrected solution for the flow in the corner region that removes the vorticity singularity was proposed by JZ87.

10.4.5 Howard's Scaling for High-Ra Convection

While the preceding solution for cellular convection is illuminating, it is known that convection rolls at high Rayleigh number are unstable to small perturbations (§ 10.3.2). In reality, convection at high Rayleigh number $Ra > 10^6$ is a quasi-periodic process in which the TBLs grow by thermal diffusion, become unstable and then empty rapidly into plumes, at which point the cycle begins again. Figure 10.5 shows laboratory images of four stages of this process in a layer of sugar syrup heated suddenly from below. The evolving shape of the plumes is visualized using thermochromic liquid crystals. The thinning of the TBL as it empties into the plumes can be traced via the evolution of the height of the 24.35°C isotherm on either side of the central plume.

The characteristic time scale of the TBL instability can be estimated via a simple scaling argument (Howard, 1964). The thickness δ of the TBL initially increases by thermal diffusion as $\delta \approx (\pi \kappa t)^{1/2}$. Instability sets in when the Rayleigh number $Ra(\delta) \equiv g\alpha\Delta T\delta^3/\nu\kappa$ based on the TBL thickness attains a critical value

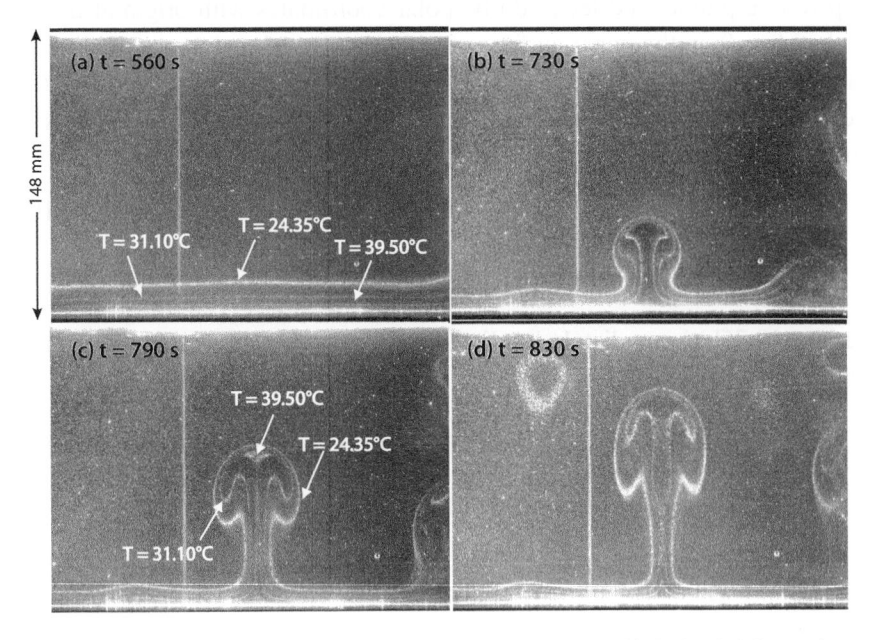

Figure 10.5 Evolution of thermal plumes in convection at high Rayleigh number. A layer of sugar syrup with initial temperature $T_0 = 21.8°C$ is suddenly heated from below to a temperature $T_{\text{hot}} = 46.9°C$ at $t = 0$. The Rayleigh number based on the temperature difference $\Delta T = T_{\text{hot}} - T_0$ and the viscosity ν_0 at the temperature T_0 is $Ra = 8.6 \times 10^5$. Three isotherms with the temperatures indicated are visualized using thermochromic liquid crystals. The time since the beginning of heating is indicated at the upper left of each image. Images from Androvandi (2009), courtesy of S. Androvandi.

$Ra_c \approx 1100$, which is the critical Rayleigh number for the onset of convection in a fluid layer with one rigid and one free-slip boundary (Chandrasekhar, 1981). The instability occurs after a time

$$t_c \approx \frac{1}{\pi\kappa}\left(\frac{\kappa\nu Ra_c}{g\alpha\Delta T}\right)^{2/3} \qquad (10.79)$$

that is independent of the layer depth. The phenomenology underlying Howard's model has been amply confirmed by laboratory experiments (e.g., Sparrow et al., 1970; Davaille and Vatteville, 2005). More quantitatively, Le Bars and Davaille (2004) showed that (10.79) is valid for $Ra_c = 1300 \pm 500$. An alternative way of deriving (10.79) is the subject of Exercise 10.5.

Exercises

10.1 [SM] Consider a two-dimensional system comprising a horizontal layer of fluid with viscosity η_1 sandwiched between two infinite half-spaces of fluid with viscosity η_2 (Figure 10.6). The base flow in both fluids is pure shear with a strain rate $\dot\epsilon > 0$, so that the horizontal and vertical velocities are $u = -\dot\epsilon x$ and $w = \dot\epsilon z$ (heavy arrows in Figure 10.6). In the absence of perturbations, the interfaces are horizontal (dashed lines in Figure 10.6) and the layer has thickness $2h_0(t) = 2h(0)\exp(\dot\epsilon t)$, where $2h(0)$ is the thickness at some initial time. Examine the stability of this system to small sinusoidal perturbations that have the form of folding of the layer.

10.2 [SM] Consider a layer $z \in [-1/2, 1/2]$ of fluid in which an imposed shear velocity $\mathbf{u} = -U\sin\pi z\, \mathbf{e}_x$ is driven by a fictitious force in the momentum equation, the (dimensionless) $\mathbf{e}_x$-component of which is

$$0 = -\partial_x p + \nabla^2 u - \pi^2 U \sin\pi z. \qquad (10.80)$$

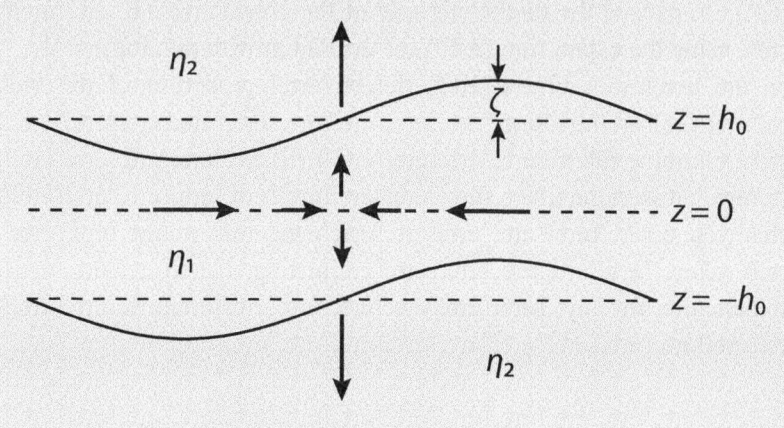

Figure 10.6 Definition sketch for Exercise 10.1.

Suppose further that two sets of convection rolls exist in the fluid: one with axes perpendicular to the shear direction, and another with axes parallel to it. The dimensionless amplitudes (= maximum vertical velocity) of these rolls are A and B, respectively, and are functions of the slow time $T = \epsilon^2 t$ where $\epsilon^2 = (Ra - Ra_0)/Ra_0$, Ra is the Rayleigh number and $Ra_0 = (k^2 + \pi^2)^3/k^2$. The boundaries $z = \pm 1/2$ of the layer are free-slip. At order ϵ, the poloidal scalar for the total flow (two sets of rolls plus imposed shear) is

$$\mathcal{P}_1 = -\frac{1}{k^2}[A(T)\cos kx + B(T)\cos ky]\cos \pi z - \frac{U}{\pi}x\cos \pi z. \quad (10.81)$$

Using a procedure similar to that of § 10.3.1, derive the equations governing the temporal evolution of the amplitudes $A(T)$ and $B(T)$.

10.3 [SM] Derive the envelope equation (10.55) by determining the solvability condition at order ϵ^3.

10.4 [SM] The envelope equation (10.55) admits a steady finite-amplitude solution

$$W = 2\sqrt{2}\epsilon \left(\frac{3\pi^2}{2} - \frac{4}{3}K^2\right)^{1/2} e^{iKX},$$

where $\epsilon^2 = (Ra - Ra_c)/Ra_c$. Perform a LSA of this solution. Hint: look for a solution of the linearized perturbation equation in the form

$$u = e^{iKX}e^{\lambda T}\left[Ae^{i(MX+NY)} + Be^{-i(MX+NY)}\right],$$

where M and N are wavenumbers and λ is the growth rate.

10.5 The form of the scaling law (10.79) of Howard (1964) can be determined in an alternative way by estimating the time required for the growth rate of a RTI to exceed the thickening rate of the conductive TBL. Demonstrate this, using the expression (3.20) for the RTI growth rate for $\gamma = 1$.

10.6 In the images of Figure 10.5, the physical properties of the fluid are $\nu_0 \equiv \nu(T_0) = 0.0046$ m^2 s^{-1}, $\kappa = 1.16 \times 10^{-7}$ and $\alpha = 5.73 \times 10^{-4}$. (a) Determine the time t_c defined by (10.79) assuming $Ra_c = 1100$, and compare it with the times shown on the images of Figure 10.5. (b) Suppose that you don't know the time at which the first image of Figure 10.5 was taken. Devise a procedure for estimating that time, and suggest an explanation for any discrepancy between your estimate and the true time $t = 560$ s.

11
Convection in More Realistic Systems

As the simplest and most widely studied example of thermal convection, the Rayleigh–Bénard configuration is the source of most of what has been learned about thermal convection during the past century. From a geodynamic point of view, however, it lacks many crucial features that are important in Earth's mantle: compressibility; the variation of viscosity as a function of pressure, temperature and stress; density variations associated with differences in chemical composition; the presence of solid-state phase changes; internal production of heat by radioactive decay; and a host of other factors such as rheologic anisotropy, grain-size evolution, the effect of volatile content, etc. Because theoretical methods thrive on simple model problems, the addition of each new complexity reduces their room for manoeuvre; but many important results have been obtained nevertheless. Here I briefly review some of these.

11.1 Compressible Convection and the Anelastic Liquid Equations

In the previous chapter, the equations governing convection were written using the Boussinesq approximation, wherein the density of the fluid is taken to be constant everywhere in the governing equations except in the body-force term in the equation of motion. While the Boussinesq approximation is valid for many convecting systems including most laboratory models, it breaks down for deep layers such as planetary interiors in which large hydrostatic pressures lead to significant compression of the fluid.

The principal tool for studying compressible convection under mantle-like conditions is the so-called anelastic liquid equations. The following derivation of these equations is based on Schubert et al. (2001) (henceforth STO01), whose careful treatment avoids the thermodynamically inconsistent assumption that the various thermodynamic parameters of the model are constants. From the point of view of this book, the derivation is a good example of how scaling arguments and nondimensionalization can be used to simplify a set of complicated governing equations.

Our starting point is the complete set of equations describing time-dependent laminar convection, which are

$$\frac{\partial \rho}{\partial t} + \frac{\partial}{\partial x_j}(\rho u_j), \tag{11.1a}$$

$$\rho \frac{D u_i}{Dt} = -\frac{\partial p}{\partial x_i} + \frac{\partial}{\partial x_j}\tau_{ij} + \rho g_i, \tag{11.1b}$$

$$\rho c_p \frac{DT}{Dt} - \alpha T \frac{DP}{Dt} = \frac{\partial}{\partial x_j}\left(k\frac{\partial T}{\partial x_j}\right) + \Phi + \rho H, \tag{11.1c}$$

$$\tau_{ij} = \eta\left(\frac{\partial u_i}{\partial x_j} + \frac{\partial u_j}{\partial x_i} - \frac{2}{3}\delta_{ij}\frac{\partial u_k}{\partial x_k}\right), \tag{11.1d}$$

$$\Phi = \tau_{ij}\frac{\partial u_i}{\partial x_j}, \tag{11.1e}$$

where all the symbols have their usual meanings. The quantity Φ is the rate of viscous dissipation of energy, and H is the rate of internal heat production per unit mass. The vertical coordinate $x_3 \equiv z$ increases downward.

The first step is to write the variables T, p and ρ as

$$T = \overline{T} + T', \quad p = \overline{p} + p', \quad \rho = \overline{\rho}(\overline{T},\overline{p}) + \rho' \tag{11.2}$$

where overbars denote a hydrostatic, motionless and time-independent reference state and primes denote departures (not necessarily small) from this state. Because the reference state is motionless, the velocity is by definition a departure from its value in the reference state and will be written without a prime. The reference state pressure satisfies the usual hydrostatic relation

$$\frac{d\overline{p}}{dz} = \overline{\rho g}, \tag{11.3}$$

where z is depth and $\overline{g}$ is the (depth-dependent) gravitational acceleration. Now because density changes due to the pressure and temperature variations associated with convection are small compared to $\overline{\rho}$, we can linearize the equation of state as follows:

$$\rho = \overline{\rho} + \overline{\left(\frac{\partial \rho}{\partial p}\right)_T}p' + \overline{\left(\frac{\partial \rho}{\partial T}\right)_p}T' = \overline{\rho}(1 + \overline{\chi}_T p' - \overline{\alpha}T'), \tag{11.4}$$

where $\overline{\chi}_T$ is the isothermal compressibility in the reference state. Next, we define dimensionless (starred) variables and thermodynamic parameters using the depth d of the convecting layer and other characteristic scales denoted by hats:

11.1 Compressible Convection

$$T'^* = T'/\Delta \hat{T}, \quad p'^* = d^2 p'/\hat{\kappa}\hat{\eta}, \quad \rho^* = \rho/\bar{\rho}, \quad u_i^* = u_i d/\hat{\kappa},$$

$$x_i^* = x_i/d, \quad t^* = t\hat{\kappa}/d^2, \quad \overline{\chi}_T^* = \overline{\chi}_T/\hat{\chi}_T, \quad \bar{\alpha}^* = \bar{\alpha}/\hat{\alpha},$$

where $\hat{\kappa}$ and $\hat{\eta}$ are characteristic values of the thermal diffusivity and viscosity, respectively. The scales chosen for p, $\mathbf{u}$ and t correspond to the commonly used thermal diffusion scaling. The dimensionless form of (11.4) is

$$\rho^* = \bar{\rho}^* \left(1 + M^2 Pr \overline{\chi}_T^* p'^* - \epsilon \bar{\alpha}^* T'^*\right), \tag{11.5}$$

where

$$M = \frac{\hat{\kappa}/d}{\sqrt{1/\hat{\rho}\hat{\chi}_S}} \sqrt{\frac{\hat{\chi}_T}{\hat{\chi}_S}}, \quad Pr = \frac{\hat{\nu}}{\hat{\kappa}}, \quad \epsilon = \hat{\alpha}\Delta \hat{T}, \tag{11.6}$$

$\hat{\chi}_S$ is the characteristic isentropic compressibility, and $\hat{\nu} = \hat{\eta}/\hat{\rho}$ is the kinematic viscosity. Apart from the factor $\sqrt{\hat{\chi}_T/\hat{\chi}_S}$ of order unity, M is the ratio of the characteristic velocity $\hat{\kappa}/d$ to the bulk sound speed $\sqrt{1/\hat{\rho}\hat{\chi}_S}$ in the fluid. M will be called the Mach number by analogy to the similar parameter in gas dynamics; STO01 estimate $M^2 \approx 10^{-33}$ in Earth. Given that the Prandtl number $Pr \approx 10^{23}$, $M^2 Pr \ll 1$. Using $\Delta \hat{T} = 1000$ K and $\hat{\alpha} = 3 \times 10^{-5}$ K^{-1}, we find that $\epsilon \approx 0.03$ is also small.

Using (11.5), we can write the dimensionless form of the continuity equation (11.1a) as

$$\bar{\rho}^* \left(M^2 Pr \overline{\chi}_T^* \frac{\partial p'^*}{\partial t^*} - \epsilon \bar{\alpha}^* \frac{\partial T'^*}{\partial t^*}\right)$$

$$+ \frac{\partial}{\partial x_j} \left\{\bar{\rho}^* \left(1 + M^2 Pr \overline{\chi}_T^* p'^* - \epsilon \bar{\alpha}^* T'^*\right) u_j^*\right\} = 0. \tag{11.7}$$

Taking the limit $M^2 Pr \to 0$ and $\epsilon \to 0$ of (11.7) and redimensionalizing the result, we obtain

$$\frac{\partial}{\partial x_j}(\bar{\rho} u_j) = 0, \tag{11.8}$$

which is known as the anelastic continuity equation. The disappearance of the term $\partial \rho/\partial t$ from the original form (11.1a) of the continuity equation means that sound waves cannot propagate in the fluid, whence the qualifier 'anelastic'. This makes physical sense because the characteristic time scale of mantle convection (millions of years) exceeds by many orders of magnitude the time scale for seismic waves and free oscillations (seconds to hours).

We now determine the conditions under which the anelastic continuity equation (11.8) reduces to the more familiar continuity equation $\nabla \cdot \mathbf{u} = 0$. If the reference state is adiabatic, then

$$\frac{\overline{d\rho}}{dz} = \overline{\left(\frac{\partial \rho}{\partial p}\right)_s \frac{d\overline{p}}{dz}} = \overline{\rho} \chi_s \frac{d\overline{p}}{dz} = \overline{\rho}^2 \chi_s \overline{g}, \tag{11.9}$$

where the hydrostatic relation (11.3) has been used in the last step. Now the local scale height for the reference density is

$$\overline{h}_\rho = \left(\frac{1}{\overline{\rho}} \left|\frac{d\overline{\rho}}{dz}\right|\right)^{-1} = (\overline{\rho} \chi_s \overline{g})^{-1} = \frac{\overline{\gamma c_p}}{\overline{\alpha g}}, \tag{11.10}$$

where

$$\overline{\gamma} \equiv \frac{\overline{\alpha}}{\overline{\rho c_v \chi_T}} = \frac{\overline{\alpha}}{\overline{\rho c_p \chi_S}} \tag{11.11}$$

is the Grüneisen parameter. Replacing the parameters on the right-hand side (RHS) of (11.10) by their characteristic (hatted) values, we see that the change in $\overline{\rho}$ across the fluid layer will be small if

$$\frac{d}{\hat{h}_\rho} = \frac{1}{\hat{\gamma}} \frac{\hat{\alpha} \hat{g} d}{\hat{c}_p} \ll 1. \tag{11.12}$$

The quantity $\hat{\alpha}\hat{g}d/\hat{c}_p$ is known as the dissipation number Di:

$$Di \equiv \frac{\hat{\alpha} \hat{g} d}{\hat{c}_p}. \tag{11.13}$$

The dissipation number is the ratio of the layer depth to the vertical distance over which a parcel of fluid must be moved in order to change its temperature by adiabatic expansion or compression by a factor e. For the whole mantle of Earth, $Di \approx 0.5$-0.6.

We turn now to the momentum equation (11.1b). Its scaled and dimensionless form is

$$\frac{1}{Pr}\overline{\rho}^* \left(1 + M^2 Pr \overline{\chi}_T^* p'^* - \epsilon \overline{\alpha}^* T'^*\right) \frac{Du_i^*}{Dt^*} = -\frac{\partial p'^*}{\partial x_i^*} + \overline{g}_i^* \overline{\rho}^* \overline{\chi}_T^* \frac{Di \hat{c}_p}{\hat{\gamma} \hat{c}_v} p'^*$$

$$-\overline{g}_i^* \overline{\rho}^* \overline{\alpha}^* Ra T'^* + \frac{\partial}{\partial x_j^*}\left[\eta^*\left(\frac{\partial u_i^*}{\partial x_j^*} + \frac{\partial u_j^*}{\partial x_i^*} - \frac{2}{3}\delta_{ij}\frac{\partial u_k^*}{\partial x_k^*}\right)\right], \tag{11.14}$$

In (11.14), the variations of the gravitational acceleration due to convection have been assumed to be negligible, and the reference state pressure gradient has been eliminated using (11.3). The Rayleigh number Ra is defined as

$$Ra \equiv \frac{\hat{g}\hat{\alpha}\Delta\hat{T}d^3}{\hat{\nu}\hat{\kappa}}. \tag{11.15}$$

Because $Pr \gg 1$, the left-hand side (LHS) of (11.14) is negligible. Redimensionalizing the remaining terms, we obtain

11.1 Compressible Convection

$$0 = -\frac{\partial p'}{\partial x_i} + \overline{\rho} g_i \left(\overline{\chi}_T p' - \overline{\alpha} T' \right) + \frac{\partial}{\partial x_j} \left[\eta \left(\frac{\partial u_i}{\partial x_j} + \frac{\partial u_j}{\partial x_i} - \frac{2}{3} \delta_{ij} \frac{\partial u_k}{\partial x_k} \right) \right]. \quad (11.16)$$

We have seen that the incompressible form $\nabla \cdot \mathbf{u} = 0$ of the continuity equation used in the Boussinesq approximation is valid in the limit $Di/\hat{\gamma} \to 0$. Taking this limit together with $Pr \to \infty$ in (11.14), setting $\partial u_k^*/\partial x_k^* = 0$ and $\overline{\rho}^* = \overline{\alpha}^* = 1$ and redimensionalizing, we recover the Boussinesq form of the momentum equation, viz.,

$$0 = -\frac{\partial p'}{\partial x_i} - \overline{\rho} g_i \overline{\alpha} T' + \frac{\partial}{\partial x_j} \left[\eta \left(\frac{\partial u_i}{\partial x_j} + \frac{\partial u_j}{\partial x_i} \right) \right]. \quad (11.17)$$

Finally, we consider the energy equation (11.1c), the dimensionless form of which is

$$\overline{\rho}^* (1 + M^2 Pr \overline{\chi}_T^* p'^* - \epsilon \overline{\alpha}^* T'^*) \overline{c}_p^* \frac{D}{Dt^*} (\overline{T}^* + T'^*)$$

$$- \overline{\alpha}^* (\overline{T}^* + T'^*) \left[\frac{Dp'^*}{Dt^*} \frac{\epsilon Di}{Ra} + \overline{g}_j^* u_j^* \overline{\rho}^* Di \right]$$

$$= \frac{\partial}{\partial x_j^*} \left[\overline{k}^* \frac{\partial}{\partial x_j^*} (\overline{T}^* + T'^*) \right] + \Phi^* \frac{Di}{Ra}$$

$$+ \overline{\rho}^* (1 + M^2 Pr \overline{\chi}_T^* p'^* - \epsilon \overline{\alpha}^* T'^*) H^* \left(\frac{d^2 \hat{H} \hat{\rho}}{\hat{k} \Delta \hat{T}} \right). \quad (11.18)$$

The quantity $d^2 \hat{H} \hat{\rho}/\hat{k} \Delta \hat{T}$ that appears in the last term is the ratio of a characteristic temperature difference $d^2 \hat{H} \hat{\rho}/\hat{k}$ associated with internal heat production to the applied temperature difference $\Delta \hat{T}$. The product of that ratio and the Rayleigh number Ra is the internal heating Rayleigh number Ra_H given by

$$Ra_H = \frac{\hat{g} \hat{\alpha} \hat{\rho} \hat{H} d^5}{\hat{\nu} \hat{\kappa} \hat{k}}, \quad (11.19)$$

which measures the contribution of internal heating to the overall convective vigor.

To obtain the anelastic form of the energy equation we take the limits $M^2 Pr \to 0$ and $\epsilon \to 0$ of (11.18), which becomes

$$\overline{\rho}^* \overline{c}_p^* \frac{D}{Dt^*} (\overline{T}^* + T'^*) - \overline{\alpha}^* (\overline{T}^* + T'^*) \overline{g}_j^* u_j^* \overline{\rho}^* Di$$

$$= \frac{\partial}{\partial x_j^*} \left[\overline{k}^* \frac{\partial}{\partial x_j^*} (\overline{T}^* + T'^*) \right] + \Phi^* \frac{Di}{Ra} + \overline{\rho}^* H^* \frac{Ra_H}{Ra}. \quad (11.20)$$

Equation (11.20) can be simplified further by using

$$\overline{\rho}^* \overline{c}_p^* \frac{D\overline{T}^*}{Dt^*} = \overline{\rho}^* \overline{c}_p^* u_j^* \frac{\partial \overline{T}^*}{\partial x_j^*} = \overline{\rho}^* \overline{\alpha}^* \overline{T}^* Di \overline{g}_j^* u_j^*. \tag{11.21}$$

The first step in (11.21) uses the fact that $\overline{T}^*$ is time-independent, while the second step uses the dimensionless equation for the adiabatic gradient, viz., $\partial \overline{T}^* / \partial x_j^* = \overline{\alpha}^* \overline{T}^* Di \overline{g}_j^* / \overline{c}_p^*$. The same relation can be used to eliminate $\partial \overline{T}^* / \partial x_j^*$ in the first term on the RHS of (11.20), which represents conduction of heat along the adiabatic gradient. Equation (11.20) now becomes

$$\overline{\rho}^* \overline{c}_p^* \frac{DT'^*}{Dt^*} - T'^* \overline{\alpha}^* \overline{g}_j^* u_j^* \overline{\rho}^* Di$$

$$= \frac{\partial}{\partial x_j^*} \left[\overline{k}^* \left(Di\overline{\alpha}^* \overline{g}_j^* \overline{T}^* / \overline{c}_p^* + \frac{\partial T'^*}{\partial x_j^*} \right) \right] + \Phi^* \frac{Di}{Ra} + \overline{\rho}^* H^* \frac{Ra_H}{Ra}. \tag{11.22}$$

Now in the special case when $\overline{k}^*$, $\overline{\alpha}^*$, $\overline{c}_p^*$ and $\overline{g}_j^*$ are constants, we may use the adiabatic relation $\partial \overline{T}^* / \partial x_j^* = \overline{\alpha}^* \overline{T}^* Di \overline{g}_j^* / \overline{c}_p^*$ again to write

$$\frac{\partial}{\partial x_j^*} \left[\overline{k}^* Di\overline{\alpha}^* \overline{g}_j^* \overline{T}^* / \overline{c}_p^* \right] = Di^2 \overline{k}^* \left(\frac{\overline{\alpha}^* \overline{g}^*}{\overline{c}_p^*} \right)^2 \overline{T}^*, \tag{11.23}$$

where $\overline{g}^* = (\overline{g}_j^* \overline{g}_j^*)^{1/2}$. The first term in square brackets on the RHS of (11.22) is therefore inherently of order Di^2. Redimensionalizing (11.22), we obtain

$$\overline{\rho} \overline{c}_p \frac{DT'}{Dt} - \overline{\alpha} \overline{\rho} g_j u_j T' = \frac{\partial}{\partial x_j} \left[\overline{k} \left(\frac{\overline{\alpha} g_j}{\overline{c}_p} \overline{T} + \frac{\partial T'}{\partial x_j} \right) \right] + \Phi + \overline{\rho} H. \tag{11.24}$$

To obtain the form of the energy equation in the Boussinesq approximation, we take the limit $Di \to 0$ of (11.22). The result after redimensionalization is

$$\overline{\rho} \overline{c}_p \frac{DT'}{Dt} = \frac{\partial}{\partial x_j} \left(\overline{k} \frac{\partial T'}{\partial x_j} \right) + \overline{\rho} H. \tag{11.25}$$

The linear stability of a layer of compressible fluid heated from below was first studied by Jeffreys (1930). He showed that the criterion for convective onset is the same as for an incompressible fluid if the temperature gradient in the latter is interpreted as a temperature gradient in excess of the adiabatic gradient. However, as pointed out by Jarvis and McKenzie (1980) (henceforth JM80), Jeffreys's approach is primarily useful for gases, for which $\alpha \approx 1/T$ and the adiabatic gradient $-g/c_p$ is constant. In a liquid with constant α, by contrast, the adiabatic gradient $\propto T$ and increases from top to bottom across the layer. The Rayleigh number is therefore

11.1 Compressible Convection 225

not uniquely defined. Accordingly, JM80 use an alternative parameter R_0 that takes into account the variation of the density across the layer. Its definition is

$$R_0 = \frac{g\alpha\beta_0 d^4}{\kappa_0 \nu_0} \left(\frac{\overline{\rho}}{\rho_0}\right)^2, \qquad (11.26)$$

where β_0 is the imposed temperature gradient, $\overline{\rho}$ is the mean density of the layer and κ_0, ν_0 and ρ_0 are the thermal diffusivity, kinematic viscosity and density, respectively, at the top of the layer.

JM80's formulation of the linear stability problem is similar to the one already introduced for RBC (§ 10.2.1), and so the details are omitted here. JM80 assume free-slip top and bottom boundaries, but use a condition of constant heat flux on the bottom boundary rather than constant temperature. For these boundary conditions, the critical value of R_0 (denoted by R_{0c}) for the onset of convection is 385 in the Boussinesq limit $Di \to 0$. Figure 11.1 shows R_{0c} (circles, left scale) and the critical wavenumber k_c (squares, right scale) as functions of the dissipation number Di for a Grüneisen parameter $\gamma = 1.1$. R_{0c} increases strongly as Di increases. This can be understood by noting that for each value of $Di > 0$, the fluid layer comprises a gravitationally unstable zone at the top where the temperature gradient is superadiabatic and a gravitationally stable zone at the bottom with a subadiabatic temperature gradient. Convection begins in the form of cells that are confined mostly to the upper layer; for large Di, these drive weaker countercells in the lower layer by viscous coupling. However, with increasing Di the thickness of the upper gravitationally unstable layer decreases, strongly decreasing the effective value of

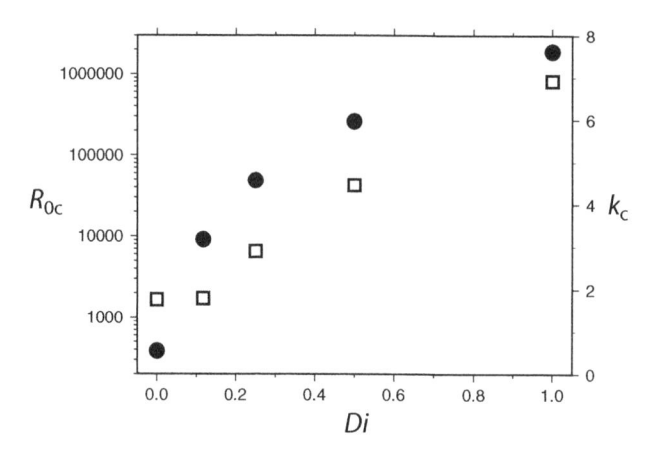

Figure 11.1 Results of a LSA for convection in a compressible fluid (Jarvis and McKenzie, 1980, Table 2). Shown as functions of the dissipation number Di are the critical value R_{0c} of R_0 (circles, left scale) and the critical wavenumber k_c (squares, right scale) for the onset of convection. The parameter R_0 is defined by (11.26), and the Grüneisen parameter $\gamma = 1.1$. See § 11.1 for discussion.

R_0 (which varies as the fourth power of the layer depth; see (11.26)). Overcoming this decrease requires a correspondingly higher 'true' value of R_0. The decrease in the thickness of the upper layer also explains why k_c is an increasing function of Di: the horizontal wavelength of convection is generally proportional to the layer depth.

While most subsequent studies of compressible convection have been numerical, some recent semianalytical studies are worth noting. Liu and Zhong (2013) used a propagator matrix method (PMM) to carry out a linear stability analysis (LSA) of a compressible fluid layer with constant physical properties and constant-temperature boundary conditions. Comparing their results with those of Jarvis and McKenzie (1980), they concluded that the bottom thermal boundary condition has a strong effect on the dependence of the critical Rayleigh number on Di. Kameyama et al. (2015) performed an LSA for a fluid layer with a depth-dependent thermal expansivity, using a truncated anelastic liquid approximation that neglects the effect of pressure fluctuations on the density in the buoyancy term of the momentum equation (second term on the RHS of (11.16)). Kameyama (2016) extended the results of Kameyama et al. (2015) to depth-dependent thermal conductivity and temperature-dependent viscosity.

11.2 Convection with Temperature-Dependent Viscosity

The viscosity of mantle rocks depends strongly on temperature. The key parameter here is the rheological temperature scale $\Delta T_r \equiv -\eta/(\partial \eta/\partial T)$, defined as the temperature difference required to change the viscosity by a factor e. For a viscosity that depends on temperature according to the Arrhenius law (6.80), $\Delta T_r = RT^2/H$, where T is the absolute temperature, H is the activation enthalpy and $R = 8.3$ J mol^{-1} K^{-1} is the universal gas constant. For representative upper mantle values $H \approx 5 \times 10^5$ J mol^{-1} (Bai et al., 1991) and $T \approx 1700$ K, $\Delta T_r \approx 50$ K.

The dynamic significance of the rheological temperature scale can be understood by considering the viscous force term in the Navier–Stokes equations, viz., $\nabla \cdot (2\eta \mathbf{e})$, where $\eta(T)$ is the viscosity and $\mathbf{e}$ is the strain rate tensor. Expanding the derivative and using the definition of the strain rate tensor, we find

$$\nabla \cdot (2\eta \mathbf{e}) = \eta \nabla^2 \mathbf{u} + 2 \frac{d\eta}{dT} \nabla T \cdot \mathbf{e}. \tag{11.27}$$

Now suppose that typical scales for the variables are $\mathbf{x} \sim d$, $\mathbf{u} \sim U$, $\mathbf{e} \sim U/d$ and $T \sim \Delta T$. The ratio of the two terms on the RHS of (11.27) is then

$$\frac{2(d\eta/dT)\nabla T \cdot \mathbf{e}}{\eta \nabla^2 \mathbf{u}} \sim \frac{\Delta T}{\Delta T_r}. \tag{11.28}$$

11.2 Temperature-Dependent Viscosity

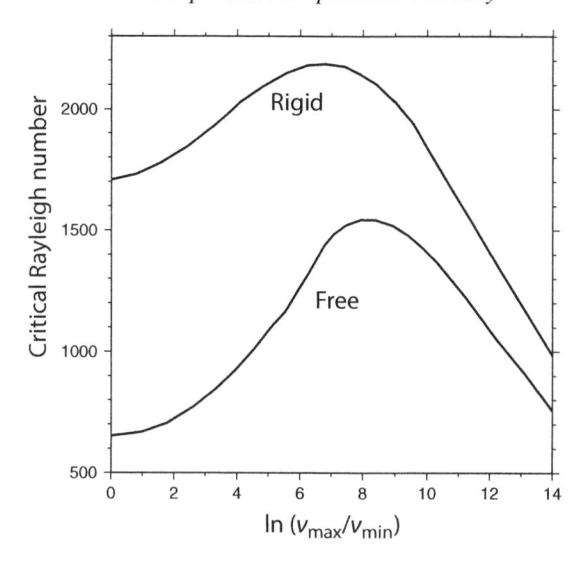

Figure 11.2 Critical Rayleigh number (based on the viscosity at the mean of the boundary temperatures) for convection with exponentially temperature-dependent viscosity, as a function of the total viscosity contrast across the layer. Curves for both free-slip and rigid boundaries are shown. For discussion, see § 11.2. Figure redrawn from Figure 2 of Stengel et al. (1982) by permission of Cambridge University Press.

Viscosity variations due to temperature differences are therefore dynamically important if the rheological temperature scale ΔT_r is comparable to or smaller than the typical temperature difference ΔT in the fluid.

A variety of important theoretical results have been obtained for convection in fluids with temperature-dependent viscosity. Stengel et al. (1982) performed a LSA of convection in a fluid whose viscosity $\nu(T)$ varies exponentially with temperature by up to factor 1.2×10^6. The critical Rayleigh number (based on the viscosity at the mean of the boundary temperatures) first increases and then decreases as the total viscosity contrast across the layer increases (Figure 11.2). The decrease reflects the fact that convection begins first in a low-viscosity sublayer near the hot bottom boundary when the viscosity contrast is large. A nonlinear stability analysis of convection in a fluid with a weak linear dependence of viscosity on temperature was carried out by Palm (1960), using a method like that described in § 10.3.1. Because the variation of the viscosity breaks the symmetry between upwelling and downwelling motions, the amplitude equation analogous to (10.43) contains a quadratic term that permits transcritical bifurcation. Subsequent work (e.g., Busse, 1967b; Palm et al., 1967) showed that the stable planforms are hexagons, hexagons and 2-D rolls, or 2-D rolls alone, depending on the Rayleigh number. Busse and Frick (1985) used a Galerkin method (§ 10.3.2) to study convection in a fluid whose viscosity depends strongly (but linearly) on temperature and found that a square planform becomes stable when the viscosity contrast is sufficiently large. McKenzie (1988) used group theoretic methods to classify the symmetries of convective transitions in fluids with strongly temperature-dependent viscosity and was able to recover all the major planforms and transitions between them observed in laboratory experiments by White (1981, 1988).

Convection at high Rayleigh number in fluids with strongly variable viscosity is a particularly difficult analytical challenge, but some noteworthy results have been obtained. Morris and Canright (1984) and Fowler (1985a) used boundary-layer analyses similar to that described in § 10.4 to determine the Nusselt number in the stagnant lid limit $\Delta T/\Delta T_r \to \infty$ for convection in a fluid whose viscosity depends on temperature as $\nu = \nu_0 \exp[-(T - T_0)/\Delta T_r]$. Both studies find the Nusselt number to be $Nu \sim (\Delta T/\Delta T_r)^{-1} Ra_r^{1/5}$, where $Ra_r = \alpha g \Delta T_r d^3 / \nu_0 \kappa$ is the Rayleigh number based on the rheological temperature scale ΔT_r and the viscosity ν_0 at the hot bottom boundary. Fowler et al. (2016) extended the asymptotic analysis of Fowler (1985a) to the case of strongly temperature- and pressure-dependent viscosity. Solomatov (1995) presented a scaling analysis for convection in fluids with temperature- and stress-dependent viscosity, using arguments based on the balance between dissipation and the rate of mechanical work. He found that three different dynamic regimes occur as the total viscosity contrast R_η across the layer increases: (I) quasi-isoviscous convection, (II) a transitional regime with a mobile cold upper BL and (III) a stagnant-lid regime in which the upper BL is motionless and convection is confined beneath it. Figure 11.3 shows the domains of these three regimes in the space Ra_1-R_η, where Ra_1 is the Rayleigh number based on the viscosity at the bottom of the convecting layer.

The studies cited in the foregoing paragraph do not predict the same exponent n in the Nusselt number–Rayleigh number scaling relationship for the stagnant lid regime. Morris and Canright (1984) and Fowler (1985a) predict $n = 1/5$, which is consistent with the value determined in § 10.4.1 by scaling analysis of isoviscous RBC between rigid boundaries. By contrast, the scaling analysis of Solomatov (1995) predicts $n = 1/3$. The latter value is more consistent with the fact that high Rayleigh-number convection with temperature-dependent viscosity, like its constant-viscosity counterpart, is time-dependent and disordered, taking the form of plumes generated quasi-periodically by Howard's (1964) mechanism (§ 10.4.5). That mechanism implies that both the heat flux $q \propto (k\Delta T/d) Ra_r^{1/3}$ (k is the thermal conductivity) and the characteristic period of plume formation $t_c \propto (d^2/\kappa) Ra_r^{-2/3}$ are independent of the layer depth. Manga et al. (2001) demonstrated experimentally that Howard's mechanism is operative in stagnant-lid convection and verified that t_c is indeed independent of the layer depth.

Another convective phenomenon in which strongly temperature-dependent viscosity plays a role is small-scale convection beneath the cooling oceanic

11.2 Temperature-Dependent Viscosity

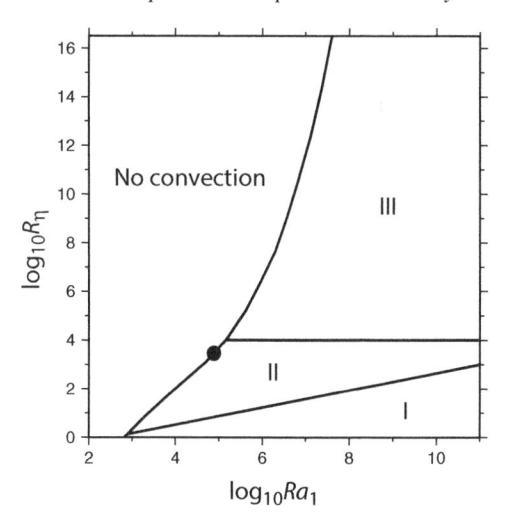

Figure 11.3 Regimes of convection in a Newtonian fluid with strongly temperature-dependent viscosity. Ra_1 is the Rayleigh number based on the viscosity at the bottom of the convecting layer, and R_η is the ratio of the viscosities at the top and the bottom of the layer. I: small viscosity-contrast regime; II: transitional (sluggish lid) regime; III: stagnant-lid regime. The solid circle separates the neutral stability curve into a lower portion where instability occurs in the entire layer from an upper portion where it occurs in a low-viscosity sublayer. See § 11.2 for discussion. Redrawn from Figure 4 of Solomatov (1995) by permission of AIP Publishing LLC.

lithosphere. In an early study, Parsons and McKenzie (1978) assumed constant viscosity and invoked the scaling argument of Howard (1964) to estimate an onset time of the form (10.79) with a critical Rayleigh number $Ra_c = 473$. Davaille and Jaupart (1994) extended that analysis to a fluid with temperature-dependent viscosity and emphasized the importance of the rheological temperature scale $\Delta T_r(T = T_0)$ evaluated at the interior temperature T_0 below the unstable TBL. Korenaga and Jordan (2003a) proposed an alternative scaling analysis based on the concept of available buoyancy (§ 10.1). The predictions of the onset time by Davaille and Jaupart (1994) and Korenaga and Jordan (2003a) agree with each other and with laboratory experiments once Davaille and Jaupart's (1994) extrapolation to larger values of $\Delta T/\Delta T_r$ is corrected (Davaille and Limare, 2015). Using an energy stability analysis in conjunction with a similarity transformation, Kim and Choi (2006) recovered the same trend of dimensionless onset time versus $\Delta T/\Delta T_r$ as the two previous studies. Finally, Landuyt and Ierley (2012) performed a LSA for the onset of 2-D sublithospheric convection in a fluid with temperature-dependent viscosity subject to shear. For small plate velocities, the analysis reproduces the dependence of the onset time on activation energy and Rayleigh number predicted by Korenaga and Jordan (2003a). However, a significant deviation from this behaviour is observed at large plate velocities, for which the onset of small-scale convection is delayed.

11.3 Convection in a Compositionally Layered Mantle

Convection in a compositionally layered mantle was first studied by Richter and Johnson (1974), who performed a linear stability analysis of a system comprising two superposed fluid layers of equal depth and viscosity but different densities. The critical Rayleigh number Ra_c depends on the value of a second Rayleigh number $Ra_\rho = g\Delta\rho d^3/\nu\kappa$ proportional to the magnitude of the stabilizing density difference $\Delta\rho$ between the layers. Three distinct modes of instability are possible: convection over the entire depth of the fluid with advection of the interface; separate convection within each layer; and standing waves on the interface, corresponding to an imaginary growth rate at marginal stability. Busse (1981) extended these results to the case of layers of different thicknesses. Renardy and Joseph (1985) and Renardy and Renardy (1985) considered the case of two fluids with nearly the same properties. Rasenat et al. (1989) examined a larger region of the parameter space and demonstrated the possibility of an oscillatory two-layer regime with no interface deformation, in which the coupling between the layers oscillates between thermal and mechanical. Le Bars and Davaille (2002) mapped out the linear stability of the system as a function of the interlayer viscosity contrast and the layer depth ratio. Jaupart et al. (2007) examined the linear stability of a system with a negligible temperature gradient in one of the layers, with application to the stability of continental lithosphere. The following development is that of Le Bars and Davaille (2002) with some small changes of notation.

Figure 11.4 is a schematic of the two-layer system studied. The lower layer (fluid 1) has thickness d_1, intrinsic (compositional) density ρ_{10} and kinematic viscosity ν_1, while the upper layer (fluid 2) has thickness d_2, density ρ_{20} and viscosity ν_2. The density stratification is intrinsically stable, i.e., $\rho_{10} > \rho_{20}$. The two fluids are perfectly miscible, but diffusion across their interface is assumed to be sufficiently slow that it can be neglected. The temperature at the upper boundary is T_2, and that at the lower boundary is $T_1 = T_2 + \Delta T$.

To nondimensionalize the problem, we define dimensionless (primed) variables

$$\mathbf{x}' = \frac{\mathbf{x}}{d}, \quad t' = \frac{g\alpha\Delta T d}{\nu_2}t, \quad \mathbf{u}' = \frac{\nu_2}{g\alpha\Delta T d^2}\mathbf{u},$$

$$T' = \frac{T - T_2}{\Delta T}, \quad p' = \frac{p}{\rho_{20}g\alpha\Delta T d}, \qquad (11.29)$$

where $d = d_1 + d_2$, α is the thermal expansivity and g is the gravitational acceleration. From this point on all variables are dimensionless, but the primes are dropped to simplify the notation.

11.3 A Compositionally Layered Mantle

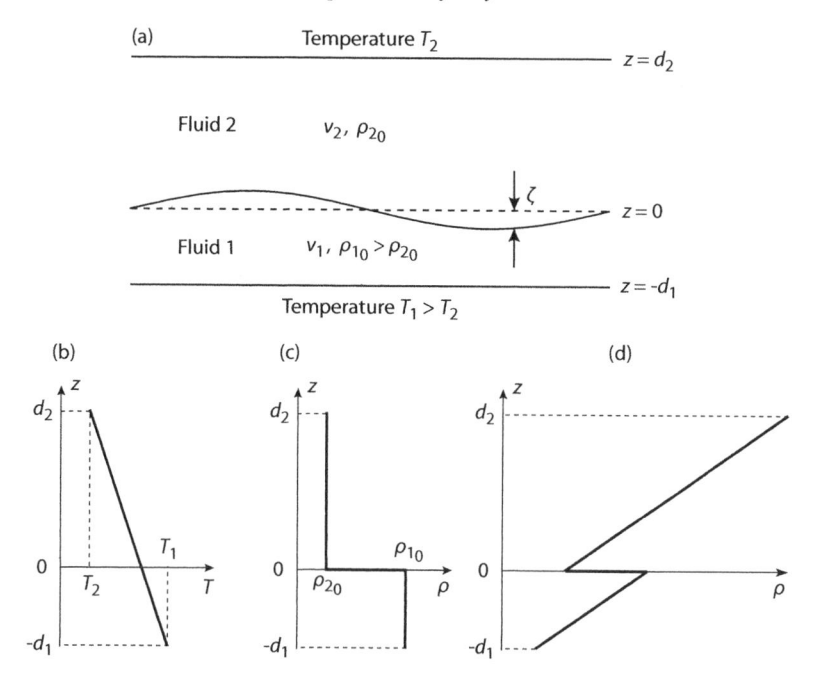

Figure 11.4 (a) Definition sketch of two-layer convection, (b) initial linear temperature profile, (c) compositional density profile and (d) effective density profile accounting for both thermal and compositional effects. See § 11.3 for discussion. Figure redrawn from Figure 1 of Le Bars and Davaille (2002) by permission of Cambridge University Press.

We study the linear stability of the static solution, which is characterized by a linear temperature profile

$$T = 1 - a - z, \quad a = \frac{d_1}{d}. \tag{11.30}$$

The equation of state for the density of each layer is

$$\rho_n(T) = \rho_{n0}(1 - \alpha \Delta T\, T). \tag{11.31}$$

Let θ_n and p_n be the perturbations of the temperature and pressure in fluid n, respectively, relative to their static distributions, and let $\mathbf{u}_n$ be the velocity vector in fluid n. Making the Boussinesq approximation and assuming that inertia is negligible, we may write the dimensionless equations governing the flow in layer n as

$$\nabla \cdot \mathbf{u}_n = 0, \tag{11.32a}$$

$$-\nabla p_n + \theta_n \mathbf{k} + \frac{\nu_n}{\nu_2}\nabla^2 \mathbf{u}_n = 0, \tag{11.32b}$$

$$Ra\left[\left(\frac{\partial}{\partial t} + \mathbf{u}_n \cdot \nabla\right)\theta_n - \mathbf{u}_n \cdot \mathbf{k}\right] = \nabla^2 \theta_n, \tag{11.32c}$$

where the vertical unit vector **k** points upward and the Rayleigh number is

$$Ra = \frac{\alpha g \Delta T d^3}{\kappa \nu_2}, \tag{11.33}$$

where κ is the thermal diffusivity. Moreover, because we are only concerned with infinitesimal perturbations of the static reference state, the nonlinear term $\mathbf{u}_n \cdot \nabla \theta_n$ in (11.32c) can be neglected. Now take the curl of (11.32b) twice and use (11.32c) to eliminate the temperature, which yields the following equations for the vertical velocity $w_n \equiv \mathbf{u}_n \cdot \mathbf{k}$:

$$\left(Ra\partial_t - \nabla^2\right) \nabla^4 w_1 = -\frac{Ra}{\gamma} \partial_{xx}^2 w_1, \tag{11.34a}$$

$$\left(Ra\partial_t - \nabla^2\right) \nabla^4 w_2 = -Ra\partial_{xx}^2 w_2, \tag{11.34b}$$

where $\gamma = \nu_1/\nu_2$.

Solution of equations (11.34) requires three boundary conditions at each of the boundaries $z = -a$ and $z = 1 - a$, the kinematic condition for the interface, and six matching conditions at the interface. The conditions at the boundaries are

$$w = \partial_z w = 0 \quad \text{for a rigid boundary}, \tag{11.35a}$$

$$w = \nabla^2 w = 0 \quad \text{for a free-slip boundary}, \tag{11.35b}$$

and $\theta = 0$, which is equivalent to

$$\nabla^4 w = 0. \tag{11.35c}$$

Because the deformation $\zeta(x, t)$ of the interface is assumed to be infinitesimal, the kinematic and matching conditions there can be linearized about the undeformed position $z = 0$ of the interface, as discussed previously in connection with the RTI (§ 10.1). The kinematic condition required by the material character of the interface is

$$w_1 = \partial_t \zeta. \tag{11.36}$$

Continuity of the two components of the velocity requires

$$w_1 = w_2, \tag{11.37}$$

$$\partial_z w_1 = \partial_z w_2. \tag{11.38}$$

Continuity of the shear stress requires

$$\gamma(\partial_{zz}^2 - \partial_{xx}^2)w_1 = (\partial_{zz}^2 - \partial_{xx}^2)w_2. \tag{11.39}$$

11.3 A Compositionally Layered Mantle

Continuity of the normal stress requires

$$p_1 - 2\gamma \partial_z w_1 = p_2 - 2\partial_z w_2 + B\zeta, \tag{11.40a}$$

where

$$B = \frac{\rho_{10} - \rho_{20}}{\alpha \rho_0 \Delta T} \tag{11.40b}$$

is the buoyancy number, the ratio of the stabilizing compositional density anomaly to the destabilizing thermal density anomaly. To write (11.40a) in terms of w_1 and w_2 alone we take ∂_{txx}^3 of that equation and eliminate p_n using (11.32b) and ζ using (11.36), obtaining

$$\partial_{tz}^2 \left(\nabla^2 + 2\partial_{xx}^2 \right) (\gamma w_1 - w_2) = -B \partial_{xx}^2 w_1. \tag{11.40c}$$

Continuity of temperature requires

$$\theta_1 = \theta_2 \quad \Rightarrow \quad \gamma \nabla^4 w_1 = \nabla^4 w_2. \tag{11.41}$$

Finally, continuity of heat flux requires

$$\partial_z \theta_1 = \partial_z \theta_2 \quad \Rightarrow \quad \gamma \nabla^4 \partial_z w_1 = \nabla^4 \partial_z w_2. \tag{11.42}$$

Analyzing the problem in terms of normal modes, we seek solutions of the form

$$w(x, z, t) = W(z) \exp(ikx + st) \quad \text{with } s = \sigma + i\omega, \tag{11.43}$$

where σ and ω are the real and imaginary parts of the growth rate, respectively. The amplitudes W_1 and W_2 satisfy the equations

$$(sRa + k^2 - D^2)(D^2 - k^2)^2 W_1 = k^2 \frac{Ra}{\gamma} W_1, \tag{11.44a}$$

$$(sRa + k^2 - D^2)(D^2 - k^2)^2 W_2 = k^2 Ra W_2, \tag{11.44b}$$

where $D = d/dz$. The general solution of (11.44) is

$$W_1 = \sum_{m=1}^{3} \left[A_{1m} \exp(q_{1m}(a + z)) + B_{1m} \exp(-q_{1m}(a + z)) \right], \tag{11.45a}$$

$$W_2 = \sum_{m=1}^{3} \left[A_{2m} \exp(q_{2m}(1 - a - z)) + B_{2m} \exp(-q_{2m}(1 - a - z)) \right], \tag{11.45b}$$

where the coefficients q_{nm} are solutions of the equations

$$(sRa + k^2 - q_{1m}^2)(q_{1m}^2 - k^2)^2 = \frac{Ra}{\gamma} k^2, \tag{11.46a}$$

$$(sRa + k^2 - q_{2m}^2)(q_{2m}^2 - k^2)^2 = Ra k^2. \tag{11.46b}$$

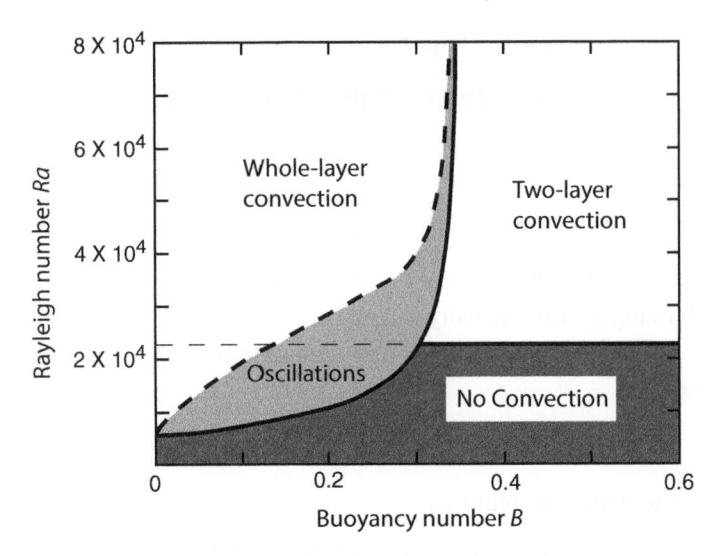

Figure 11.5 Modes of convection in a fluid comprising two layers with different intrinsic densities, as a function of the Rayleigh number Ra and the buoyancy number B. The layer depth ratio is $a = 0.5$, and the viscosity ratio is $\gamma = 6.7$. The upper boundary of the oscillatory field is determined by fixing the wavenumber k to its marginal stability value and increasing Ra until the imaginary part of the growth rate becomes zero. See § 11.3 for discussion. Figure redrawn from Figures 2 and 10 of Le Bars and Davaille (2002) by permission of Cambridge University Press.

The twelve constants A_{nm} and B_{nm} satisfy a set of twelve homogeneous equations represented by the six outer boundary conditions (11.35) and the six interfacial matching conditions (11.37)–(11.42). The determinant of the coefficient matrix must vanish for a nontrivial solution to exist; Le Bars and Davaille (2002) give that matrix for the case of rigid boundaries in their Appendix. The condition for the vanishing of the determinant is a transcendental equation that links the growth rate s and the parameters a, γ, B, Ra and k, and that must be solved numerically. Because the problem is not self-adjoint, the growth rate s is complex in general.

The essential dynamics of the system are illustrated in Figure 11.5, which shows the mode of convection as a function of the Rayleigh number Ra and the buoyancy number B for equal layer depths ($a = 0.5$), a viscosity ratio $\gamma = 6.7$ and rigid boundaries. Three distinct types of convection are possible: whole-layer, two-layer and oscillatory. Whole-layer convection occurs for small values of the buoyancy number B, while two-layer convection occurs for large values. In two-layer convection, the interface is undeformed and convection occurs in one of the layers, the flow in the other (passive) layer being driven by viscous coupling to the active layer. For $a = 0.5$ and $\gamma = 6.7$ the critical Rayleigh number for two-layer convection is

22400, independent of the buoyancy number B (horizontal line in Figure 11.5). The fields for whole-layer and two-layer convection are separated by a field of oscillatory convection in the form of travelling waves. The oscillatory mode sets in because the density at the bottom of the lower layer is smaller than the density of the upper layer despite the stabilizing density jump across the interface (Figure 11.4d). The oscillatory mode disappears as $Ra \to \infty$ because the compositional stratification ($B|_{Ra\to\infty} = 0.352$) becomes too large to be reversed by any thermal effect.

11.4 Convection with a Phase Transition

Another convective phenomenon amenable to analytic treatment is convection through a univariant phase transition. The first LSA of this situation was that of Busse and Schubert (1971), who focussed on the instability due to the phase transition alone. Schubert and Turcotte (1971) (henceforth ST71) took the next step by considering the combined effects of a phase transition and the ordinary Rayleigh instability due to thermal expansion of the fluid. ST71's treatment was limited to the case of two layers having the same viscosity and thickness. Their results were extended to the case of arbitrary layer depth ratio and viscosity ratio by Sotin and Parmentier (1989) (henceforth SP89), and to spherical geometry by Peltier et al. (1989). Roberts et al. (2007) generalized the model of Schubert and Turcotte (1971) to a phase change at arbitrary depth and presented a perturbation analysis for a phase change close to the bottom of the fluid layer (e.g., the perovskite/post-perovskite phase change on Earth).

The equations describing convection through a phase transition share many features with those for two-layer convection. The following discussion therefore focusses on the differences between the two cases, using as much as possible the notation of SP89 to facilitate reference to that paper, but with some small changes for greater consistency with the notation of Le Bars and Davaille (2002). One difference between the two sets of notation is that SP89 denote the upper and lower layers by the indices 1 and 2, respectively, which is opposite to the convention of Le Bars and Davaille (2002). SP89 also nondimensionalize their equations using thermal diffusion scaling wherein the velocity scale is κ/d, where κ is the thermal diffusivity and d is the sum of the depths of the two layers. Finally, SP89 define two layer-specific Rayleigh numbers

$$Ra_j = \frac{g\alpha(\beta - \beta_{ad})d^4}{\kappa \nu_j} \qquad (11.47)$$

where $j = 1$ or 2 and β_{ad} is the adiabatic temperature gradient.

Relative to two-layer convection, there are three novel elements in the phase-change problem: (1) flow through the phase boundary can occur because that interface is not a material surface, (2) the phase boundary must lie on the Clapeyron slope and (3) the condition of continuity of heat flux across the interface involves an additional term representing the latent heat released or absorbed by the phase change. Let us consider each of these in turn.

(1) Because the interface is not a material surface, the kinematic condition (11.36) does not apply. Moreover, because the mass flux ρw across the interface must be continuous, the condition (11.37) is replaced by

$$\rho_1 w_1 = \rho_2 w_2. \tag{11.48}$$

(2) The basic form of the normal stress continuity condition is the same as for two-layer convection, viz.,

$$-p_1 + 2\eta_1 \partial_z w_1 - g\zeta \Delta\rho = -p_2 + 2\eta_2 \partial_z w_2, \tag{11.49}$$

where $\Delta\rho > 0$ is the density difference between the phases. However, the interface distortion ζ is no longer advected by the flow, but is instead fixed by the requirement that it lie on the Clapeyron slope of the phase transition. This condition is (ST71)

$$\zeta = \frac{p_1 - \chi\theta_1}{\rho_1 g - \beta\chi} = \frac{p_2 - \chi\theta_2}{\rho_2 g - \beta\chi} \tag{11.50}$$

where χ is the Clapeyron slope, θ_j is the temperature perturbation in layer j and β is the temperature gradient. The pressure and temperature perturbations in (11.50) are evaluated at $z = 0$ because the displacement of the phase boundary is infinitesimal. Equation (11.50) can be better understood when written in the form

$$\chi = \left(\frac{p_1 - \rho_1 g \zeta}{\theta_1 - \beta\zeta}\right)_{z=0} = \left(\frac{p_2 - \rho_2 g \zeta}{\theta_2 - \beta\zeta}\right)_{z=0}. \tag{11.51}$$

The numerators in (11.51) are the differences in pressure between a point located at $z = \zeta$ in the perturbed state and a point located at $z = 0$ in the unperturbed state. The denominators are similar temperature differences.

Now substitute either of the expressions (11.50) into (11.49), and assume $g\Delta\rho/(g\rho_j - \beta\chi) \ll 1$ for $j = 1$ or 2. The result is

$$p_2 - p_1 = -\frac{g\Delta\rho \, \chi}{g\rho - \beta\chi}\theta + 2(\eta_2 - \eta_1)\partial_z w_1. \tag{11.52}$$

In (11.52), ρ_j has been replaced by the mean density ρ in the denominator of the first term on the RHS. Also, the temperature perturbation θ_j has been replaced by θ because $\theta_1 = \theta_2$ at the interface. Equation (11.52) can be rewritten in terms of w_1 and w_2 alone by differentiating it twice with respect to x and applying the horizontal force balance equation $\partial_x p_j = \eta_j \nabla^2 u_j$, the continuity equation $\partial_x u_j + \partial_z w_j = 0$ and the continuity of $\partial_z w$ across the interface. In addition, the temperature perturbation θ is eliminated using the equation

$$\nu_j \nabla^4 w_j = -g\alpha \partial_{xx}^2 \theta_j. \tag{11.53}$$

Finally, the matching condition is nondimensionalized using d as the length scale and κ/d as the velocity scale. The result is

$$\partial_{zzz}^3 (R_\eta w_2 - w_1) = -S\nabla^4 w_1 - 3(R_\eta - 1)\partial_{xxz}^3 w_1, \tag{11.54}$$

where $R_\eta = \eta_2/\eta_1$ and

$$S = \frac{\Delta \rho}{\alpha \rho d\,(\rho g/\chi - \beta)}. \tag{11.55}$$

Equation (11.54) corrects two misprints in eqn. (A12) of SP89.

(3) The third new condition is that the difference in the heat flux on the two sides of the interface is equal to the rate of latent heat release by the phase transition. In dimensional form this condition is

$$\rho_1 w_1 Q = -k\,(\partial_z \theta_2 - \partial_z \theta_1) \tag{11.56}$$

where Q is the latent heat per unit mass of transformation of the lower phase into the upper phase. To write (11.56) in terms of w_1 and w_2 alone, we first differentiate it twice with respect to x and then use (11.53) to eliminate the temperature perturbations. Setting $k = \rho_1 c_p \kappa$ and nondimensionalizing the result, we obtain

$$Ra_1 R' \partial_{xx}^2 w_1 = \partial_z \left(\frac{R_\eta}{R_\rho} \nabla^4 w_2 - \nabla^4 w_1 \right) \tag{11.57}$$

where $R_\rho = \rho_2/\rho_1$ and

$$R' = \frac{Q}{c_p d(\beta - \beta_{\text{ad}})}. \tag{11.58}$$

With all the matching conditions in hand, we now turn to the dimensionless equations governing the vertical velocity in each layer, which are

$$(\partial_t - \nabla^2)\nabla^4 w_j = -Ra_j \partial_{xx}^2 w_j. \tag{11.59}$$

Introducing the normal mode decomposition (11.43), we find that $W_j(z)$ satisfy

$$(s + k^2 - D^2)(D^2 - k^2)^2 W_j = k^2 Ra_j W_j. \tag{11.60}$$

Because the problem is not self-adjoint, the growth rate s can be complex. However, SP89 show that s is real for values of R' and S corresponding to natural exothermic

($R' > 0, S > 0$) and endothermic ($R' < 0, S < 0$) phase transitions. The neutral stability curve is therefore obtained by setting $s = 0$.

We now follow ST71 and SP89 in considering only the case of free-slip boundaries. The general solutions of (11.60) with $s = 0$ that satisfy the boundary conditions (11.35b) and (11.35c) are

$$W_1 = \sum_{l=1}^{3} A_{1l} \sinh[g_{1l}(a - z)] \quad \text{for} \quad a > z > 0, \tag{11.61a}$$

$$W_2 = \sum_{l=1}^{3} A_{2l} \sinh[g_{2l}(1 - a + z)] \quad \text{for} \quad 0 > z > a - 1, \tag{11.61b}$$

where $a = d_1/d$ is the dimensionless thickness of the upper layer and

$$g_{jl}^2 = k^2 + (Ra_j k^2)^{1/3} \exp[i(\pi/3)(2l - 1)] \tag{11.62}$$

for $j = 1, 2$ and $l = 1, 2, 3$. Equation (11.62) corrects two misprints in eqn. (9) of SP89.

The six constants A_{jl} are determined by substituting $w_j = W_j(z) \exp ikx$ with W_j given by (11.61) into the six matching conditions (11.38), (11.39), (11.41), (11.48), (11.54) and (11.57). Setting the determinant of the 6 × 6 coefficient matrix equal to zero in the usual way, we obtain an equation that links the seven dimensionless parameters $k, Ra_1, R', S, a, R_\eta$ and R_ρ. For given values of a, R_η and R_ρ, the resulting equation can be solved numerically to determine the critical wavenumber k_{crit} and critical Rayleigh number $Ra_{\text{crit}} \equiv Ra_1(k_{\text{crit}})$ for the onset of convection as functions of R' and S.

The essential physics are well illustrated by the simplest model: two equal layers ($a = 1/2$) of equal viscosity ($R_\eta = 1$) and (nearly) equal densities ($R_\rho = 1$). Figure 11.6 shows Ra_{crit} as a function of R' and S for this case, for an exothermic phase transition with $R' > 0$ and $S > 0$. Ra_{crit} for an endothermic phase transition can be obtained via the transformation $Ra_{\text{crit}}(S, R') = Ra_{\text{crit}}(-2R', -S/2)$, which is implicit in equations (13) and (20) of SP89. The parameters R' and S have opposite effects on the stability of the system: increasing R' promotes stability (i.e., increases Ra_{crit}) while increasing S promotes instability. This behaviour can be understood by considering a fluid element that moves downward through an exothermic phase transition (ST71). Owing to the zeroth-order temperature gradient, the element approaching the boundary will be cooler than undisturbed fluid at the boundary. Because the Clapeyron slope is positive, the phase boundary will therefore be displaced upward to lower hydrostatic pressure. The heavier material below the boundary will then provide a hydrostatic pressure head tending to drive downward flow, promoting instability. On the other hand, the downward flow of fluid through

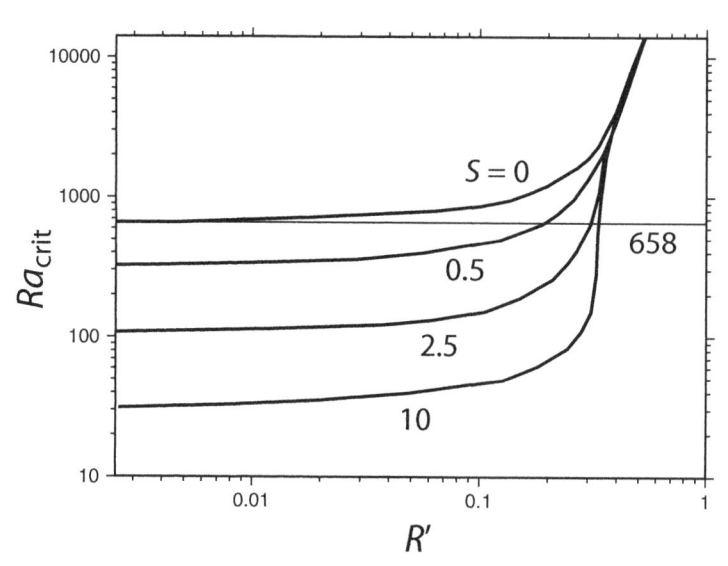

Figure 11.6 Critical Rayleigh number Ra_{crit} for convection with an exothermic phase transition, as a function of R' (defined by (11.58)) for several values of S (defined by (11.55)). The phase transition occurs at mid-depth in the fluid layer, and the viscosities above and below the transition are equal. The horizontal line indicates the critical Rayleigh number (= 658) for convection without a phase transition. For discussion, see § 11.4.

the boundary releases heat, which tends to warm the fluid and return the phase boundary to its unperturbed depth, promoting stability.

The results discussed earlier have been extended to the more realistic case of a divariant phase transition by Schubert et al. (1975).

In conclusion, I note that Buffett et al. (1994) have presented a general PMM for analyzing the stability of a system comprising multiple flat homogeneous layers heated from below. Effects that can be handled include multiple phase transitions, rheological layering (§ 4.9.2) and mobile surface plates.

Exercises

11.1 Referring to the notation for the problem of convection in a compositionally layered mantle (§ 11.3), show that the constant-temperature boundary condition $\theta = 0$ is equivalent to $\nabla^4 w = 0$.

11.2 [SM] Consider convection in a compositionally layered mantle comprising two equal layers of depth $d/2$ with the same viscosity. Suppose further that the upper and lower boundaries of the system are free-slip. Carry out a linear stability analysis of this situation, up to the point where you require the determinant of a coefficient matrix to vanish. Verify that a solution is $s = 0.009i$, $k = 2.2$, $B = 0.36737$ and $Ra = 2705.7$.

11.3 Consider steady, two-dimensional, high Rayleigh-number convection in a fluid with constant physical properties heated uniformly from within at a rate H per unit mass (J kg^{-1} s^{-1}). Let the depth of the fluid layer be d, and assume that the typical width of a convection cell $\sim d$. The top and bottom boundaries are free-slip. The temperature at the top boundary is zero, and the vertical gradient of the temperature vanishes on the bottom boundary (= zero heat flux). Each convection cell comprises a uniform interior with temperature $\sim \Delta T$ and vertical velocity W, a TBL of thickness δ adjacent to the top boundary and a cold downwelling plume with thickness δ and vertical velocity $-w$. Perform a scaling analysis of this situation to determine w, W, δ and ΔT.

12
Solutions to Exercises

Exercise 2.1. The parameters g and $\Delta\rho$ must appear in the combination $g\Delta\rho$. There are six parameters, the SI units of which are $[W] = \text{m}$, $[x] = \text{m}$, $[Q] = \text{m}^3\,\text{s}^{-1}$, $[U] = \text{m s}^{-1}$, $[\eta_p] = \text{kg m}^{-1}\,\text{s}^{-1}$ and $[g\Delta\rho] = \text{kg m}^{-2}\,\text{s}^{-2}$. Three of these parameters have independent dimensions, and so there are three independent dimensionless groups. A natural length scale is $(Q/U)^{1/2}$, whence two groups are

$$\Pi_1 = W\left(\frac{U}{Q}\right)^{1/2}, \quad \Pi_2 = x\left(\frac{U}{Q}\right)^{1/2}.$$

The third group must contain η_p and $g\Delta\rho$. These must appear in the combination $g\Delta\rho/\eta_p$ to remove the units of kg. Using Q and U to remove the remaining units of m and s, we find

$$\Pi_3 = \frac{Qg\Delta\rho}{U^2\eta_p}.$$

We therefore have

$$W = \left(\frac{Q}{U}\right)^{1/2} \text{fct}\left(x\left(\frac{U}{Q}\right)^{1/2}, \frac{Qg\Delta\rho}{U^2\eta_p}\right).$$

Exercise 2.2. Define dimensionless (primed) variables as follows:

$$\mathbf{x} = d\mathbf{x}', \quad \mathbf{u} = \frac{\kappa}{d}\mathbf{u}', \quad t = \frac{d^2}{\kappa}t',$$

$$p = \frac{\kappa\eta_0}{d^2}p', \quad T = T_0 + \frac{d^2 H}{c_p\kappa}T', \quad \eta = \eta_0\eta'.$$

The scales chosen for $\mathbf{u}$, t and p are known as thermal diffusion scaling. The scale for p is chosen to balance the pressure gradient and the viscous term in the momentum equation (2.27b). The temperature scale is chosen to balance the

241

thermal diffusion and heat production terms on the RHS of (2.27b). Rewriting the equations in terms of the dimensionless variables and dropping the primes, we obtain

$$\eta = \exp(-\beta T),$$

$$\nabla \cdot \mathbf{u} = 0,$$

$$0 = -\nabla p - Ra_H T \mathbf{e}_z + \nabla \cdot \{\eta [\nabla \mathbf{u} + (\nabla \mathbf{u})^T]\},$$

$$\frac{DT}{Dt} = \nabla^2 T + 1,$$

where

$$\beta = \frac{d^2 H}{c_p \kappa \Delta T_r}, \quad Ra_H = \frac{\rho_0 g \alpha d^5 H}{c_p \kappa^2 \eta_0}.$$

The dimensionless forms of the boundary conditions are

$$w(0) = w(1) = \partial_z u(0) = \partial_z u(1) = \partial_z T(0) = T(1) = 0,$$

where u and w are the horizontal and vertical components of the velocity, respectively, and the arguments indicate the (dimensionless) value of z. Because no dimensionless groups appear in the boundary conditions, the only ones in the problem are β and Ra_H. β is the ratio of the characteristic temperature scale $\Delta T_H \equiv d^2 H/c_p \kappa$ to the rheological temperature scale ΔT_r. Ra_H is the Rayleigh number based on the temperature scale ΔT_H:

$$Ra_H = \frac{\rho_0 g \alpha d^3 \Delta T_H}{\kappa \eta_0}.$$

Exercise 2.3. The inward radial velocity at the surface of the sphere due to the suction is $-v_0 = -Q/4\pi a^2 < 0$. The presence of this additional radial velocity means that we must balance radial advection and diffusion in the boundary-layer heat equation (2.22b). For a no-slip surface, the surface-tangential velocity in the boundary layer given by Stokes's solution is $u \sim \zeta U/a$ where $\zeta = r - a$. The continuity equation (2.22a) then implies that the radial velocity associated with Stokes's solution is $v \sim \zeta^2 U/a^2$. Balancing radial advection and diffusion at the edge $\zeta \sim \delta$ of the BL, we obtain

$$\max\left(\frac{\delta^2 U}{a^2}, v_0\right) \frac{\Delta T}{\delta} \sim \kappa \frac{\Delta T}{\delta^2}. \tag{12.1}$$

Defining $Nu \sim a/\delta$, $Pe = Ua/\kappa$ and $\epsilon = v_0/U$, we can write (12.1) as

$$\max\left(\frac{1}{Nu^2}, \epsilon\right) \sim \frac{Nu}{Pe}. \tag{12.2}$$

The low-suction limit obtains when $\epsilon \ll Nu^{-2}$, and is $Nu \sim Pe^{1/3}$, in agreement with our previous result for $v_0 = 0$. Substituting this expression for Nu into the condition $\epsilon \ll Nu^{-2}$, we find that the low-suction limit obtains when $\epsilon Pe^{2/3} \ll 1$. Summarizing, we have

$$Nu \sim Pe^{1/3} \quad \text{when} \quad \epsilon Pe^{2/3} \ll 1 \quad \text{(low-suction limit)}. \tag{12.3}$$

At the opposite extreme, the high-suction limit obtains when $\epsilon \gg Nu^{-2}$. Proceeding as before, we obtain

$$Nu \sim \epsilon Pe \quad \text{when} \quad (\epsilon Pe^{2/3})^3 \gg 1 \quad \text{(high-suction limit)}. \tag{12.4}$$

The dimensionless parameter that controls the transition between the two limits is evidently $\epsilon Pe^{2/3}$.

Exercise 3.1. The equation governing F_2 is

$$2F_2'' + \eta F_2' + 2F_2 = 0, \tag{12.5}$$

which has the general solution

$$F_2 = C_1 G_1(\eta) + C_2 G_2(\eta), \tag{12.6a}$$

$$G_1(\eta) = \eta \exp\left(-\frac{\eta^2}{4}\right), \tag{12.6b}$$

$$G_2(\eta) = 1 - \frac{1}{2}\sqrt{\pi}\,\eta \exp\left(-\frac{\eta^2}{4}\right) \text{erfi}\left(\frac{\eta}{2}\right), \tag{12.6c}$$

where erfi is the imaginary error function and C_1 and C_2 are arbitrary constants. Because $G_1(0) = 0$ and $G_2(0) = 1$, we must choose $C_2 = 0$ to satisfy the boundary condition $F_2(0) = 0$. The function G_1 then satisfies the other boundary condition $F_2(\infty) = 0$. The value of C_1 is determined by the temperature moment constraint (3.9), which in terms of F_2 is

$$\int_0^\infty \eta F_2 d\eta = 1. \tag{12.7}$$

Substituting (12.6) into (12.7), we find $C_1 = 1/(2\sqrt{\pi})$. Finally substituting the preceding results into (3.10), we obtain (3.11).

Exercise 3.2. For clarity, let ∇_x and ∇_X be the gradient operators relative to $\mathbf{x}$ and $\mathbf{X}$, respectively. Then

$$\partial_t T = \partial_t \left[\frac{B}{\alpha g (4\kappa t)^{3/2}} \Theta(\mathbf{X}) \right]$$

$$= \frac{B}{\alpha g (4\kappa)^{3/2}} \left[-\frac{3}{2} t^{-5/2} \Theta + t^{-3/2} \nabla_X \Theta \cdot \partial_t \mathbf{X} \right]$$

$$= \frac{B}{\alpha g(4\kappa)^{3/2}} \left[-\frac{3}{2} t^{-5/2} \Theta - \frac{1}{2} t^{-5/2} \frac{1}{(4\kappa t)^{1/2}} \nabla_X \Theta \cdot \mathbf{x} \right]$$
$$= \frac{B}{\alpha g(4\kappa)^{3/2}} \left[-\frac{3}{2} t^{-5/2} \Theta - \frac{1}{2} t^{-5/2} \nabla_X \Theta \cdot \mathbf{X} \right]. \tag{12.8}$$

$$\mathbf{u} \cdot \nabla_x T = \left(\frac{\kappa}{4t}\right)^{1/2} \mathbf{U} \cdot \nabla_x T = \left(\frac{\kappa}{4t}\right)^{1/2} \frac{1}{(4\kappa t)^{1/2}} \mathbf{U} \cdot \nabla_X T$$
$$= \left(\frac{\kappa}{4t}\right)^{1/2} \frac{1}{(4\kappa t)^{1/2}} \frac{B}{\alpha g(4\kappa t)^{3/2}} \mathbf{U} \cdot \nabla_X \Theta$$
$$= \frac{B}{4\alpha g(4\kappa)^{3/2}} t^{-5/2} \mathbf{U} \cdot \nabla_X \Theta. \tag{12.9}$$

$$\kappa \nabla_x^2 T = \kappa \frac{1}{4\kappa t} \nabla_X^2 T = \kappa \frac{1}{4\kappa t} \frac{B}{\alpha g(4\kappa t)^{3/2}} \nabla_X^2 \Theta$$
$$= \frac{B}{4\alpha g(4\kappa)^{3/2}} t^{-5/2} \nabla_X^2 \Theta. \tag{12.10}$$

Now substitute (12.8), (12.9) and (12.10) into (3.12c) to obtain

$$-6\Theta - 2\nabla_X \Theta \cdot \mathbf{X} + \mathbf{U} \cdot \nabla_X \Theta = \nabla_X^2 \Theta. \tag{12.11}$$

To rewrite the LHS of (12.11), note that $\nabla_X \cdot \mathbf{X} = 3$, whence

$$-6\Theta - 2\nabla_X \Theta \cdot \mathbf{X} + \mathbf{U} \cdot \nabla_X \Theta = -2(\nabla_X \cdot \mathbf{X})\Theta - 2\nabla_X \Theta \cdot \mathbf{X} + \mathbf{U} \cdot \nabla_X \Theta$$
$$= -2\nabla_X \cdot (\mathbf{X}\Theta) + \nabla_X \cdot (\mathbf{U}\Theta) - \Theta \nabla_X \cdot \mathbf{U}$$
$$= -2\nabla_X \cdot (\mathbf{X}\Theta) + \nabla_X \cdot (\mathbf{U}\Theta)$$
$$= \nabla_X \cdot [(\mathbf{U} - 2\mathbf{X})\Theta].$$

Exercise 3.3. Substituting the expansions (3.25) into the energy equation (3.15c) and retaining only the lowest-order terms, we obtain

$$-2\nabla \cdot \left(\mathbf{X}\Theta^{(0)}\right) = \nabla^2 \Theta^{(0)}. \tag{12.12}$$

Now set $\Theta^{(0)} = F(r)$ and $\mathbf{X} = r\mathbf{e}_r$, where $\mathbf{e}_r$ is the radial unit normal vector. Then (12.12) takes the form

$$-2\frac{1}{r^2} \partial_r(r^3 F) = \frac{1}{r^2} \partial_r(r^2 \partial_r F)$$

or

$$rF'' + 2(1+r^2)F' + 6rF = 0.$$

The general solution is

$$F = C_1 e^{-r^2} + C_2 \left(\frac{1}{r} - \sqrt{\pi} e^{-r^2} \operatorname{erfi} r\right),$$

where erfi is the imaginary error function and C_1 and C_2 are arbitrary constants. To make the temperature finite at the origin, we must choose $C_2 = 0$. The constant C_1 is then determined from the normalization condition (3.15d), yielding

$$\Theta^{(0)} = \pi^{-3/2} e^{-r^2}.$$

Exercise 3.4. Write the long-wavelength growth rate as

$$\frac{\epsilon\gamma(2\epsilon + 3\gamma)}{2(2\epsilon^3 + 3\gamma + 6\epsilon\gamma^2)} = \frac{N_1 + N_2}{D_1 + D_2 + D_3}. \tag{12.13}$$

Each intermediate asymptotic limit corresponds to a pair of dominant terms in (12.13), one each in the numerator and the denominator. There are six possible pairs, as follows:

1. (N_1, D_1).

   - $N_1 \gg N_2 \to \gamma \ll \epsilon$
   - $D_1 \gg D_2 \to \gamma \ll \epsilon^3$
   - $D_1 \gg D_3 \to \gamma \ll \epsilon$

   The limits on γ are consistent for $\gamma \ll \epsilon^3$.

2. (N_1, D_2)

   - $N_1 \gg N_2 \to \gamma \ll \epsilon$
   - $D_2 \gg D_1 \to \gamma \gg \epsilon^3$
   - $D_2 \gg D_3 \to \gamma \ll \epsilon^{-1}$

   The limits are consistent for $\epsilon^3 \ll \gamma \ll \epsilon$.

3. (N_1, D_3)

   - $N_1 \gg N_2 \to \gamma \ll \epsilon$
   - $D_3 \gg D_1 \to \gamma \gg \epsilon$
   - $D_3 \gg D_2 \to \gamma \gg \epsilon^{-1}$

   The limits are contradictory, so (N_1, D_3) is not an allowable pair.

4. (N_2, D_1)

   - $N_2 \gg N_1 \to \gamma \gg \epsilon$
   - $D_1 \gg D_2 \to \gamma \ll \epsilon^3$
   - $D_1 \gg D_3 \to \gamma \ll \epsilon$

   The limits are contradictory, so (N_2, D_1) is not an allowable pair.

Table 12.1. *Rayleigh–Taylor instability: pressures*

Limit	W/U	$\hat{u}/U$	p_0	p_1	p_1/p_0
$\gamma \ll \epsilon^3$	$\epsilon^2/\gamma \gg 1$	$\epsilon/\gamma \gg 1$	$\eta_0 kW$	$\eta_1 \hat{u}/h_0^2 k$	$\gamma/\epsilon^3 \ll 1$
$\epsilon^3 \ll \gamma \ll \epsilon$	$\epsilon^2/\gamma \sim 1$	$\epsilon/\gamma \gg 1$	$\eta_0 kW$	$\eta_1 \hat{u}/h_0^2 k$	$\gamma/\epsilon^3 \gg 1$
$\epsilon \ll \gamma \ll \epsilon^{-1}$	$\epsilon \ll 1$	$\epsilon/\gamma \sim 1$	$\eta_0 kU$	$\eta_1 \hat{u}/h_0^2 k$	$\epsilon^{-1} \gg 1$
$\epsilon^{-1} \ll \gamma$	$\epsilon \ll 1$	$\epsilon^2 \ll 1$	$\eta_0 kU$	$\eta_1 kU$	$\gamma \gg 1$

5. (N_2, D_2)

- $N_2 \gg N_1 \rightarrow \gamma \gg \epsilon$
- $D_2 \gg D_1 \rightarrow \gamma \gg \epsilon^3$
- $D_2 \gg D_3 \rightarrow \gamma \ll \epsilon^{-1}$

The limits are consistent for $\epsilon \ll \gamma \ll \epsilon^{-1}$.

6. (N_2, D_3)

- $N_2 \gg N_1 \rightarrow \gamma \gg \epsilon$
- $D_3 \gg D_1 \rightarrow \gamma \gg \epsilon$
- $D_3 \gg D_2 \rightarrow \gamma \gg \epsilon^{-1}$

The limits are consistent for $\gamma \gg \epsilon^{-1}$.

Exercise 3.5. The solution is given in tabular form in Table 12.1.

Exercise 4.1. Suppose for definiteness that the particle tends to move towards the wall (Figure 12.1). Reversal of the force (i.e., the applied pressure gradient) and subsequent reflection across a vertical plane leads to a contradiction. The particle therefore has no tendency to migrate sideways.

Exercise 4.2. (a) Using Cartesian tensor notation, write the operator $\mathbf{e}_z \times \nabla$ as $\epsilon_{i3k}\partial_k$. Then

$$\epsilon_{i3k}\partial_k(-\partial_i p + \eta\nabla^2 u_i - g\rho\delta_{i3}) = 0.$$

Now $\epsilon_{i3k}\partial_{ki}^2 p = 0$ because the contraction of a symmetric and an antisymmetric tensor vanishes, and $\epsilon_{i3k}\partial_k(g\rho\delta_{i3}) = 0$ because $\epsilon_{33k} = 0$. The remaining term is

$$\epsilon_{i3k}\partial_k(\eta\nabla^2 u_i) = \eta\nabla^2(\epsilon_{123}\partial_2 u_1 + \epsilon_{231}\partial_1 u_2) = \eta\nabla^2(-\partial_2 u_1 + \partial_1 u_2)$$
$$= \eta\nabla^2[-\partial_2(-\partial_{13}^2\mathcal{P} - \partial_2\mathcal{T}) + \partial_1(-\partial_{23}^2\mathcal{P} + \partial_1\mathcal{T})] = \nabla_h^2\nabla^2\mathcal{T} = 0.$$

(b) Write the operator $\nabla \times (\mathbf{e}_z \times \nabla)$ as

$$\epsilon_{ijk}\partial_j(\epsilon_{k3m}\partial_m) = \epsilon_{kij}\epsilon_{k3m}\partial_{jm}^2 = (\delta_{i3}\delta_{jm} - \delta_{im}\delta_{j3})\partial_{jm}^2 = \delta_{i3}\partial_{jj}^2 - \partial_{i3}^2.$$

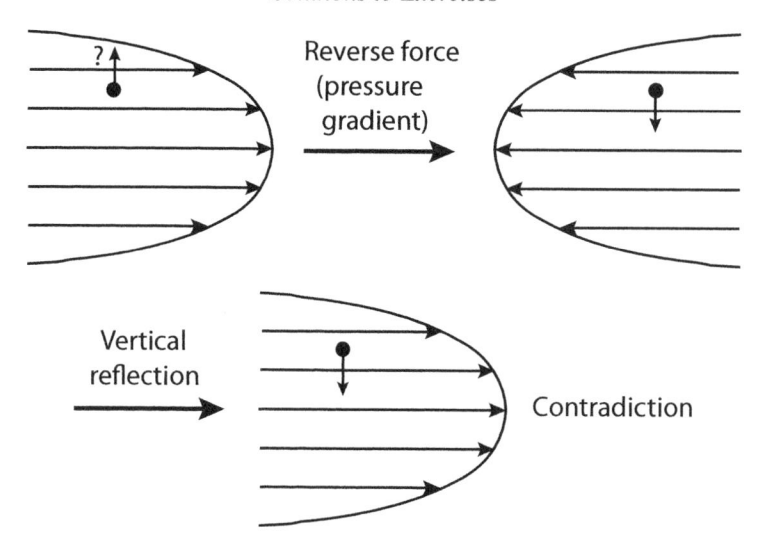

Figure 12.1 A spherical particle freely suspended in a Poiseuille flow has no tendency to migrate towards or away from the wall of the pipe.

Then
$$(\delta_{i3}\partial_{jj}^2 - \partial_{i3}^2)(-\partial_i p + \eta\nabla^2 u_i - g\rho\delta_{i3}) = 0.$$

Now
$$(\delta_{i3}\partial_{jj}^2 - \partial_{i3}^2)\partial_i p = (\partial_{jj3}^3 - \partial_{ii3}^3)p = 0,$$
$$(\delta_{i3}\partial_{jj}^2 - \partial_{i3}^2)(g\rho\delta_{i3}) = g(\partial_{jj}^2 - \partial_{33}^2)\rho = g\nabla_h^2 \rho,$$
$$(\delta_{i3}\partial_{jj}^2 - \partial_{i3}^2)(\eta\nabla^2 u_i) = \eta\nabla^2(\partial_{jj}^2 u_3 - \partial_3 \partial_i u_i) = \eta\nabla^2 \partial_{jj}^2 u_3 = \eta\nabla_h^2 \nabla^4 \mathcal{P}.$$

The poloidal scalar therefore satisfies $\eta\nabla_h^2 \nabla^4 \mathcal{P} = g\nabla_h^2 \rho$.

Exercise 4.3. Take the curl of the momentum equations by taking ∂_θ of (4.152a), multiplying (4.152b) by r and taking ∂_r of the result and forming the difference of the two resulting equations to obtain

$$\partial_r(r\partial_r \tau) + 2\partial_r \tau - \frac{1}{r}\partial_{\theta\theta}^2 \tau = 0. \tag{12.14}$$

Now because (12.14) is linear and homogeneous in τ, the constant prefactor in the expression for τ cancels and can be ignored. The viscosity is

$$\eta \propto (e_{r\theta})^{\frac{1}{n}-1} \propto \left(\frac{F'' + F}{r}\right)^{\frac{1}{n}-1}.$$

The stress τ is therefore

$$\tau \equiv 2\eta e_{r\theta} \propto \left(\frac{F'' + F}{r}\right)^{\frac{1}{n}}. \tag{12.15}$$

Substitution of (12.15) into (12.14) yields

$$r^{-\frac{1}{n}-1}\left(\frac{d^2}{d\phi^2} + \frac{2n-1}{n^2}\right)\left[\left(\frac{d^2F}{d\phi^2} + F\right)^{1/n}\right] = 0.$$

Exercise 4.4. We must prove that

$$\Re\left[(a+ib)\left(\frac{r}{r_0}\right)^{iq_1}\right] \equiv \Re(\Psi) = \gamma \sin\left(q_1 \ln \frac{r}{r_0} + \epsilon\right).$$

Now

$$\Psi = (a+ib)\exp\left[\ln\left(\frac{r}{r_0}\right)^{iq_1}\right] = (a+ib)\exp\left[iq_1 \ln\left(\frac{r}{r_0}\right)\right]$$

$$= (a+ib)\left[\cos\left(q_1 \ln \frac{r}{r_0}\right) + i\sin\left(q_1 \ln \frac{r}{r_0}\right)\right].$$

Therefore

$$\Re(\Psi) = a\cos\left(q_1 \ln \frac{r}{r_0}\right) - b\sin\left(q_1 \ln \frac{r}{r_0}\right).$$

Now define

$$\sin \epsilon = \frac{a}{\sqrt{a^2+b^2}} \equiv \frac{a}{\gamma}, \quad \cos \epsilon = -\frac{b}{\sqrt{a^2+b^2}} \equiv -\frac{b}{\gamma}.$$

Then

$$\Re(\Psi) = \gamma\left[\sin \epsilon \cos\left(q_1 \ln \frac{r}{r_0}\right) + \cos \epsilon \sin\left(q_1 \ln \frac{r}{r_0}\right)\right]$$

$$= \gamma \sin\left(q_1 \ln \frac{r}{r_0} + \epsilon\right).$$

Exercise 4.5. The biharmonic equation is

$$F_{xxxx} + 2F_{xxyy} + F_{yyyy} = 0.$$

Working out the derivatives, we find

$$F_{xxxx} = 4if'''(z) + (y+ix)f''''(z),$$

$$2F_{xxyy} = -2(y+ix)f''''(z),$$

$$F_{yyyy} = -4if'''(z) + (y+ix)f''''(z).$$

Exercise 4.6. (a) We must show that

$$\partial_x \psi + i\partial_y \psi = \phi(z) + \overline{z\phi'(z)} + \overline{\chi'(z)}.$$

Now

$$\partial_x \psi = \Re\left[\phi + \bar{z}\phi' + \chi'\right],$$
$$\partial_y \psi = \Re\left[-i\phi + i\bar{z}\phi' + i\chi'\right] = \Im\left[\phi - \bar{z}\phi' - \chi'\right].$$

Therefore

$$\partial_x \psi + i\partial_y \psi = \Re\phi + i\Im\phi + \Re\left(\bar{z}\phi'\right) - i\Im\left(\bar{z}\phi'\right) + \Re\chi' - i\Im\chi' = \phi + \overline{z\phi'} + \overline{\chi'}.$$

(b) In terms of ψ, σ_{xx} is

$$\sigma_{xx} = -p + 2\eta\partial_x u = -p - 2\eta\partial^2_{xy}\psi.$$

From (4.74),

$$p = 4\eta\Im\left[\phi'(z)\right].$$

Differentiating with respect to x our earlier expression for $\partial_y\psi$, we obtain

$$\partial^2_{xy}\psi = -\Im\left[\bar{z}\phi'' + \chi''\right],$$

whence

$$\sigma_{xx} = -p - 2\eta\partial^2_{xy}\psi = -2\eta\Im\left[2\phi' - \bar{z}\phi'' - \chi''\right].$$

Turning to $\sigma_{yy} \equiv -p + 2\eta\partial^2_{xy}\psi$, we note that this expression differs from the expression for σ_{xx} only by the sign of the second term. The expression (4.75b) therefore follows by inspection. Finally,

$$\sigma_{xy} = \eta\left(\partial^2_{xx}\psi - \partial^2_{yy}\psi\right)$$
$$= \eta\Re\left[2\phi' + \bar{z}\phi'' + \chi'' - \left(2\phi' - \bar{z}\phi'' - \chi''\right)\right] = 2\eta\Re\left(\bar{z}\phi'' + \chi''\right).$$

Exercise 4.7. The definition of the stress tensor is

$$\sigma_{ik} = -p\delta_{ik} + \eta(\partial_i u_k + \partial_k u_i).$$

Substituting into this the expressions (4.78) and dropping the arbitrary vector F_j, we obtain

$$K_{ijk} = -\Pi_j\delta_{ik} + \partial_i J_{kj} + \partial_k J_{ij}.$$

Work out the derivatives by noting that $\partial_i r_k = \delta_{ik}$ and $\partial_k r = r_k/r$.

Exercise 4.8. From (4.89b), the stress due to a line force is

$$\sigma_{ik} = -\frac{1}{\pi}\frac{r_i r_j r_k}{r^4}F_j.$$

Now the normal stress on the surface is σ_{22}, so $i = k = 2$. Moreover, the line force acts vertically, so $j = 2$. To simplify the notation, let $F = F_2$, $x = r_1$ and $y = r_2$. Summing the contributions of the line force at $y = d$ and its reflected image at $y = -d$, we have

$$\sigma_{22} = FK_{222}^+ - FK_{222}^- = F\left(-\frac{y_+^3}{\pi r_+^4} + \frac{y_-^3}{\pi r_-^4}\right)$$

where the super/subscripts $+$ and $-$ refer to the line forces above and below the wall, respectively. Now

$$y_+ = d, \quad y_- = -d, \quad r_+ = r_- = (x^2 + d^2)^{1/2}.$$

Then

$$\sigma_{22} = -\frac{2d^3 F}{\pi(x^2 + d^2)^2} \quad \text{whence} \quad \int_{-\infty}^{\infty}\sigma_{22}\mathrm{d}x = -F.$$

Exercise 4.9. The poloidal scalar $\mathcal{P}$ satisfies

$$\eta\nabla^4\mathcal{P} = f\delta(x)\delta(y)\delta(z - z'),$$

where f is the magnitude of the point force. Nondimensionalizing $\mathbf{x}$ by d and $\mathcal{P}$ by fd/η, we obtain the dimensionless equation

$$\nabla^4\mathcal{P} = \delta(x)\delta(y)\delta(z - z'). \tag{12.16}$$

Define the 2-D Fourier transform pair

$$\overline{\mathcal{P}}(\mathbf{k}) = \frac{1}{2\pi}\int \mathcal{P}(\mathbf{x})\exp(-i\mathbf{k}\cdot\mathbf{x})\mathrm{d}\mathbf{x}, \tag{12.17a}$$

$$\mathcal{P}(\mathbf{x}) = \frac{1}{2\pi}\int \overline{\mathcal{P}}(\mathbf{k})\exp(i\mathbf{x}\cdot\mathbf{k})\mathrm{d}\mathbf{k}, \tag{12.17b}$$

where an overbar denotes the Fourier transform, $\mathbf{k}$ is the 2-D wavevector nondimensionalized by d^{-1} and the integrals are over the whole $x - y$ and $k_x - k_y$ planes, respectively. The Fourier transform of (12.16) is

$$(D^2 - k^2)^2\overline{\mathcal{P}} = \frac{1}{2\pi}\delta(z - z'), \tag{12.18}$$

where $D = \mathrm{d}/\mathrm{d}z$ and $k = |\mathbf{k}|$. The general solutions of (12.18) for $z \neq z'$ that satisfy the free-slip boundary conditions are

$$\overline{\mathcal{P}}_1 = A_1 \sinh kz + B_1 kz \cosh kz \quad (0 \leq z < z'), \tag{12.19a}$$

$$\overline{\mathcal{P}}_2 = A_2 \sinh k(z-1) + B_2 k(z-1) \cosh k(z-1) \quad (z' < z \leq 1). \tag{12.19b}$$

Continuity of the velocity and the shear stress at $z = z'$ requires

$$[\overline{\mathcal{P}}] = [D\overline{\mathcal{P}}] = [D^2\overline{\mathcal{P}}] = 0,$$

where $[\ldots]$ is the jump in the enclosed quantity moving upward across $z = z'$. The fourth matching condition is obtained by integrating (12.18) across $z = z'$ and is

$$[D^3\overline{\mathcal{P}}] = \frac{1}{2\pi}.$$

Substitution of (12.19) into the matching conditions yields four simultaneous linear equations for A_1, B_1, A_2 and B_2. The solutions are

$$A_1 = \frac{\cosh k(z'-2) - kz' \sinh k(z'-2) - \cosh kz' + k(z'-2) \sinh kz'}{8\pi k^3 \sinh^2 k},$$

$$B_1 = \frac{\sinh k(z'-1)}{4\pi k^3 \sinh k},$$

$$A_2 = \frac{kz' \cosh kz' - (1 + k \coth k) \sinh kz'}{4\pi k^3 \sinh k},$$

$$B_2 = \frac{\sin kz'}{4\pi k^3 \sinh k}.$$

The poloidal scalar is now obtained by inverse Fourier transformation using (12.17b). The integral must be evaluated numerically.

Exercise 4.10 (Pozrikidis, 1992, § 7.3). The source quadrupole Green function is

$$G^{SQ}_{ijk} \equiv -\frac{\partial G^{SD}_{ij}}{\partial r_k} = \frac{1}{4\pi}\left(15\frac{r_i r_j r_k}{r^7} - 3\frac{\delta_{ij} r_k + \delta_{jk} r_i + \delta_{ki} r_j}{r^5}\right).$$

Write the total flow as

$$u_i = A_{ij} r_j + B_{kl} G^{FD}_{ikl} + C_{kl} G^{SQ}_{ikl}. \tag{12.20}$$

The no-slip boundary condition is then

$$0 = A_{ij} r_j + B_{kl} G^{FD}_{ikl}|_{r=a} + C_{kl} G^{SQ}_{ikl}|_{r=a}$$

where the point r_j is on the surface of the sphere. Cancellation of the terms $\propto r_i r_j r_k$ requires

$$\mathbf{B} = -\frac{10}{a^2}\mathbf{C}.$$

The remaining terms in the boundary condition are then

$$0 = A_{ij}r_j - \frac{1}{2\pi a^5}C_{kl}(4\delta_{ik}r_l - \delta_{il}r_k - \delta_{kl}r_i)$$

$$= A_{ij}r_j - \frac{1}{2\pi a^5}(4C_{il}r_l - C_{ki}r_k - C_{kk}r_i)$$

$$= A_{ij}r_j - \frac{1}{2\pi a^5}(4C_{ij}r_j - C_{ji}r_j - C_{kk}r_j\delta_{ij})$$

$$= \left[A_{ij} - \frac{1}{2\pi a^5}(4C_{ij} - C_{ji} - C_{kk}\delta_{ij})\right]r_j.$$

Because the vector r_j is an arbitrary point on the sphere, the terms in square brackets must vanish, yielding

$$\mathbf{A} = \frac{1}{2\pi a^5}[4\mathbf{C} - \mathbf{C}^T - \mathbf{I}\,\text{tr}(\mathbf{C})],$$

where ^{T} denotes the transpose, $\mathbf{I}$ is the identity tensor and tr denotes the trace. The next step is to invert the preceding equation to determine $\mathbf{C}$ in terms of $\mathbf{A}$. The easiest way to do this is to assume $\text{tr}(\mathbf{C}) = 0$ and to verify this assumption a posteriori. Splitting $\mathbf{A}$ and $\mathbf{C}$ into symmetric (superscript S) and antisymmetric (superscript A) parts, we have

$$\mathbf{A}^S + \mathbf{A}^A = \frac{1}{2\pi a^5}\left[4(\mathbf{C}^S + \mathbf{C}^A) - (\mathbf{C}^S - \mathbf{C}^A)\right] = \frac{1}{2\pi a^5}(3\mathbf{C}^S + 5\mathbf{C}^A).$$

The symmetric and antisymmetric parts of the preceding equation are $\mathbf{C}^S = 2\pi a^5 \mathbf{A}^S/3$ and $\mathbf{C}^A = 2\pi a^5 \mathbf{A}^A/5$, whence

$$\mathbf{C} = \frac{2\pi a^5}{15}(5\mathbf{A}^S + 3\mathbf{A}^A) = \frac{2\pi a^5}{15}(4\mathbf{A} + \mathbf{A}^T).$$

Because $\text{tr}(\mathbf{A}) = 0$, $\text{tr}(\mathbf{C}) = 0$ also, validating our earlier assumption.

To make the contour plot, set $B_{kl} = -20\pi a^3 A_{kl}/3$ and $C_{kl} = 2\pi a^5 A_{kl}/3$ in (12.20). Using spherical coordinates (r, θ, ϕ) we write

$$r_1 = r\cos\phi\sin\theta, \quad r_2 = r\sin\phi\sin\theta, \quad r_3 = r\cos\theta.$$

The radial velocity is

$$v = u_1\cos\phi\sin\theta + u_2\sin\phi\sin\theta + u_3\cos\theta$$

$$= \dot{\epsilon}(r-a)^2\frac{3a^3 + 6a^2r + 4ar^2 + 2r^3}{8r^4}(1 + 3\cos 2\theta).$$

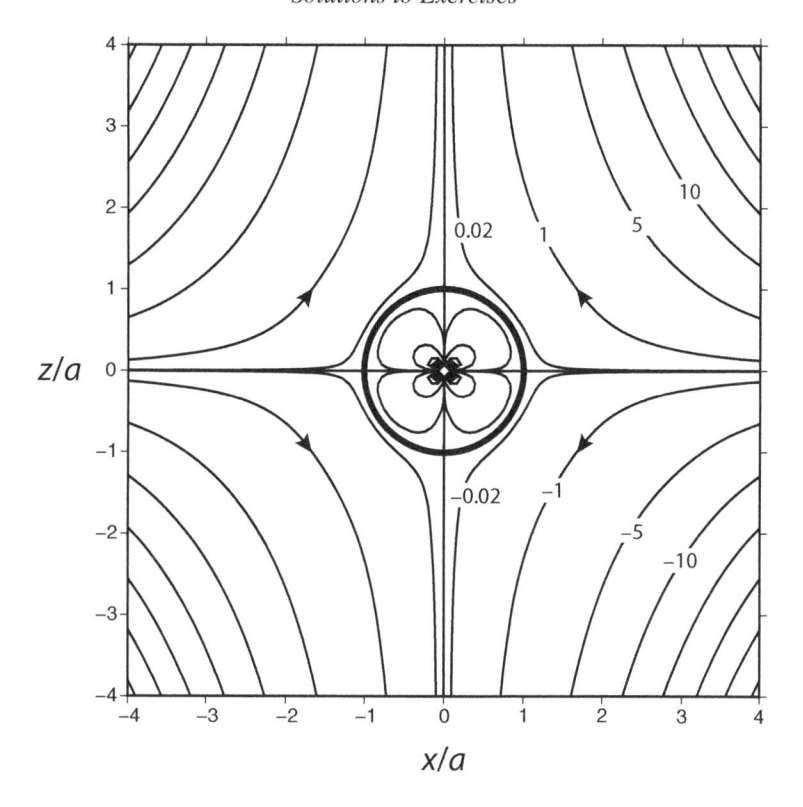

Figure 12.2 Contour plot of the streamfunction for the flow around a sphere embedded in an axisymmetric pure straining flow. The line $x = 0$ is the symmetry axis, and extension is in the vertical direction. Values of the contours are in units of $\dot\epsilon a^3$.

The streamfunction is obtained by integrating the definition $\partial_\theta \psi = vr^2 \sin\theta$, which yields

$$\psi = \dot\epsilon (r-a)^2 \frac{3a^3 + 6a^2 r + 4ar^2 + 2r^3}{4r^4} \cos\theta \sin^2\theta.$$

Plotting is facilitated by using Cartesian coordinates (x, z) such that $\cos\theta = z/r$, $\sin\theta = x/r$ and $r = (x^2 + z^2)^{1/2}$. Figure 12.2 is a contour plot of the streamlines in a plane containing the symmetry axis $x = 0$. The streamfunction is singular at the origin, which is inside the sphere.

Exercise 4.11. Form the dot product of (4.112) and **t** and solve the resulting equation for $\mathbf{F} \cdot \mathbf{t}$ to obtain

$$\mathbf{F} \cdot \mathbf{t} = -4\pi \eta \epsilon L \mathbf{U} \cdot \mathbf{t}.$$

Use the preceding equation to eliminate $\mathbf{F} \cdot \mathbf{t}$ from (4.112) and then solve the resulting equation for **F**.

Exercise 4.12. We consider an axisymmetric ($m = 0$) flow without loss of generality. Define

$$U = y_1 Y_l^0, \quad V = y_2 \partial_\theta Y_l^0, \quad \Sigma_{rr} = \frac{\eta_0 y_3}{r} Y_l^0, \quad \Sigma_{r\theta} = \frac{\eta_0 y_4}{r} \partial_\theta Y_l^0. \quad (12.21)$$

In the following, $D = d/dr$ and primes denote derivatives with respect to $z = \ln r$.

The first row of **A** corresponds to the continuity equation. The divergence of the velocity is

$$\nabla \cdot \mathbf{u} = \frac{1}{r^2} \partial_r (r^2 U) + \frac{1}{r \sin \theta} \partial_\theta (V \sin \theta)$$

$$= \frac{1}{r}(2y_1 - Ly_2 + rDy_1)Y_l^0 = \frac{1}{r}(2y_1 - Ly_2 + y_1')Y_l^0, \quad (12.22)$$

where the relation $\partial_{\theta\theta}^2 Y_l^0 = -LY_l^0 - \cot\theta \partial_\theta Y_l^0$ has been used. The continuity equation is therefore $y_1' = -2y_1 + Ly_2$, which gives $A_{11} = -2$ and $A_{12} = L$.

The second row of **A** corresponds to the definition of the shear stress

$$\Sigma_{r\theta} = \eta_n [r \partial_r (V/r) + (1/r) \partial_\theta U]. \quad (12.23)$$

Substitution of (12.21) into (12.23) yields

$$\frac{\eta_0 y_4}{r} \partial_\theta Y_l^0 = \eta_n \left[r \partial_r \left(\frac{y_2}{r} \right) + \frac{1}{r} y_1 \right] \partial_\theta Y_l^0, \quad (12.24)$$

which is equivalent to

$$y_4 = \eta^* (rDy_2 - y_2 + y_1) = \eta^* (y_2' - y_2 + y_1).$$

Rearranging, we obtain $y_2' = -y_1 + y_2 + y_4/\eta^*$, which implies $A_{21} = -1, A_{22} = 1$ and $A_{24} = 1/\eta^*$.

The third row of **A** corresponds to the radial component of the momentum equation, which is

$$(\nabla \cdot \boldsymbol{\Sigma})_r - g_0 \hat{\rho} Y_l^0 - \rho_n \partial_r \hat{\chi} Y_l^0 = 0, \quad (12.25)$$

where $(\nabla \cdot \boldsymbol{\Sigma})_r$ is the $\mathbf{e}_r$-component of the divergence of the stress tensor. Relative to general orthogonal curvilinear coordinates (q_1, q_2, q_3),

$$(\nabla \cdot \boldsymbol{\Sigma})_1 = \frac{1}{h_1 h_2 h_3} \left[\frac{\partial}{\partial q_1}(\Sigma_{11} h_2 h_3) + \frac{\partial}{\partial q_2}(\Sigma_{12} h_3 h_1) + \frac{\partial}{\partial q_3}(\Sigma_{13} h_1 h_2) \right]$$

$$+ \frac{\Sigma_{12}}{h_1 h_2} \frac{\partial h_1}{\partial q_2} + \frac{\Sigma_{31}}{h_3 h_1} \frac{\partial h_1}{\partial q_3} - \frac{\Sigma_{22}}{h_1 h_2} \frac{\partial h_2}{\partial q_1} - \frac{\Sigma_{33}}{h_3 h_1} \frac{\partial h_3}{\partial q_1}, \quad (12.26)$$

where (h_1, h_2, h_3) are the scale factors for the coordinates. The expressions for $(\nabla \cdot \boldsymbol{\Sigma})_2$ and $(\nabla \cdot \boldsymbol{\Sigma})_3$ are obtained from (12.26) by cyclic permutation of indices. The radial component of the divergence for our axisymmetric problem is obtained by setting $q_1 = r$, $q_2 = \theta$, $h_1 = 1$, $h_2 = r$, $h_3 = r\sin\theta$, $\partial/\partial q_3 = 0$, $\Sigma_{11} = \Sigma_{rr}$, $\Sigma_{12} = \Sigma_{r\theta}$, $\Sigma_{22} = -\hat{p}Y_l^0 + (2\eta_n/r)(\partial_\theta V + U)$, $\Sigma_{33} = -\hat{p}Y_l^0 + (2\eta_n/r)(V\cot\theta + U)$ and $\Sigma_{13} = 0$. The result is

$$(\nabla \cdot \mathbf{\Sigma})_r = \left(\frac{2\hat{p}}{r} + \frac{2\hat{\sigma}_{rr}}{r} - \frac{L\hat{\sigma}_{r\theta}}{r} - \frac{4\eta_n}{r^2}\hat{u} + \frac{2\eta_n L}{r^2}\hat{v} + D\hat{\sigma}_{rr}\right)Y_l^0. \qquad (12.27)$$

Now eliminate the pressure $\hat{p} = 2\eta_n D\hat{u} - \hat{\sigma}_{rr}$ from (12.27) and substitute the result into (12.25) to obtain

$$\frac{4\eta_n}{r}D\hat{u} - \frac{L\hat{\sigma}_{r\theta}}{r} - \frac{4\eta_n}{r^2}\hat{u} + \frac{2\eta_n L}{r^2}\hat{v} + D\hat{\sigma}_{rr} - g_0\hat{\rho} - \rho_n\partial_r\hat{\chi} = 0. \qquad (12.28)$$

Substituting into (12.28) the definitions (4.141) of the variables y_1–y_6, we obtain

$$\frac{4\eta_n}{r}Dy_1 - \frac{L\eta_0}{r^2}y_4 - \frac{4\eta_n}{r^2}y_1 + \frac{2L\eta_n}{r^2}y_2 + \eta_0 D\left(\frac{y_3}{r}\right) - \frac{\rho_n\eta_0}{\rho_0 r^2}y_6 - g_0\hat{\rho} = 0. \qquad (12.29)$$

Now multiply (12.29) by r^2/η_0 and transform the radial derivatives to derivatives with respect to $z = \ln r$. Solving the resulting equation for y_3' and eliminating $y_1' = -2y_1 + Ly_2$, we obtain

$$y_3' = 12\eta^* y_1 - 6L\eta^* y_2 + y_3 + Ly_4 + \rho^* y_6 + r^2 g_0\hat{\rho}/\eta_0.$$

Thus $A_{31} = 12\eta^*$, $A_{32} = -6L\eta^*$, $A_{33} = 1$, $A_{34} = L$ and $A_{36} = \rho^*$.

Exercise 4.13. The flow satisfies the Stokes momentum equation

$$0 = \nabla \cdot \hat{\sigma} - \hat{\rho}g_0\mathbf{e}_z,$$

where $\hat{\sigma}$ is the perturbation stress tensor and $\mathbf{e}_z$ is an upward-pointing vertical unit vector. Now integrate the preceding equation over the volume V of the mantle and use the divergence theorem to obtain

$$0 = \int_{S_1+S_2} \hat{\sigma} \cdot \mathbf{n} dS - \mathbf{e}_z \int_V \hat{\rho}g_0 dV,$$

where S_1 is the upper surface of the layer and S_2 is the lower surface. The vertical component of the preceding equation is

$$0 = \int_{S_1} \hat{\sigma}_{zz}(d)dS - \int_{S_2} \hat{\sigma}_{zz}(0)dS - \int_V \hat{\rho}g_0 dV, \qquad (12.30)$$

where the arguments of $\hat{\sigma}_{zz}$ are values of z. The sign of the second integral on the RHS is negative because the outward normal $\mathbf{n}$ to the lower surface points down. By analogy to (4.133),

$$\hat{\sigma}_{zz}(d) = -g_0(\rho_0 - \rho_w)\hat{\zeta}_1, \quad \hat{\sigma}_{zz}(0) = -g_0(\rho_0 - \rho_c)\hat{\zeta}_2,$$

where $\hat{\zeta}_j$ is the deflection of surface S_j. Thus the integral balance (12.30) becomes

$$\int_V g_0 \hat{\rho} dV = -\int_{S_1} g_0(\rho_0 - \rho_w)\hat{\zeta}_1 dS - \int_{S_2} g_0(\rho_c - \rho_0)\hat{\zeta}_2 dS,$$

which was to be proved.

Exercise 5.1. The general laterally periodic solution for χ is

$$\chi = \left[(A + Bkz)e^{kz} + (C + Dkz)e^{-kz}\right] \cos kx.$$

The constants A–D are determined by the boundary conditions

$$-\partial_{xz}^2 \chi|_{z=-T/2} = -\partial_{xz}^2 \chi|_{z=T/2} = \partial_{xx}^2 \chi|_{z=-T/2} = 0$$

and

$$\partial_{xx}^2 \chi|_{z=T/2} = -[g\Delta(\rho_c - \rho_w) + g\xi(\rho_m - \rho_c)]\cos kx,$$

and are

$$A = \frac{[(2+k')\exp(2k') - 2k'^2 + 3k' - 2]\Phi}{4\exp(k'/2)},$$

$$B = -\frac{[\exp(2k') + 2k' - 1]\Phi}{2\exp(k'/2)},$$

$$C = -\frac{[(2k'^2 + 3k' + 2)\exp(2k') + k' - 2]\Phi}{4\exp(3k'/2)},$$

$$D = -\frac{[(2k'+1)\exp(2k') - 1]\Phi}{2\exp(3k'/2)},$$

where $k' = kT$ and

$$\Phi = \frac{g[\Delta(\rho_c - \rho_w) + \xi(\rho_m - \rho_c)]}{k^2(\cosh 2k' - 2k'^2 - 1)}.$$

The displacements $\zeta_1(z) \sin kx$ and $\zeta_3(z) \cos kx$ within the plate are then determined from the constitutive relations

$$\sigma_{xz} \equiv -\partial_{xz}^2 \chi = \mu(\zeta_1' - k\zeta_3) \sin kx$$

and

$$\sigma_{zz} \equiv \partial_{xx}^2 \chi = \left[2\mu\zeta_3' + \left(K - \frac{2}{3}\mu\right)(\zeta_3' + k\zeta_1)\right] \cos kx,$$

where primes denote differentiation with respect to z. The preceding constitutive relations are two coupled first-order ODEs for $\zeta_1(z)$ and $\zeta_3(z)$, of which only the particular solutions are relevant. The particular solution for $\zeta_3(z)$ is

$$\zeta_3 = \frac{k}{2\mu(3K+\mu)} \left\{ e^{kz}[3\mu B - (3K+\mu)(A+Bkz)] \right.$$
$$\left. + e^{-kz}[3\mu D + (3K+\mu)(C+Dkz)] \right\}.$$

Now by definition, $\xi = \zeta_3|_{z=T/2}$. Injecting into this equation the expressions for A–D and solving for Δ/ξ, we obtain

$$\frac{\Delta}{\xi} = -\frac{\rho_m - \rho_c}{\rho_c - \rho_w} - \frac{1}{2(1-\nu)} \frac{\mu F(k')}{gT(\rho_c - \rho_w)}$$

where ν is Poisson's ratio and F is given by (5.15). The preceding solution agrees with (5.14) in the incompressible limit $\nu = 1/2$.

Exercise 5.2. The displacement vector is $\hat{\zeta} = y_1 \mathbf{e}_r + y_2 \mathbf{e}_\theta$. Define

$$U = y_1 P_l(\theta), \quad V = y_2 P'_l(\theta), \quad \Sigma_{rr} = y_3 P_l(\theta),$$
$$\Sigma_{r\theta} = y_4 P'_l(\theta), \quad X = y_5 P_l(\theta), \quad R = \hat{\rho} P_l(\theta). \quad (12.31)$$

The dilatation $\Delta = \nabla \cdot \hat{\zeta}$ is

$$\Delta = \frac{1}{r^2} \partial_r(r^2 U) + \frac{1}{r \sin\theta} \partial_\theta(V \sin\theta) \equiv \frac{1}{r}(2y_1 - Ly_2 + ry'_1) P_l(\theta), \quad (12.32)$$

where the relation $P''_l(\theta) = -LP_l(\theta) - \cot\theta\, P'_l(\theta)$ has been used.

The first row of $\mathbf{A}$ corresponds to the definition of the radial normal stress,

$$\Sigma_{rr} = (K - 2\overline{\mu}/3)\Delta + 2\overline{\mu} \partial_r U. \quad (12.33)$$

Substituting (12.31) and (12.32) into (12.33) and solving for y'_1, we obtain the first row of (5.22).

The second row of $\mathbf{A}$ corresponds to the definition of the shear stress,

$$\Sigma_{r\theta} = \overline{\mu}[r\partial_r(V/r) + (1/r)\partial_\theta U]. \quad (12.34)$$

Substituting (12.31) into (12.34) and solving for y'_2, we obtain the second row of (5.22).

The third row of $\mathbf{A}$ corresponds to the radial component of the momentum equation. To write this down in terms of the variables y_1–y_6, we need the radial component of the divergence of the stress tensor Σ. Relative to general orthogonal curvilinear coordinates (q_1, q_2, q_3),

$$\nabla \cdot \mathbf{\Sigma} \cdot \mathbf{e}_1 = \frac{1}{h_1 h_2 h_3} \left[\frac{\partial}{\partial q_1}(\Sigma_{11} h_2 h_3) + \frac{\partial}{\partial q_2}(\Sigma_{12} h_3 h_1) + \frac{\partial}{\partial q_3}(\Sigma_{13} h_1 h_2) \right]$$
$$+ \frac{\Sigma_{12}}{h_1 h_2} \frac{\partial h_1}{\partial q_2} + \frac{\Sigma_{31}}{h_3 h_1} \frac{\partial h_1}{\partial q_3} - \frac{\Sigma_{22}}{h_1 h_2} \frac{\partial h_2}{\partial q_1} - \frac{\Sigma_{33}}{h_3 h_1} \frac{\partial h_3}{\partial q_1}, \tag{12.35}$$

where (h_1, h_2, h_3) are the scale factors for the coordinates. The remaining components of $\nabla \cdot \mathbf{\Sigma}$ are obtained from (12.35) by cyclic permutation of indices. The radial component of the divergence of the stress tensor for our axisymmetric problem is obtained by setting $q_1 = r$, $q_2 = \theta$, $h_1 = 1$, $h_2 = r$, $h_3 = r \sin \theta$, $\partial/\partial q_3 = 0$, $\Sigma_{11} = \Sigma_{rr}$, $\Sigma_{12} = \Sigma_{r\theta}$, $\Sigma_{22} = (K - 2\bar{\mu}/3)\Delta + (2\bar{\mu}/r)(\partial_\theta V + U)$, $\Sigma_{33} = (K - 2\bar{\mu}/3)\Delta + (2\bar{\mu}/r)(V \cot \theta + U)$ and $\Sigma_{13} = 0$. We thereby find

$$\nabla \cdot \mathbf{\Sigma} \cdot \mathbf{e}_r = \left[y_3' + \frac{2}{r} y_3 - \frac{L}{r} y_4 + \frac{2(K + \bar{\mu}/3)}{r^2}(Ly_2 - 2y_1) - \frac{2\lambda}{r} y_1' \right] P_l(\theta).$$

The radial momentum equation is

$$0 = \nabla \cdot \mathbf{\Sigma} \cdot \mathbf{e}_r - \rho_0 \partial_r X - g_0 R - \partial_r(\rho_0 g_0 U). \tag{12.36}$$

where

$$R = -\nabla \cdot (\rho_0 \mathbf{U}) \equiv \left[\frac{\rho_0}{r}(Ly_2 - 2y_1) - \rho_0' y_1 - \rho_0 y_1' \right] P_l(\theta).$$

Now substitute into (12.36) the definitions of $\nabla \cdot \mathbf{\Sigma} \cdot \mathbf{e}_r$, U, X and R and the definitions (5.21) of the variables y_1–y_6. The result is

$$y_3' + \frac{2}{r} y_3 - \frac{L}{r} y_4 + \frac{2(K + \bar{\mu}/3)}{r^2}(Ly_2 - 2y_1) - \frac{2\lambda}{r} y_1'$$
$$- \rho_0 \left(y_6 - \frac{M}{r} y_5 - 4\pi G \rho_0 y_1 \right) - \frac{\rho_0 g_0}{r}(Ly_2 - 2y_1) - \rho_0 g_0' y_1 = 0.$$

Next, eliminate g_0' using $g_0' = -2g_0/r + 4\pi G \rho_0$, and eliminate y_1' using the expression from the first row of (5.22). Solving the resulting equation for y_3', we obtain the third row of (5.22).

Exercise 6.1. The governing equation in the boundary-layer approximation is

$$w T_r + \frac{u}{a} T_\theta = \kappa T_{rr},$$

where the velocity $w\mathbf{e}_r + u\mathbf{e}_\theta$ is given by the Stokes solution. The boundary conditions are

$$T(a, \theta) - T_0 = T(\infty, \theta) - T_\infty = 0.$$

Now define dimensionless (primed) variables $r' = r/a$, $\mathbf{u}' = \mathbf{u}/U$ and $T' = (T - T_\infty)/(T_0 - T_\infty)$. Dropping the primes, we may write the dimensionless governing equation and boundary conditions as

$$Pe(wT_r + uT_\theta) = T_{rr},$$

$$T(1,\theta) - 1 = T(\infty, \theta) = 0.$$

Now the velocity components are defined in terms of the Stokes streamfunction ψ as

$$(w, u) = \left(\frac{1}{r^2 \sin\theta} \psi_\theta, -\frac{1}{r \sin\theta} \psi_r \right).$$

But $r \approx 1$ in the BL, so

$$(w, u) = \left(\frac{1}{\sin\theta} \psi_\theta, -\frac{1}{\sin\theta} \psi_r \right).$$

We now apply the von Mises transformation to change the independent variables from (r, θ) to (ψ, θ). Using the chain rule, we find

$$T_r = \psi_r T_\psi \equiv -u \sin\theta \, T_\psi,$$

$$T_{rr} = \psi_r (\psi_r T_\psi)_\psi \equiv u \sin^2\theta (uT_\psi)_\psi.$$

Finally, note that T_θ in the BL equation is actually the partial derivative $T_\theta|_r$ at constant r and must be rewritten in terms of the derivative $T_\theta|_\psi$ at constant ψ. The chain rule gives

$$T_\theta|_r = T_\theta|_\psi + T_\psi|_\theta \, \psi_\theta|_r = T_\theta|_\psi + w \sin\theta \, T_\psi.$$

Substitution of the preceding expressions into the BL equation yields

$$PeT_\theta = \sin^2\theta \, (uT_\psi)_\psi.$$

Now within the BL, the Stokes solution is

$$\psi \approx -\frac{3}{4} \zeta^2 \sin^2\theta, \quad u \approx \frac{3}{2} \zeta \sin\theta,$$

where $\zeta = r - 1$ is the (dimensionless) distance from the sphere's surface. The preceding expressions imply

$$u = \sqrt{-3\psi}.$$

The BL equation therefore becomes

$$PeT_\theta = \sqrt{3} \sin^2\theta \, (\sqrt{-\psi} T_\psi)_\psi.$$

Now define a new variable

$$t = \frac{\sqrt{3}}{Pe}\int_0^\theta \sin^2\theta \, d\theta = \frac{\sqrt{3}}{2Pe}\left(\theta - \frac{1}{2}\sin 2\theta\right),$$

in terms of which the BL equation becomes

$$T_t = (\sqrt{-\psi}\, T_\psi)_\psi.$$

This is a linear diffusion equation with a diffusivity that depends on the 'position' ψ; t is a time-like variable that is zero at the front stagnation point and increases along the BL.

Because the BL is thin, it sees the sphere locally as a flat surface, and so a is not a length scale for the solution, which is self-similar. To find the similarity variable, we use a different method than the one used in the rest of the book: we look for a change of scale $t \to \mu t$ and $\psi \to \lambda\psi$ that leaves the diffusion equation and its boundary conditions unchanged. The rescaled equation is

$$\frac{1}{\mu}T_t = \frac{1}{\lambda^{3/2}}(\sqrt{-\psi}\, T_\psi)_\psi,$$

which is the same as the original equation if $\mu = \lambda^{3/2}$. The similarity variable is therefore

$$\xi = \frac{-\psi}{t^{2/3}},$$

which is the only combination of ψ and t that remains invariant under the transformation $t \to \lambda^{3/2}t$, $\psi \to \lambda\psi$. In terms of ξ, the diffusion equation takes the form

$$-\frac{2}{3}\xi T' = (\sqrt{\xi}\, T')',$$

where primes denote differentiation with respect to ξ. Now set $z = \sqrt{\xi}$, whereupon the diffusion equation becomes

$$T_{zz} + \frac{4}{3}z^2 T_z = 0.$$

The solution that satisfies the boundary conditions $T(z=0) = 1$, $T(z=\infty) = 0$ is

$$T = A\int_0^z \exp\left(-\frac{4}{9}z^3\right)dz + B,$$

$$A = -\left[\int_0^\infty \exp\left(-\frac{4}{9}z^3\right)dz\right]^{-1} = -\frac{3}{(9/4)^{1/3}\Gamma(1/3)} \approx -0.855,$$

$$B = 1.$$

The variable z is related to the original variables (ζ, θ) by

$$z = \left[\frac{3Pe}{4(\theta - \sin\theta \cos\theta)}\right]^{1/3} \zeta \sin\theta.$$

The dimensionless heat flux per unit area from the sphere is

$$q \equiv -T_\zeta|_{\zeta=0} = \frac{9^{1/3}}{\Gamma(1/3)} \frac{\sin\theta}{(\theta - \sin\theta \cos\theta)^{1/3}} Pe^{1/3}.$$

The total heat flux (Nusselt number) is obtained by integrating q over the sphere:

$$Nu = 2\pi \int_0^\pi q \sin\theta \, d\theta = \frac{(3\pi)^{5/3}}{2\Gamma(1/3)} Pe^{1/3} \approx 7.85 Pe^{1/3}.$$

Exercise 6.2. Written in scalar form, (4.114b) is

$$F_{\text{longitudinal}} = -\frac{4\pi \eta L U}{\ln(2L/R)}.$$

Now $F_{\text{longitudinal}}$ is the total force on a rod of length $2L$, whereas ρF is a force per unit length of the plume. Now make the substitutions $2L \to z$, $R \to a$, $U \to w$ and $F_{\text{longitudinal}} \to -2L\rho F$, where the negative sign is present because F is the force exerted by the plume on the surrounding fluid. Rearranging, we obtain (6.33).

Exercise 6.3. Scaling the Stokes equation (6.30b), we find

$$\frac{\nu \tilde{w}}{a^2} \sim b \quad \Rightarrow \quad \tilde{w} \sim \frac{ba^2}{\nu} \sim \frac{F}{\nu} \sim \left(\frac{B}{\nu}\right)^{1/2} \left[\ln\left(\frac{z}{a}\right)\right]^{-1/2}.$$

Now the scaling for w_0 is given by (6.33). Thus

$$\frac{\tilde{w}}{w_0} \sim \left[\ln\left(\frac{z}{a}\right)\right]^{-1} \ll 1.$$

Exercise 6.4. The boundary-layer momentum equation is obtained from the vertical component of (6.30b). Neglecting the vertical derivatives in the Laplacian and the vertical derivative of the modified pressure, we have

$$\frac{\nu}{s} \frac{\partial}{\partial s}\left(s \frac{\partial w}{\partial s}\right) = -b.$$

Inserting the expressions (6.37) and (6.38) for b and w and the definition (6.36) of ξ, we obtain (6.39a).

The boundary-layer form of the buoyancy transport equation is

$$u\frac{\partial b}{\partial s} + w\frac{\partial b}{\partial z} = \kappa \frac{1}{s}\frac{\partial}{\partial s}\left(s\frac{\partial b}{\partial s}\right). \tag{12.37}$$

Now from (6.35) and (6.37a), $u = -4\kappa(zf)_z/s$. Now substitute this expression together with (6.38) and (6.37b) into (12.37), and use (6.36) to transform s to ξ. The result is (6.39b).

Finally, (6.31) is transformed into (6.39c) by straightforward substitution of (6.36), (6.37b) and (6.38).

Exercise 6.5. Denote the scales for x, y, u, v and T by X, Y, U, V and T, respectively. Scaling the continuity equation (6.2a), we obtain $U/X \sim V/Y$. Scaling of the boundary-layer momentum equation (6.57) yields $\eta_0 U/Y^2 \sim \rho_0 g\alpha T$. Scaling of the heat equation (6.2b) gives $UT/X \sim \kappa T/Y^2$. Finally, scaling of the definition (6.60) of the heat flux gives $Q \sim \rho_0 c_p TUY^2$. Setting $T \sim \gamma^{-1}$ and solving the four coupled equations, we obtain

$$X = \frac{\gamma Q}{\kappa c_p \rho_0} \equiv \frac{\gamma Q}{k}, \quad Y = \frac{X}{Ra^{1/4}}, \quad U = \frac{\kappa}{X}Ra^{1/2}, \quad V = \frac{\kappa}{X}Ra^{1/4}.$$

Exercise 6.6 (Nayfeh, 1973) Begin with the outer solution, which we write as $y = h(x, \epsilon)$. The outer solution is governed by a balance between the second and third terms in (6.86a), implying that $h \sim O(1)$. Thus we seek an expansion of the form

$$h \sim h_0 + \mu_1(\epsilon)h_1 + \mu_2(\epsilon)h_2 + \cdots$$

where μ_n are unknown gauge functions. Substituting this expansion into (6.86a) we obtain

$$\epsilon(h_0'' + \mu_1 h_1'' + \mu_2 h_2'') + h_0' + \mu_1 h_1' + \mu_2 h_2' + h_0 + \mu_1 h_1 + \mu_2 h_2 = 0.$$

Now the $O(1)$ terms in the preceding equation are

$$h_0' + h_0 = 0.$$

The terms of next lowest order are not yet obvious and depend on the relative magnitudes of ϵ and μ_1. If $\epsilon/\mu_1 \to 0$, we find $h_n' + h_n = 0$. Then all h_n satisfy the same equation, and the asymptotic expansion is not a proper one. If on the other hand $\epsilon/\mu_1 \to \infty$, we find $h_0'' = 0$, which contradicts the preceding equation for h_0. Thus the only possible choice is $\mu_1 = \epsilon$ and $\mu_n = \epsilon^n$. The higher-order problems then become

$$h_1' + h_1 = -h_0'', \quad h_2' + h_2 = -h_1''.$$

The preceding equations are first-order, and so each can satisfy only one boundary condition. To find out which one, we must consider the scaling of the inner solution, which is defined by one of the two remaining balances of the terms in (6.86a). Supposing that the balance is $\epsilon y'' \sim y$, we find $d/dx \sim \epsilon^{-1/2}$, which implies $y' \sim \epsilon^{-1/2} \gg \epsilon y''$ and $\gg y$. The assumed balance is therefore contradictory.

If however we assume $\epsilon y'' \sim y'$, then $d/dx \sim \epsilon^{-1}$, and $y \ll \epsilon y''$ and $y \ll y'$. The assumed balance is therefore consistent. Now the equation $\epsilon y'' + y' = 0$ has exponential solutions that decay to the right. The BL must therefore be at the left end ($x = 0$) of the interval, which implies that the outer solutions must satisfy the boundary condition $y(1) = e^{-1}$ at the right end of the interval. The appropriate solutions for h_0 and h_1 are therefore

$$h_0 = e^{-x}, \quad h_1 = (1-x)e^{-x}$$

where $h_1(1) = 0$ because h_0 satisfies the boundary condition by itself.

We turn now to the inner solution. Our scaling analysis has already shown that $d/dx \sim \epsilon^{-1}$ in the BL. Thus we define a new stretched variable $\zeta = x/\epsilon$ that is of order unity there, and we denote the inner solution by $y = g(\zeta, \epsilon)$. Substituting this into the governing equation, we find

$$g'' + g' + \epsilon g = 0$$

where the primes now denote $d/d\zeta$. We now expand g in an asymptotic series in unknown gauge functions $\nu_n(\epsilon)$. Arguments like those used for the outer solution show that $\nu_n = \epsilon^n$, so that

$$g = g_0 + \epsilon g_1 + \cdots$$

Substituting this expansion into the differential equation for g and gathering terms proportional to like powers of ϵ, we obtain

$$g_0'' + g_0' = 0,$$
$$g_1'' + g_1' = -g_0.$$

The solutions that satisfy the boundary condition $g(0) = 0$ are

$$g_0 = A(1 - e^{-\zeta}),$$
$$g_1 = B(1 - e^{-\zeta}) - A\zeta(1 + e^{-\zeta}).$$

The constants A and B will now be determined using the IMP. The first step is to define a variable $\tilde{x}$ that is intermediate between x and ζ, viz.,

$$\tilde{x} = \frac{x}{\xi(\epsilon)}, \quad \epsilon \ll \xi \ll 1.$$

The IMP is then

$$\lim_{\epsilon \to 0} \frac{1}{\epsilon^R} \left[\sum_{n=0}^{P} h_n(\eta \tilde{x}) \epsilon^n - \sum_{n=0}^{Q} g_n(\eta \tilde{x}/\epsilon) \epsilon^n \right] = 0,$$

where the limit $\epsilon \to 0$ is taken with $\tilde{x}$ fixed. The order of the matching is given by the integer R. The IMP states that the difference between the (P-term) outer expansion and the (Q-term) inner expansion must be smaller than the Rth gauge function, which is ϵ^R in this case. For each $R = 0, 1, 2$ one must choose the values of P and Q that yield a consistent matching. In most cases, $P = Q = R$.

Starting with the lowest-order matching, we set $R = 0$ and assume that $P = Q = 0$. Then the IMP takes the form

$$\lim_{\epsilon \to 0}[h_0(\eta\tilde{x}) - g_0(\eta\tilde{x}/\epsilon)] = 0$$

Substituting in the explicit expressions for h_0 and g_0, we obtain

$$\lim_{\epsilon \to 0}[e^{-\eta\tilde{x}} - A(1 - e^{-\eta\tilde{x}/\epsilon})] = 0.$$

Now because $\eta/\epsilon \to \infty$, the term $e^{-\eta\tilde{x}/\epsilon}$ is transcendentally small, whereas $e^{-\eta\tilde{x}} \to 1$. We therefore find $A = 1$.

Proceeding now to the next order, we set $R = 1$ and assume that $P = Q = 1$. The IMP then becomes

$$\lim_{\epsilon \to 0} \frac{1}{\epsilon} \left\{ e^{-\eta\tilde{x}} + \epsilon(1 - \eta\tilde{x})e^{-\eta\tilde{x}} - (1 - e^{-\eta\tilde{x}/\epsilon}) \right.$$
$$\left. - \epsilon \left[B(1 - e^{-\eta\tilde{x}/\epsilon}) - \frac{\eta\tilde{x}}{\epsilon}(1 + e^{-\eta\tilde{x}/\epsilon}) \right] \right\} = 0.$$

Now drop the transcendentally small term $e^{-\eta\tilde{x}/\epsilon}$ and set $e^{-\eta\tilde{x}} \approx 1 - \eta\tilde{x}$, whereupon the IMP simplifies to

$$\lim_{\epsilon \to 0}[1 - B - 2\eta\tilde{x} + O(\eta^2)] = 0.$$

This requires $B = 1$.

To determine the composite expansion, we write both the outer and inner expansions in terms of the intermediate variable:

$$h \sim e^{-\eta\tilde{x}} + \epsilon(1 - \eta\tilde{x})e^{-\eta\tilde{x}},$$

$$g \sim 1 - e^{-\eta\tilde{x}/\epsilon} + \epsilon \left[1 - e^{-\eta\tilde{x}/\epsilon} - \frac{\eta\tilde{x}}{\epsilon}(1 + e^{-\eta\tilde{x}/\epsilon}) \right].$$

In the limit $\epsilon \to 0$, these become

$$h \sim 1 + \epsilon - \eta\tilde{x} + O(\epsilon\eta),$$

$$g \sim 1 + \epsilon - \eta\tilde{x} + \text{TST},$$

where TST indicates transcendentally small terms. Thus the common part is

$$\text{CP} = 1 + \epsilon - \eta\tilde{x}.$$

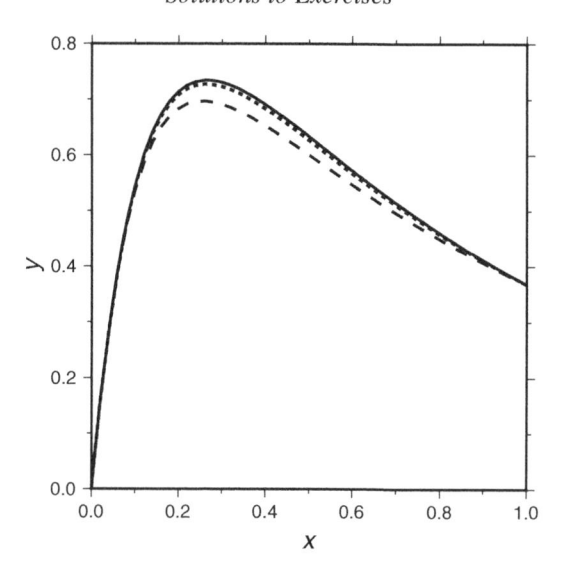

Figure 12.3 Solution of (6.86) with $\epsilon = 0.1$ using matched asymptotic expansions. Solid line: exact analytical solution (6.87). Dashed line: one-term composite expansion. Dotted line: two-term composite expansion.

The composite expansion is $y = h + g - \text{CP}$, or

$$y = e^{-x} - (x+1)e^{-x/\epsilon} + \epsilon[(1-x)e^{-x} - e^{-x/\epsilon}].$$

Figure 12.3 shows the one-term composite expansion (dashed line), the two-term composite expansion (dotted line) and the exact analytical solution (6.87) (solid line) for $\epsilon = 0.1$.

Exercise 6.7. (Morris, 1982) The scaling analysis of § 6.5.2 shows that the following variables are of order unity in the outer region:

$$Z = \frac{z}{a}, \quad F = \frac{(1+\Delta)f}{U\Delta}, \quad \theta = \gamma(T - T_\infty).$$

Substituting these definitions into $[\nu(T)f'']'' = 0$ yields

$$(e^{t-\theta}F_{ZZ})_{ZZ} = 0,$$

where $t = \gamma \Delta T$ and the subscripts indicate differentiation with respect to Z. Now for $t \gg 1$, $\exp(t - \theta) \approx \exp(t)$ (constant) in the outer region, so F satisfies

$$F_{ZZZZ} = 0.$$

The solution that satisfies the outer boundary conditions

$$F_Z(1) = F(1) - \frac{1+\Delta}{\Delta} = 0$$

is

$$F = \frac{1+\Delta}{\Delta}\left[1 + A_1 + 2B_1 - (2A_1 + 3B_1)Z + A_1 Z^2 + B_1 Z^3\right]$$

where the constants A_1 and B_1 will be determined later by matching.

Turning now to the inner region, we note that the following dimensionless variables are $O(1)$ in the deformation layer:

$$\zeta = tNu\frac{z}{a}, \quad G = \frac{(1+\Delta)f}{U}.$$

The length $a/(tNu)$ is the distance over which the viscosity in the deformation layer varies by a factor e, and the viscosity in the deformation layer is $\nu_0 \exp(\zeta)$. The equation satisfied by $G(\zeta)$ is therefore

$$(e^\zeta G_{\zeta\zeta})_{\zeta\zeta} = 0.$$

The solution that satisfies the conditions

$$G(0) = G_{\zeta\zeta}(0) = 0$$

on the traction-free boundary $\zeta = 0$ is

$$G = A_0[(\zeta + 2)e^{-\zeta} - 2] + B_0\zeta,$$

where the constants A_0 and B_0 must be determined by matching.

To perform the matching of the inner and outer expansions for f, we use a generalized form of Prandtl's matching principle

$$\lim_{\zeta\to\infty} f_{\text{inner}} = \lim_{Z\to 0} f_{\text{outer}}$$

wherein both the constant and linear terms are retained in the outer expansion. In terms of dimensional variables, this principle requires

$$\frac{U}{1+\Delta}\left(-2A_0 + B_0 \frac{tNu}{a}z\right) = U\left[1 + A_1 + 2B_1 - (2A_1 + 3B_1)\frac{z}{a} + O(Z^2, Z^3)\right].$$

Matching the constant and linear terms in the preceding equation, we obtain

$$-\frac{2U}{1+\Delta}A_0 = U(1 + A_1 + 2B_1)$$

and

$$\frac{U}{1+\Delta}B_0 \frac{tNu}{a}z = -U(2A_1 + 3B_1)\frac{z}{a}.$$

Next, we match the shear stress $\nu f''$. In the outer region, the shear stress is

$$\nu f''_{\text{outer}} = \nu_\infty U \left(\frac{2A_1}{a^2} + \frac{6B_1 z}{a^3} \right).$$

In the inner region,

$$\nu f''_{\text{inner}} = \nu_0 e^\zeta \frac{d^2}{dz^2} \left[\frac{A_0 U}{1+\Delta} (\zeta + 2) e^{-\zeta} \right] = \frac{\nu_0 A_0 U}{1+\Delta} \left(\frac{tNu}{a} \right)^3 z.$$

Matching the constant and linear terms in the preceding expressions gives $A_1 = 0$ and

$$\nu_0 A_0 \frac{U}{1+\Delta} \left(\frac{tNu}{a} \right)^3 z = \frac{6B_1 \nu_\infty U}{a^3} z \longrightarrow A_0 \frac{U\Delta}{1+\Delta} = 6UB_1,$$

where we have used the definition $\epsilon Nu^3 \equiv \Delta$ with $\epsilon = (\nu_0/\nu_\infty)t^3$. Solving the four simultaneous equations for A_0, B_0, A_1 and B_1, we obtain

$$A_0 = -\frac{1+\Delta}{2(1+\Delta/6)}, \quad B_0 = \frac{\Delta(1+\Delta)}{4tNu(1+\Delta/6)},$$

$$A_1 = 0, \quad B_1 = -\frac{\Delta}{12(1+\Delta/6)}.$$

The final step is to determine the composite expansion. The common part of the inner and outer expansions is

$$f_{\text{CP}} = \lim_{\zeta \to \infty} f_{\text{inner}} = \frac{U}{1+\Delta} (-2A_0 + B_0 \zeta).$$

The composite expansion is thus

$$f_{\text{composite}} = f_{\text{inner}} + f_{\text{outer}} - f_{\text{CP}}$$

$$= U \left(1 + \frac{\epsilon Nu^3}{6} \right)^{-1} \left[1 - \left(1 + \frac{\zeta}{2} \right) e^{-\zeta} - \frac{\epsilon Nu^3}{12} (Z^3 - 3Z) \right].$$

The preceding expression has two limits. In the 'lubrication' limit $\epsilon Nu^3 \ll 1$, the deformation of the fluid is confined to the deformation layer, and

$$f_{\text{lubrication}} \sim U \left[1 - \left(1 + \frac{\zeta}{2} \right) e^{-\zeta} \right].$$

The other extreme is the 'Stokes' limit $\epsilon Nu^3 \gg 1$, for which

$$f_{\text{Stokes}} \sim \frac{U}{2} (3Z - Z^3).$$

Exercise 6.8. The relevant experimental data are in Figure 7 of Ansari and Morris (1985) (henceforth AM85), which shows the Péclet number P as a function of the dimensionless drag D for several values of the modified viscosity ratio ϵ. The Péclet number corresponding to our desired value of U is $P = Ua/\kappa \approx 91$, where the value $\kappa = 0.0011$ cm^2 s^{-1} given in Table 2 of AM85 has been used. The next task is to determine $\epsilon = \theta^3 \nu_0/\nu_\infty$, where ν_0 and ν_∞ are the kinematic viscosities of the fluid at the temperatures T_0 and T_∞, respectively, and $\theta = \gamma(T_0 - T_\infty)$ where

$$\gamma = -\frac{d}{dT} \ln \nu(T) \bigg|_{T=T_0}$$

The quantity γ is just the local (at $T = 50°$ C) slope of the curve of $\ln \mu$ vs. T shown in Figure 6 of AM85, where μ is the dynamic viscosity. Drawing a tangent to the curve by hand, we find $\gamma \approx 0.137$, whence $\theta \approx 6.9$. Also from Figure 6, $\log_{10} \mu(T = 0°) \approx 8.0$ and $\log_{10} \mu(T = 50°) \approx 3.07$, whence $\nu_\infty/\nu_0 \approx 85000$. Thus $\epsilon \approx 0.0038$. Turning to Figure 7, we move rightward along the horizontal line $Pe = 91$ until we reach a point corresponding to $\epsilon \approx 0.0038$, which lies between the data points for $\epsilon = 0.0041$ and $\epsilon = 0.0026$. The abscissa of this point is $D \approx 34000$. The dimensional drag is $D' = \frac{2}{3}\pi\mu_0\kappa\theta^3 D$, where $\mu_0 \approx 1170$ g cm^{-1} s^{-1} is determined from Figure 6, yielding $D' \approx 3.0 \times 10^7$ g cm s^{-2}. This drag must be equal to the weight Mg of the load, whence $M \approx 30$ kg. This value is within the range mentioned on p. 472 of AM85.

Exercise 7.1. (a) The limiting forms of the decay rate are

$$s(kd \gg 1) = -\frac{g}{2k\nu},$$

$$s(kd \ll 1) = -\frac{gd^3k^2}{3\nu} \equiv -\frac{g}{2k\nu}\left[\frac{2}{3}(kd)^3\right].$$

The surface relaxes more slowly for a shallower layer because the no-slip condition on the rigid bottom boundary inhibits the flow.

(b) The lubrication equations are

$$p_x = \eta u_{zz}, \quad (12.38a)$$

$$p_z = -\rho g, \quad (12.38b)$$

$$u_x + w_z = 0. \quad (12.38c)$$

Integrate (12.38b) subject to $p(\zeta) = 0$ to obtain

$$p = -\rho g(z - \zeta). \quad (12.39)$$

Substitute (12.39) into (12.38a) and integrate twice subject to $u_z(z=0) = u(z=-d) = 0$ to obtain

$$u = \frac{g}{\nu}\zeta_x \left[\frac{z^2-d^2}{2} - \zeta(z+d)\right] \quad (12.40)$$

where $\nu = \eta/\rho$ is the kinematic viscosity. Integration of (12.38c) over the layer yields

$$\zeta_t = \partial_x \int_{-d}^{\zeta} u\,dz = 0. \quad (12.41)$$

Substitution of (12.40) into (12.41) yields

$$\zeta_t = \frac{g}{3\nu}\left[(\zeta+d)^3 \zeta_x\right]_x. \quad (12.42)$$

In the small-amplitude limit $\zeta \ll d$, (12.42) takes the form

$$\zeta_t = \frac{gd^3}{3\nu}\zeta_{xx},$$

which has the solution

$$\zeta = \zeta_0 \sin kx \exp st, \quad s = -\frac{gd^3 k^2}{3\nu},$$

which is identical to $s(kd \ll 1)$ from part a).

Exercise 7.2. Work in Cartesian coordinates (x,z), where x increases to the left starting from the trench and z increases downward from the upper surface. The lubrication equations are (7.10). Equation (7.10c) implies $p = p(x)$. Integrating (7.10b) twice subject to the boundary conditions $u(x,0) = u(x,\alpha x) - U_0 = 0$ yields

$$u = \frac{p_x}{2\eta}z(z-\alpha x) + \frac{U_0 z}{\alpha x}.$$

Integration of (7.10a) with respect to z subject to $w(x,0) = 0$ yields

$$w = \frac{z^2}{2\eta}\left(\frac{\alpha x}{2} - \frac{z}{3}\right)p_{xx} + \frac{\alpha z^2}{4\eta}p_x + \frac{U_0 z^2}{2\alpha x^2} = 0.$$

The remaining boundary condition $w(0,\alpha x) = U_0 \sin\alpha \approx U_0 \alpha$ then implies, after some rearrangement,

$$\frac{\alpha^2}{6\eta}(x^3 p_x)_x = U_0,$$

which has the solution
$$p = -\frac{6\eta U_0}{\alpha^2 x}.$$

Expanding the exact solution (4.37), (4.42) in powers of α, we find
$$p_{\text{exact}} = -\frac{6\eta U_0}{\alpha^2 r}\cos\phi + O(\alpha^{-1}).$$

However, $\cos\phi \approx 1$ and $r \approx x$ to within errors of order α^2. The two expressions for the pressure therefore agree.

Exercise 7.3. (Morris, 2008) Define a new variable $z = y + 1$ that ranges from 0 at the bottom of the sublayer to 1 at the top. The sine transform and its inverse are defined by (7.94). By a standard property of the sine transform, the transforms of the relevant derivatives of ψ are

$$\overline{\psi_{xx}} = -k^2\overline{\psi}(k, z) + \sqrt{\frac{2}{\pi}}k\psi(0, z) = -k^2\overline{\psi}(k, z),$$

$$\overline{\psi_{xxxx}} = k^4\overline{\psi}(k, z) - \sqrt{\frac{2}{\pi}}k^3\psi(0, z) + \sqrt{\frac{2}{\pi}}k\psi_{xx}(0, z) = k^4\overline{\psi}(k, z) + \sqrt{\frac{2}{\pi}}k,$$

where the boundary conditions (7.72e) have been used. The sine transform of the biharmonic equation is

$$\overline{\psi}_{zzzz} - 2k^2\overline{\psi}_{zz} + k^4\overline{\psi} = -\sqrt{\frac{2}{\pi}}k,$$

the general solution of which is

$$\overline{\psi} = (A + Bz)\cosh kz + (C + Dz)\sinh kz - \sqrt{\frac{2}{\pi}}\frac{1}{k^3}$$

where A–D are constants determined by the boundary conditions. The sine-transformed boundary conditions are

$$\overline{\psi}(0) = \overline{\psi}_{zz}(0) = 0,$$

$$\overline{\psi}_{zz}(1) + k^2\overline{\psi}(1) = \frac{1}{3}\overline{\psi}_{zzz}(1) - k^2\overline{\psi}_z(1) - \epsilon\overline{\psi}(1) = 0.$$

The solution for $\overline{\psi}$ that satisfies the preceding boundary conditions is

$$\overline{\psi} = -\sqrt{\frac{2}{\pi}}\frac{1}{k^3}\left[1 + \left(a + \frac{1}{2}kz\right)\sinh kz + (bz - 1)\cosh kz\right],$$

$$a = (2\sinh k - k)\left(k^4 + k^3\sinh k + 3\epsilon\sinh^2\frac{k}{2}\right)\chi^{-1},$$

$$b = -k\sinh^2\frac{k}{2}\left[2k^3(2+\cosh k) + 3\epsilon\sinh k\right]\chi^{-1},$$

$$\chi = 2k^4 + k^3\sinh 2k + 3\epsilon\sinh^2 k.$$

The inverse transform of $\overline{\psi}$ must be evaluated numerically.

Exercise 7.4. The dimensionless BVP to be solved is

$$\nabla^4\psi = 0,$$
$$\psi(x,-1) = \psi_{yy}(x,-1) = 0,$$
$$\psi(x,0) = \psi_{yy}(x,0) = 0,$$
$$\psi(0,y) = \psi_{xx}(0,y) - 1 = 0.$$

By symmetry, ψ is an even function of the coordinate $y + 1/2$ that has its origin at the midpoint of the layer. The even solution of the biharmonic equation that satisfies the boundary conditions at $y = -1$ and $y = 0$ and that decays as $x \to \infty$ is

$$\psi = \sum_{n=0}^{\infty}(A_n + B_n x)e^{-(2n+1)\pi x}\sin(2n+1)\pi y.$$

The boundary condition $\psi(0,y) = 0$ requires $A_n = 0$. The boundary condition $\psi_{xx}(0,y) = 1$ is then

$$-2\pi\sum_{n=0}^{\infty}(2n+1)B_n\sin(2n+1)\pi y = 1.$$

Multiplying by $\sin(2m+1)\pi y$ and integrating over the layer, we obtain

$$B_n = \frac{2}{\pi^2(2n+1)^2},$$

whence the streamfunction is

$$\psi = \frac{2}{\pi^2}x\sum_{n=0}^{\infty}\frac{1}{(2n+1)^2}e^{-(2n+1)\pi x}\sin(2n+1)\pi y.$$

For large x, the dominant (most slowly decaying) term is the one with $n = 0$, which has a dimensionless decay length π^{-1}. Recalling that the decay length in the presence of a low-viscosity channel is $2/\sqrt{3\epsilon}$, we see that the presence of the channel increases the horizontal length scale of the flow by a factor $2\pi/\sqrt{3\epsilon} \approx 3.63\epsilon^{-1/2}$.

Exercise 7.5. Multiply (7.85) by $(2/A)dA/dy$ to obtain

$$c\frac{2}{A}\frac{dA}{dy}\frac{d}{dy}\left(\frac{1}{A}\frac{dA}{dy}\right) = \frac{2c}{A^2}\frac{dA}{dy} + \frac{2Q_1}{A^3}\frac{dA}{dy} - \frac{2}{A}\frac{dA}{dy}.$$

Rewrite each term as an exact differential:

$$c\frac{d}{dy}\left(\frac{1}{A}\frac{dA}{dy}\right)^2 = -2c\frac{d}{dy}A^{-1} - Q_1\frac{d}{dy}A^{-2} - 2\frac{d}{dy}\ln A.$$

Integrate once to obtain

$$c\left(\frac{1}{A}\frac{dA}{dy}\right)^2 = -\frac{2c}{A} - \frac{Q_1}{A^2} - 2\ln A + c_1.$$

Finally, multiply through by A^2 to obtain (7.86).

Exercise 8.1. To illustrate the procedure, we consider only the products $\mathbf{g}_1 \cdot \mathbf{g}^1$ and $\mathbf{g}_1 \cdot \mathbf{g}^2$. For the former, we have

$$\begin{aligned}\mathbf{g}_1 \cdot \mathbf{g}^1 &= \mu_1^\lambda \mathbf{a}_\lambda \cdot \left[h^{-1}(\mu_\rho^p \delta_\gamma^1 - \mu_\gamma^1)\mathbf{a}^\gamma\right] \\ &= h^{-1}\mu_1^\lambda(\mu_\rho^p \delta_\gamma^1 - \mu_\gamma^1)\delta_\lambda^\gamma \quad (\text{using } \mathbf{a}_\lambda \cdot \mathbf{a}^\gamma = \delta_\lambda^\gamma) \\ &= h^{-1}(\mu_1^1\mu_\rho^p - \mu_1^\lambda\mu_\lambda^1) \\ &= h^{-1}(\mu_1^1\mu_2^2 - \mu_1^2\mu_2^1) \\ &= h^{-1}\left[(1 - zb_1^1)(1 - zb_2^2) - (-zb_1^2)(-zb_2^1)\right] \\ &= h^{-1}\left[1 - z(b_1^1 + b_2^2) + z^2(b_1^1 b_2^2 - b_1^2 b_2^1)\right] \\ &= h^{-1}(1 - 2Hz + Gz^2) = 1.\end{aligned}$$

Proceeding to $\mathbf{g}_1 \cdot \mathbf{g}^2$, we have

$$\begin{aligned}\mathbf{g}_1 \cdot \mathbf{g}^2 &= \mu_1^\lambda \mathbf{a}_\lambda \cdot \left[h^{-1}(\mu_\rho^p \delta_\gamma^2 - \mu_\gamma^2)\mathbf{a}^\gamma\right] \\ &= h^{-1}\mu_1^\lambda(\mu_\rho^p \delta_\gamma^2 - \mu_\gamma^2) \\ &= h^{-1}(\mu_1^2\mu_\rho^p - \mu_1^\lambda\mu_\lambda^2) \\ &= h^{-1}[\mu_1^2(\mu_1^1 + \mu_2^2) - \mu_1^1\mu_1^2 - \mu_1^2\mu_2^2] = 0.\end{aligned}$$

Exercise 8.2. The Cartesian coordinates of a point are

$$\mathbf{r}(r,\theta,z) = r\mathbf{e}_r + z\mathbf{e}_z,$$

where $\mathbf{e}_r$ and $\mathbf{e}_z$ are unit vectors in the $r-$ and $z-$directions, respectively. The covariant basis vectors $\mathbf{g}_i \equiv \partial_i \mathbf{r}$ are

$$\mathbf{g}_1 = \mathbf{e}_r, \quad \mathbf{g}_2 = r\mathbf{e}_\theta, \quad \mathbf{g}_3 = \mathbf{e}_z,$$

where e_θ is a unit vector in the θ-direction and we have used the relation $\partial_\theta \mathbf{e}_r = \mathbf{e}_\theta$. The contravariant basis vectors $\mathbf{g}^i$ that satisfy $\mathbf{g}_i \cdot \mathbf{g}^j = \delta_i^j$ are

$$\mathbf{g}^1 = \mathbf{e}_r, \quad \mathbf{g}^2 = r^{-1}\mathbf{e}_\theta, \quad \mathbf{g}^3 = \mathbf{e}_z.$$

Now it is clear that the Christoffel symbol $\Gamma_{ij}^k = \mathbf{g}^k \cdot \mathbf{g}_{i,j}$ vanishes if $i = 3$ or $j = 3$, because none of the base vectors $\mathbf{g}_i$ depends on z and $\mathbf{g}_3 \equiv \mathbf{e}_z$ is a constant vector. We thus find

$$\mathbf{g}_{1,1} = 0, \quad \mathbf{g}_{1,2} = \mathbf{g}_{2,1} = \mathbf{e}_\theta, \quad \mathbf{g}_{2,2} = -r\mathbf{e}_r,$$

where we have used the relation $\partial_\theta \mathbf{e}_\theta = -\mathbf{e}_r$. Now the Christoffel symbol also vanishes for $k = 3$, because none of the derivatives $\mathbf{g}_{i,j}$ has a component in the $\mathbf{e}_z$-direction. A simple calculation then shows that the nonzero Christoffel symbols are

$$\Gamma_{1,2}^2 = \Gamma_{2,1}^2 = r^{-1}, \quad \Gamma_{22}^1 = -r.$$

Using the preceding, we find

$$\mathbf{v}_{,2} = (v_{1,2} - r^{-1}v_2)\mathbf{g}^1 + (v_{2,2} + rv_1)\mathbf{g}^2 + v_{3,2}\mathbf{g}^3.$$

In terms of the physical velocity components (u, v, w), the preceding expression takes the form

$$\partial_\theta \mathbf{v} = (\partial_\theta u - v)\mathbf{e}_r + (\partial_\theta v + u)\mathbf{e}_\theta + \partial_\theta w \mathbf{e}_z.$$

The same result is obtained by differentiating $\mathbf{v} = u\mathbf{e}_r + v\mathbf{e}_\theta + w\mathbf{e}_z$ and using $\partial_\theta \mathbf{e}_r = \mathbf{e}_\theta, \partial_\theta \mathbf{e}_\theta = -\mathbf{e}_r$.

Exercise 8.3.

$$\begin{aligned}g_{ij}|_r &= g_{ij,r} - \Gamma_{ir}^m g_{mj} - \Gamma_{jr}^m g_{im} \\ &= g_{ij,r} - g^{ms}g_{mj}\Gamma_{irs} - g^{ms}g_{im}\Gamma_{jrs} \\ &= g_{ij,r} - \delta_j^s \Gamma_{irs} - \delta_i^s \Gamma_{jrs} \\ &= g_{ij,r} - \Gamma_{irj} - \Gamma_{jri} \\ &= g_{ij,r} - \frac{1}{2}(g_{ij,r} + g_{rj,i} - g_{ir,j}) - \frac{1}{2}(g_{ji,r} + g_{ri,j} - g_{jr,i}) = 0,\end{aligned}$$

where the symmetry of g_{ij} has been used in the last step.

Exercise 8.4. Because $g^{13} = g^{23} = 0$ and $g^{33} = 1$, the equation $g^{ij}e_{ij} = 0$ takes the form $g^{\alpha\beta}e_{\alpha\beta} + e_{33} = 0$. Now use the definitions of $g^{\alpha\beta}$, $e_{\alpha\beta}$ and e_{33}, and note that $g^{\alpha\beta}$ has a factor of h^2 in the denominator. We therefore work with the equation $h^2\left(g^{\alpha\beta}e_{\alpha\beta} + e_{33}\right) = 0$, which takes the form

$$f_1 + f_2 u_3 + h^2 u_{3,3} = 0$$

where

$$f_1 = \frac{1}{2}\left(\mu_\rho^\rho \delta_\lambda^\alpha - \mu_\lambda^\alpha\right)\left(\mu_\gamma^\gamma \delta_\epsilon^\beta - \mu_\epsilon^\beta\right) a^{\lambda\epsilon}\left(\mu_\beta^\nu u_\nu|_\alpha + \mu_\alpha^\nu u_\nu|_\beta\right),$$

$$f_2 = -\frac{1}{2}\left(\mu_\rho^\rho \delta_\lambda^\alpha - \mu_\lambda^\alpha\right)\left(\mu_\gamma^\gamma \delta_\epsilon^\beta - \mu_\epsilon^\beta\right) a^{\lambda\epsilon}\left(\mu_\beta^\nu b_{\nu\alpha} + \mu_\alpha^\nu b_{\nu\beta}\right).$$

To simplify f_1, we expand the sums, set $a^{21} = a^{12}$, factor the result and use the definition (8.7) of the mean curvature to obtain

$$h^{-1}f_1 = \left[a^{11} + z\left(a^{1\alpha}b_\alpha^1 - 2Ha^{11}\right)\right]u_1|_1 + \left[a^{12} + z\left(a^{1\alpha}b_\alpha^2 - 2Ha^{12}\right)\right]u_1|_2$$
$$+ \left[a^{21} + z\left(a^{2\alpha}b_\alpha^1 - 2Ha^{21}\right)\right]u_2|_1 + \left[a^{22} + z\left(a^{2\alpha}b_\alpha^2 - 2Ha^{22}\right)\right]u_2|_2.$$

Now note that $a^{\beta\alpha}b_\alpha^\gamma = b^{\beta\gamma}$ and compress the sum in an obvious way to obtain

$$h^{-1}f_1 = \left[a^{\alpha\beta} + z\left(b^{\alpha\beta} - 2Ha^{\alpha\beta}\right)\right]u_\alpha|_\beta.$$

Turning to f_2, we proceed as for f_1 and find

$$h^{-1}f_2 = -a^{1\alpha}b_{1\alpha} - a^{2\alpha}b_{2\alpha}$$
$$+ z\left(a^{2\alpha}b_{2\alpha}b_1^1 - a^{1\alpha}b_{2\alpha}b_1^2 - a^{2\alpha}b_{1\alpha}b_2^1 + a^{1\alpha}b_{1\alpha}b_2^2\right).$$

Noting that $a^{\beta\alpha}b_{\gamma\alpha} = b_\gamma^\beta$ and using the definitions (8.7) of the mean and Gaussian curvatures, we find

$$h^{-1}f_2 = -2H + 2Gz \equiv h_{,3}.$$

Combining the preceding results, we obtain the continuity equation (8.19).

Exercise 8.5. From (5.12c), the load on the plate is $P_3 = -g[(\rho_c - \rho_w)\Delta + (\rho_m - \rho_c)\xi]\cos kx$. The one-dimensional form of (8.46a) is therefore

$$D\frac{d^4}{dx^4}(\xi \cos kx) = -g[(\rho_c - \rho_w)\Delta + (\rho_m - \rho_c)\xi]\cos kx,$$

where D is the flexural rigidity. The solution is

$$\frac{\Delta}{\xi} = -\frac{\rho_m - \rho_c}{\rho_c - \rho_w} - \frac{Dk^4}{g(\rho_c - \rho_w)}. \qquad (12.43)$$

Now for an incompressible plate, Poisson's ratio $\nu = 1/2$ and Young's modulus $E = 3\mu$, whence $D = \mu T^3/3$. Using (12.43) to eliminate ξ from the expression (5.16) for the free-air gravity anomaly δg, we obtain the thin-plate admittance

$$Z \equiv \frac{\delta g}{\Delta} = 2\pi G(\rho_c - \rho_w)e^{-kd}\left\{1 - \left[1 + \frac{\mu k'^4}{3gT(\rho_m - \rho_c)}\right]^{-1}e^{-kh}\right\}. \quad (12.44)$$

Comparing (12.44) with (5.17), we see that the thin-plate admittance is obtained from the thick-plate admittance by replacing $F(k')$ by $k'^4/3$, which is just the long-wavelength ($k' \to 0$) limit of $F(k')$. The thin-plate admittance is therefore simply the long-wavelength limit of the thick-plate admittance. For reasonable values of T, d, h, μ, ρ_w, ρ_c and ρ_m the two admittances are nearly indistinguishable when plotted together.

Exercise 8.6. The governing equations are (8.51), (8.44) and (8.52) with $\zeta_2 = \partial/\partial\phi = P_1 = P_2 = 0$. These conditions imply $\omega = \tau = H = S = 0$, whence the equation of equilibrium (8.51b) in the $\mathbf{e}_\phi$-direction disappears. The remaining equations admit solutions of the form $\zeta_1 = \zeta_1^{(0)}\partial_\theta Y_l^0(\theta)$, $\zeta_3 = \zeta_3^{(0)}Y_l^0(\theta)$. Substitution of these forms into the equilibrium equations yields two coupled linear equations for $\zeta_1^{(0)}$ and $\zeta_3^{(0)}$ whose solutions are

$$\zeta_1^{(0)} = \frac{[(2l^2 + 2l - 1)\epsilon^2 - 36]PR}{2\epsilon(l+2)(l-1)[(2l^2 + 2l - 1)(2l^2 + 2l - 3)\epsilon^2 - 36]\mu}$$

$$\zeta_3^{(0)} = \frac{(12 - \epsilon^2)(2l^2 + 2l - 1)PR}{2\epsilon(l+2)(l-1)[(2l^2 + 2l - 1)(2l^2 + 2l - 3)\epsilon^2 - 36]\mu}$$

where $\epsilon = h/R$ and μ is the shear modulus.

The total membrane (U_m) and bending (U_b) energies of the shell are difficult to determine directly for arbitrary l, and so we use an indirect approach. We first evaluate $U_b/(U_m + U_b)$ for two particular values $l = 2$ and 20, obtaining

$$\left.\frac{U_b}{U_m + U_b}\right|_{l=2} = \frac{128\epsilon^2}{48 + 120\epsilon^2 + 11\epsilon^4}$$

$$\left.\frac{U_b}{U_m + U_b}\right|_{l=20} = \frac{936320\epsilon^2}{48 + 936312\epsilon^2 + 78027\epsilon^4}.$$

These two examples make it obvious that the term $\propto \epsilon^4$ in the denominator is negligible when $\epsilon \ll 1$. The coefficients multiplying ϵ^2 can now be determined by assuming them to be (say) sixth-order polynomials in l. Fitting such polynomials

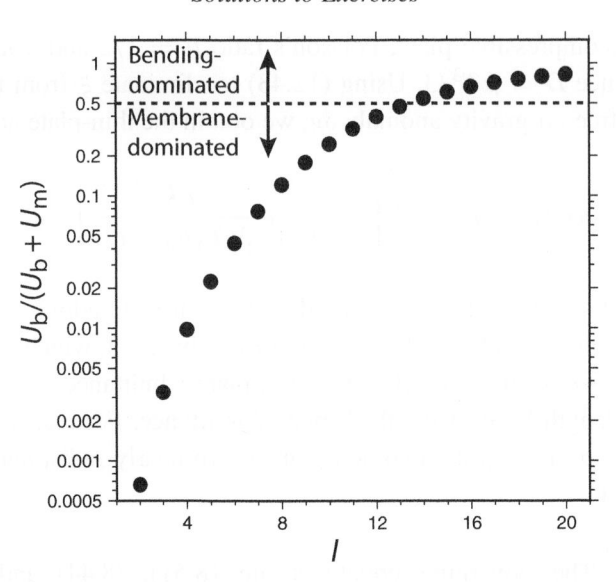

Figure 12.4 Ratio of the bending energy U_b to the total (bending plus membrane) energy $U_b + U_m$ of a complete spherical shell with $h/R = 0.0157$ deformed by a normal load $PY_l^0(\theta)$ applied to its surface, where $Y_l^0(\theta)$ is a surface spherical harmonic of degree l and order 0.

to the coefficients for $l \in [2, 8]$ shows that the coefficients are in fact fourth-order polynomials in l, and the final result is

$$\frac{U_b}{U_m + U_b} \equiv B = \frac{2(l-1)l(l+1)(l+2)\epsilon^2}{18 + (2l^4 + 4l^3 - 2l^2 - 4l - 3)\epsilon^2}.$$

Now for a given ϵ, the value of $l = l_{m \to b}$ for which $B = 1/2$ is the spherical harmonic degree separating membrane-dominated deformation ($l < l_{m \to b}$) from bending-dominated deformation ($l > l_{m \to b}$). Solving the equation $B = 1/2$ for l in the limit $l \gg 1$ gives

$$l_{m \to b} = (9/\epsilon^2)^{1/4}. \tag{12.45}$$

Figure 12.4 shows the quantity B as a function of l for $\epsilon = 0.0157$. The shell deforms primarily in membrane mode for $l \leq 13$, and primarily by bending for $l \geq 14$. For comparison, the asymptotic formula (12.45) predicts $l_{m \to b} = 13.8$.

Exercise 9.1. Regroup the terms of (9.29) as follows:

$$0 = \frac{\partial}{\partial t}\left[\phi \rho_f \varepsilon_f + (1-\phi)\rho_m \varepsilon_m\right] + \nabla \cdot \left[\phi \rho_f \varepsilon_f \mathbf{v}_f + (1-\phi)\rho_m \varepsilon_m \mathbf{v}_m\right]$$

$$+ \frac{\partial}{\partial t}(\alpha \xi_i) + \nabla \cdot (\alpha \xi_i \mathbf{v}_\omega) - \nabla \cdot (\sigma \alpha \mathbf{v}_\omega)$$

$$-\nabla \cdot \left[-\phi P_f \mathbf{v}_f - (1-\phi)P_m \mathbf{v}_m + \phi \mathbf{v}_f \cdot \boldsymbol{\tau}_f + (1-\phi)\mathbf{v}_m \cdot \boldsymbol{\tau}_m \right]$$
$$+ \phi \mathbf{v}_f \cdot (\rho_f g \mathbf{e}_z) + (1-\phi)\mathbf{v}_m \cdot (\rho_m g \mathbf{e}_z) - Q + \nabla \cdot \mathbf{q} \quad (12.46)$$

Rewrite the terms on the first line of (12.46) by expanding the derivatives:

$$\text{line 1} = \phi \rho_f \partial_t \varepsilon_f + \varepsilon_f \partial_t (\phi \rho_f) + (1-\phi)\rho_m \partial_t \varepsilon_m + \varepsilon_m \partial_t [(1-\phi)\rho_m]$$
$$+ \phi \rho_f \mathbf{v}_f \cdot \nabla \varepsilon_f + \varepsilon_f \nabla \cdot (\phi \rho_f \mathbf{v}_f)$$
$$+ (1-\phi)\rho_m \mathbf{v}_m \cdot \nabla \varepsilon_m + \varepsilon_m \nabla \cdot [(1-\phi)\rho_m \mathbf{v}_m]$$
$$= \phi \rho_f \frac{D_f \varepsilon_f}{Dt} + (1-\phi)\rho_m \frac{D_m \varepsilon_m}{Dt} + \rho_f \varepsilon_f [\partial_t \phi + \nabla \cdot (\phi \mathbf{v}_f)]$$
$$+ \rho_m \varepsilon_m \{\partial_t (1-\phi) + \nabla \cdot [(1-\phi)\mathbf{v}_m)]\}.$$

Now use the mass conservation equations (9.9) and (9.10) and the relations $d\varepsilon_j = c_j dT$ to obtain

$$\text{line 1} = \phi \rho_f c_f \frac{D_f T}{Dt} + (1-\phi)\rho_m c_m \frac{D_m T}{Dt} - \Gamma \Delta \varepsilon. \quad (12.47)$$

Turning now to line 2 of (12.46), we have

$$\text{line 2} = \frac{D_\omega}{Dt}(\alpha \xi_i) + \alpha \xi_i \nabla \cdot \mathbf{v}_\omega - \nabla \cdot (\sigma \alpha \mathbf{v}_\omega)$$
$$= \frac{D_\omega}{Dt}\left[\alpha\left(\sigma - T\frac{d\sigma}{dT}\right)\right] + \alpha\left(\sigma - T\frac{d\sigma}{dT}\right)\nabla \cdot \mathbf{v}_\omega - \nabla \cdot (\sigma \alpha \mathbf{v}_\omega)$$
$$= \frac{D_\omega}{Dt}(\sigma \alpha) - \frac{D_\omega}{Dt}\left(T\alpha \frac{d\sigma}{dT}\right) - T\alpha \frac{d\sigma}{dT}\nabla \cdot \mathbf{v}_\omega + \sigma \alpha \nabla \cdot \mathbf{v}_\omega - \nabla \cdot (\sigma \alpha \mathbf{v}_\omega)$$
$$= \frac{D_\omega}{Dt}(\sigma \alpha) - T\frac{D_\omega}{Dt}\left(\alpha \frac{d\sigma}{dT}\right) - \alpha \frac{d\sigma}{dT}\frac{D_\omega T}{Dt} - T\alpha \frac{d\sigma}{dT}\nabla \cdot \mathbf{v}_\omega - \mathbf{v}_\omega \cdot \nabla(\sigma \alpha)$$
$$= \sigma \frac{D_\omega \alpha}{Dt} - T\frac{D_\omega}{Dt}\left(\alpha \frac{d\sigma}{dT}\right) - T\alpha \frac{d\sigma}{dT}\nabla \cdot \mathbf{v}_\omega - \mathbf{v}_\omega \cdot \nabla(\sigma \alpha)$$
$$= \sigma \frac{d\alpha}{d\phi}\frac{D_\omega \phi}{Dt} - T\frac{D_\omega}{Dt}\left(\alpha \frac{d\sigma}{dT}\right) - T\alpha \frac{d\sigma}{dT}\nabla \cdot \mathbf{v}_\omega - \mathbf{v}_\omega \cdot \nabla(\sigma \alpha). \quad (12.48)$$

Finally, we turn to lines 3 and 4 of (12.46). Expanding derivatives, we obtain

$$\text{lines 3} + 4 = \phi \mathbf{v}_f \cdot \nabla P_f + P_f \nabla \cdot (\phi \mathbf{v}_f)$$
$$+ (1-\phi)\mathbf{v}_m \cdot \nabla P_m + P_m \nabla \cdot [(1-\phi)\mathbf{v}_m]$$

$$- \mathbf{v}_f \cdot \nabla \cdot (\phi \tau_f) - \phi \nabla \mathbf{v}_f : \tau_f - \mathbf{v}_m \cdot \nabla \cdot [(1-\phi)\tau_m] - (1-\phi)\nabla \mathbf{v}_m : \tau_m$$
$$+ \phi \mathbf{v}_f \cdot (\rho_f g \mathbf{e}_z) + (1-\phi)\mathbf{v}_m \cdot (\rho_m g \mathbf{e}_z).$$

Now use the momentum equations (9.23) to eliminate $\nabla \cdot (\phi \tau_f)$ and $\nabla \cdot [(1-\phi)\tau_m]$. The terms involving gravity all cancel, and we find

$$\text{lines } 3 + 4 = P_f \nabla \cdot (\phi \mathbf{v}_f) + P_m \nabla \cdot [(1-\phi)\mathbf{v}_m]$$
$$+ \mathbf{v}_\omega \cdot [\Delta P \nabla \phi + \nabla(\sigma \alpha)] - \Psi.$$

Now use the mass conservation equations (9.9) and (9.10), whereupon we obtain

$$\text{lines } 3 + 4 = \Delta P \frac{D_\omega \phi}{Dt} + \mathbf{v}_\omega \cdot \nabla(\sigma \alpha) - \Gamma \left(\frac{P_m}{\rho_m} - \frac{P_f}{\rho_f} \right) - \Psi. \qquad (12.49)$$

Substituting the expressions (12.47), (12.48) and (12.49) into (12.46), we obtain (9.32).

Exercise 9.2. (Ricard et al., 2001) The matrix velocity $w_m(z)$ satisfies (9.48) with $g\Delta\rho = 0$. The solution that satisfies the boundary conditions $w_m(0) = 0$ and $w_m(h) = -W_0$ is

$$w_m = -W_0 \frac{\sinh(z/\delta_c)}{\sinh(h/\delta_c)},$$

where δ_c is given by (9.50). The impermeability conditions $w_f(0) = w_m(0) = 0$ imply that $W(t) = 0$ in (9.45), whence

$$w_f = -\frac{(1-\phi_0)w_m}{\phi_0}$$

The nonhydrostatic part of the fluid pressure P_f satisfies (9.46a) with $\phi \to \phi_0$ and $\rho_f g \to 0$. The solution that satisfies the condition $P_f(h) = 0$ is

$$P_f = -\frac{\eta_f \delta_c W_0}{k_0 \sinh(h/\delta_c)} \left(\cosh \frac{z}{\delta_c} - \cosh \frac{h}{\delta_c} \right).$$

Finally, P_m is determined from (9.47) with $\sigma = 0$ and $\phi = \phi_0$ and is

$$P_m = -\frac{\eta_f \delta_c W_0}{k_0 \sinh(h/\delta_c)} \left(\cosh \frac{z}{\delta_c} - \cosh \frac{h}{\delta_c} \right) + \frac{K \eta_m W_0}{\phi_0 \delta_c \sinh(h/\delta_c)} \cosh \frac{z}{\delta_c}.$$

The force per unit area on horizontal planes is

$$\Sigma_{zz} = \phi_0(-P_f) + (1-\phi_0) \left[-P_m + \frac{4}{3}\eta_m w'_m \right] = -\frac{\eta_f \delta_c W_0}{k_0} \coth \frac{h}{\delta_c},$$

which is independent of depth and therefore corresponds to the force per unit area on the screen.

Exercise 10.1 (Smith, 1975). We begin by writing the different variables as sums of their values in the basic state of pure shear and a perturbation, the latter denoted by a hat:

$$h = h_0(t) + \hat{h}, \quad u = -\dot{\epsilon}x + \hat{u}, \quad w = \dot{\epsilon}z + \hat{w}, \qquad (12.50a)$$

$$\sigma_{xx}^{(i)} = -2\eta_i\dot{\epsilon} + \hat{\sigma}_{xx}, \quad \sigma_{zz}^{(i)} = 2\eta_i\dot{\epsilon} + \hat{\sigma}_{zz} \quad \sigma_{xz}^{(i)} = \hat{\sigma}_{xz}, \qquad (12.50b)$$

where the superscript (i) or the subscript i is the fluid index ($= 1$ or 2). The hatted variables are all $\sim O(\hat{h}')$, where the prime denotes d/dx. The next step is to write down the linearized matching conditions on the interface $z = h$. The continuity of the two components of velocity requires $\hat{u}_1(h) = \hat{u}_2(h)$ and $\hat{w}_1(h) = \hat{w}_2(h)$, where the arguments x and t have been suppressed to simplify the notation. These conditions are linearized by evaluating them at the undeformed surface $h_0(t)$, which yields

$$\hat{u}_1(h_0) = \hat{u}_2(h_0), \quad \hat{w}_1(h_0) = \hat{w}_2(h_0). \qquad (12.51)$$

The next condition, that of continuity of the shear stress at the interface, must be derived with some care. The general form of this condition is

$$[n_i \sigma_{ij} t_j]_h = 0, \qquad (12.52)$$

where $[\ldots]_h$ indicates the jump of the bracketed quantity from fluid 1 to fluid 2 across the interface $z = h$, σ is the stress tensor,

$$\mathbf{n} = (1 + \hat{h}'^2)^{-1/2}(\mathbf{e}_z - \hat{h}'\mathbf{e}_x), \quad \mathbf{t} = (1 + \hat{h}'^2)^{-1/2}(\mathbf{e}_x + \hat{h}'\mathbf{e}_z)$$

are unit vectors normal to ($\mathbf{n}$) and tangential to ($\mathbf{t}$) the interface and $\mathbf{e}_x$ and $\mathbf{e}_z$ are rightward-pointing and upward-pointing unit vectors, respectively. Expanding (12.52), we obtain

$$[\hat{h}'(\sigma_{zz} - \sigma_{xx}) + (1 - \hat{h}'^2)\sigma_{xz}]_h = 0. \qquad (12.53)$$

The linearized form of (12.53) is obtained by substituting in the expansions (12.50b), retaining only terms $\sim O(\hat{h}')$ and evaluating the condition at $z = h_0$, which yields

$$[\hat{\sigma}_{xz}]_{h_0} = -4\dot{\epsilon}(\eta_2 - \eta_1)\hat{h}'. \qquad (12.54)$$

The condition (12.54) implies that the instability is driven by a jump in shear stress across the (undeformed) interface that is proportional to the background strain rate, the difference of the viscosities of the two fluids and the layer slope.

The last matching condition imposes the continuity of the normal stress across the interface, viz.,

$$[n_i \sigma_{ij} n_j]_h = 0. \qquad (12.55)$$

Expanding this expression as before, we obtain

$$[\sigma_{zz} + \hat{h}'^2 \sigma_{xx} - 2\hat{h}' \sigma_{xz}]_h = 0.$$

Retaining only terms $\sim O(\hat{h}')$ and evaluating the condition at $z = h_0$, we obtain

$$[\hat{\sigma}_{zz}]_{h_0} = 0. \qquad (12.56)$$

The final condition is the kinematic condition

$$\partial_t h + u(h) \partial_x h = w(h).$$

Introducing the expansions (12.50a), we obtain

$$\partial_t h_0 + \partial_t \hat{h} + (-\dot{\epsilon} x + \hat{u}) \partial_x \hat{h} = \dot{\epsilon}(h_0 + \hat{h}) + \hat{w}(h)$$

Noting that $\partial_t h_0 = \dot{\epsilon} h_0$ and retaining only the lowest-order terms, we find

$$\partial_t \hat{h} - \dot{\epsilon} x \partial_x \hat{h} = \dot{\epsilon} \hat{h} + \hat{w}(h_0). \qquad (12.57)$$

Now write

$$\hat{h} = \zeta(t) \sin k(t) x, \quad \hat{w}(h_0) = \hat{w}_0 \sin k(t) x. \qquad (12.58)$$

Substituting (12.58) into (12.57), we obtain

$$\dot{\zeta} \sin kx + \zeta \dot{k} x \cos kx - \zeta \dot{\epsilon} \sin kx - \zeta \dot{\epsilon} x k \cos kx = \hat{w}_0 \sin kx, \qquad (12.59)$$

where dots on ζ and k denote d/dt. Now (12.59) can only be valid if the terms $\propto x \cos kx$ cancel, which requires

$$\dot{k} = \dot{\epsilon} k. \qquad (12.60)$$

Equation (12.60) says that the wavenumber of any sinusoidal disturbance increases with time due to shortening by the background flow. The remaining terms in (12.59) are

$$\dot{\zeta} = \dot{\epsilon} \zeta + \hat{w}_0. \qquad (12.61)$$

Equation (12.61) states that the rate of change of the disturbance amplitude ζ comprises two parts: a kinematic contribution $\dot{\epsilon} \zeta$ that represents the effect of shortening by the background flow; and a dynamic contribution $\hat{w}_0$ equal to the vertical velocity of the secondary flow.

The streamfunction for the secondary flow in each fluid satisfies the biharmonic equation $\nabla^4 \psi_i = 0$, where $\hat{u}_i = \partial_z \psi_i$ and $\hat{w}_i = -\partial_x \psi_i$. We seek solutions of the form $\psi_i = \phi_i(z,t)\cos kx$. To write the matching conditions (12.51), (12.54) and (12.56) in terms of ϕ_i and their derivatives, we first need to determine the pressures $\hat{p}_i$ that appear in (12.56) by integrating the momentum equations $\partial_x \hat{p}_i = \eta_i \nabla^2 \hat{u}_i$. This yields

$$\hat{p}_i = k^{-1}\eta_i(D^3\phi_i - k^2 D\phi_i)\sin kx, \tag{12.62}$$

where $D = d/dz$. The matching conditions now take the forms

$$\phi_1 = \phi_2, \tag{12.63a}$$

$$D\phi_1 = D\phi_2, \tag{12.63b}$$

$$\eta_2(D^2 + k^2)\phi_2 - \eta_1(D^2 + k^2)\phi_1 = -4\dot{\epsilon}(\eta_2 - \eta_1)\frac{\phi_1 k^2}{s}, \tag{12.63c}$$

$$\eta_1(D^3\phi_1 - 3k^2 D\phi_1) = \eta_2(D^3\phi_2 - 3k^2 D\phi_2), \tag{12.63d}$$

where $s \equiv \phi_1 k/\zeta$ is the dynamic part of the growth rate.

The next step is to write down the expressions for ϕ_i that satisfy the boundary conditions at $z = 0$ and $z = \infty$. In folding deformation, u is an odd function of z while w is even. This corresponds to an even streamfunction that satisfies $\phi_1(z) = \phi_1(-z)$ (Smith, 1975). In addition, ϕ_2 and $D\phi_2$ must vanish as $z \to \infty$. The appropriate solutions of the biharmonic equation in the region $z \geq 0$ are therefore

$$\phi_1 = A_1 \cosh kz + B_1 kz \sinh kz, \tag{12.64a}$$

$$\phi_2 = (A_2 + B_2 kz)\exp(-kz), \tag{12.64b}$$

where A_1, B_1, A_2 and B_2 are unknown constants that must be determined by the interfacial matching conditions. Substitution of (12.64) into (12.63) yields a system of four linear algebraic equations for A_1, B_1, A_2 and B_2 that have a nontrivial solution only if the determinant of the coefficient matrix vanishes. This condition yields the growth rate

$$\frac{s}{\dot{\epsilon}} = \frac{4\beta\gamma(\gamma - 1)}{(\gamma^2 + 1)\sinh 2\beta + 2\gamma\cosh 2\beta - 2\beta(\gamma^2 - 1)}, \tag{12.65}$$

where $\beta = kh_0$ and $\gamma = \eta_1/\eta_2$. Figure 12.5 shows the normalized dynamic growth rate $s/\dot{\epsilon}$ as a function of β for three values of γ. The total normalized growth rate is the sum of the dynamic and kinematic contributions, i.e., $s/\dot{\epsilon} + 1$.

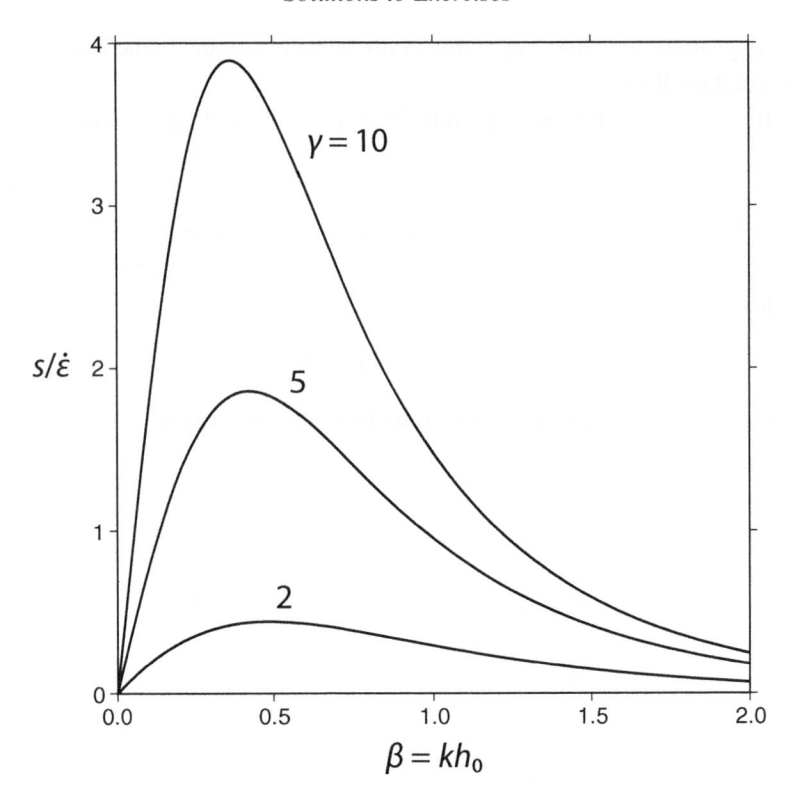

Figure 12.5 Normalized dynamic growth rate $s/\dot{\epsilon}$ defined by (12.65) as a function of $\beta = kh_0$ for three values of the viscosity ratio $\gamma = \eta_1/\eta_2$. The growth rates shown are due to the secondary flow only and do not include the kinematic contribution ($= 1$) due to background shortening.

As pointed out by Smith (1975), folding is just one of a group of four closely related instabilities, each of which is represented by a combination of a direction of shortening (parallel to or perpendicular to the layer) and the relative strength of the layer (stronger or weaker than its surroundings). The two most important of these are folding and necking, which involve a stronger layer undergoing shortening either parallel to (folding) or perpendicular to (necking) the layer. The necking instability is often called 'boudinage' or 'pinch-and-swell' instability. Early work on these instabilities considered the simplest case of a single embedded layer with either Newtonian (Smith, 1975; Fletcher, 1977) or non-Newtonian (Fletcher, 1974; Smith, 1977) rheology. An important result of this work is that necking only occurs if the rheology is non-Newtonian. Since the 1970s, study of folding and necking has generally concentrated on more realistic models incorporating multiple layers and/or laboratory-based rheological laws (e.g., Fletcher and Hallet, 1983; Ricard and Froidevaux, 1986). The reader interested in more detailed reviews of folding and necking instabilities is referred to the textbook of Johnson and Fletcher (1994) and the review articles of Cloetingh and Burov (2011) and Schmalholz and Mancktelow (2016). The latter two references include discussions of folding and necking at the lithospheric scale and are therefore of particular interest in the context of this book.

Exercise 10.2 (Richter, 1973a). The first step is to substitute the expansions (10.32) into the dimensionless governing equations

$$\nabla_h^2 \nabla^4 \mathcal{P} = -Ra \nabla_h^2 \theta, \tag{12.66a}$$

$$\partial_t \theta + \mathcal{L}_j \mathcal{P} \partial_j \theta - \nabla_h^2 \mathcal{P} = \nabla^2 \theta, \tag{12.66b}$$

where the operator $\mathcal{L}_j$ is defined by (10.22). The operator ∇_h^2 is retained in (12.66a) because the poloidal scalar is not periodic in the x-direction owing to the imposed shear. Collecting terms of order ϵ in the expanded equations, we find

$$\nabla_h^2 \nabla^4 \mathcal{P}_1 + Ra_0 \nabla_h^2 \theta_1 = 0, \tag{12.67a}$$

$$\nabla^2 \theta_1 + \nabla_h^2 \mathcal{P}_1 = 0. \tag{12.67b}$$

The solutions of (12.67) that satisfy the boundary conditions and include the imposed shear are

$$\mathcal{P}_1 = -\frac{1}{k^2}(A \cos kx + B \cos ky) \cos \pi z - \frac{U}{\pi} x \cos \pi z, \tag{12.68a}$$

$$\theta_1 = \frac{1}{\pi^2 + k^2}(A \cos kx + B \cos ky) \cos \pi z, \tag{12.68b}$$

$$Ra_0 = \frac{(\pi^2 + k^2)^3}{k^2}. \tag{12.68c}$$

The $O(\epsilon^2)$ equations are

$$\nabla^4 \mathcal{P}_2 + Ra_0 \theta_2 = 0, \tag{12.69a}$$

$$\nabla^2 \theta_2 + \nabla_h^2 \mathcal{P}_2 = F(x, y, z, \tau), \tag{12.69b}$$

where

$$F = \frac{kUA^2}{2(\pi^2 + k^2)} \sin kx \sin 2\pi z$$

$$- \frac{\pi}{2(\pi^2 + k^2)}(A^2 + B^2 + 2AB \cos kx \cos ky) \sin 2\pi z. \tag{12.69c}$$

The operator ∇_h^2 has been dropped in (12.69a) because $\mathcal{P}_2$ and θ_2, unlike $\mathcal{P}_1$, contain no components $\propto x$. The solutions of (12.69) are

$$\mathcal{P}_2 = (c_1 \cos kx \cos ky + c_2 \sin kx) \sin 2\pi z, \qquad (12.70a)$$

$$\theta_2 = (c_3 \cos kx \cos ky + c_4 \sin kx + c_5) \sin 2\pi z, \qquad (12.70b)$$

where

$$c_1 = -\frac{\pi^2(k^2+\pi^2)^2 AB}{2k^2 \phi_1}, \quad c_2 = \frac{(k^2+\pi^2)^2 UA}{2k\phi_2}, \qquad (12.71a)$$

$$c_3 = \frac{2\pi(k^2+2\pi^2)^2 AB}{(k^2+\pi^2)\phi_1}, \quad c_4 = -\frac{k(k^2+4\pi^2)^2 UA}{2(k^2+\pi^2)\phi_2}, \qquad (12.71b)$$

$$c_5 = \frac{A^2+B^2}{8\pi(k^2+\pi^2)}, \qquad (12.71c)$$

$$\phi_1 = 4(k^2+2\pi^2)^3 - (k^2+\pi^2)^3, \quad \phi_2 = (k^2+4\pi^2)^3 - (k^2+\pi^2)^3. \qquad (12.71d)$$

The $O(\epsilon^3)$ equations are

$$\nabla^4 \mathcal{P}_3 + Ra_0 \theta_3 + \chi_1 = 0, \qquad (12.72a)$$

$$\nabla^2 \theta_3 + \nabla_h^2 \mathcal{P}_3 + \chi_2 = 0, \qquad (12.72b)$$

where $\chi_1 = Ra_2(A \cos kx + B \cos ky) \cos \pi z/(k^2+\pi^2)$ and χ_2 is too complicated to write out here. Reducing (12.72) to a single equation for $\mathcal{P}_3$, we obtain

$$\nabla^6 \mathcal{P}_3 - Ra_0 \nabla_h^2 \mathcal{P}_3 = Ra_0 \chi_2 - \nabla^2 \chi_1 \equiv G. \qquad (12.73)$$

The solvability conditions are

$$\int_{-1/2}^{1/2} dz \int_{-\pi/k}^{\pi/k} dx \int_{-\pi/k}^{\pi/k} dy\, G \cos \pi z \cos kx = 0, \qquad (12.74a)$$

$$\int_{-1/2}^{1/2} dz \int_{-\pi/k}^{\pi/k} dx \int_{-\pi/k}^{\pi/k} dy\, G \cos \pi z \cos ky = 0. \qquad (12.74b)$$

Evaluating the integrals with $Ra_2 = Ra_0$, we obtain the following evolution equations for $A(T)$ and $B(T)$:

$$\dot{A} = (k^2+\pi^2)A - \frac{1}{8}A(A^2+B^2) - \beta AB^2 - \alpha U^2 A, \qquad (12.75a)$$

$$\dot{B} = (k^2+\pi^2)B - \frac{1}{8}B(B^2+A^2) - \beta BA^2, \qquad (12.75b)$$

where

$$\alpha = \frac{k^2(k^2+4\pi^2)^2}{4\phi_2} > 0, \quad \beta = \frac{\pi^2(k^2+2\pi^2)^2}{2\phi_1} > 0. \qquad (12.76)$$

The preceding expression for β corrects a misprint in the expression following eqn. (13) in Richter (1973a). The presence of the term $-\alpha U^2 A$ in (12.75a) indicates that the shear tends to suppress rolls whose axes are perpendicular to the shear velocity.

In geodynamics, convection rolls with their axes aligned with the direction of an imposed vertical shear are known as 'Richter rolls'. Korenaga and Jordan (2003b) analyzed the stability of Richter rolls to 3-D perturbations for Rayleigh numbers $Ra \leq 10^6$. For moderate Rayleigh numbers $Ra \leq 3 \times 10^5$, Korenaga's results are in excellent agreement with the experimental stability diagram of Richter and Parsons (1975) in the Pe-Ra parameter space, where Pe is the Péclet number based on the surface velocity that drives the shear flow.

Exercise 10.3. Substituting the expansion (10.51) into the governing equations (10.23), using the results (10.53) and (10.54) and collecting powers of ϵ^3, we obtain

$$\nabla^4 \mathcal{P}_3 - Ra_0 \theta_3 = \cos \pi z \left\{ \frac{Ra_0}{3\pi^2} \Psi_+ - \frac{1}{\pi^2} \partial^4_{YYYY} \Psi_+ \right.$$

$$\left. - \frac{2\sqrt{2}i}{\pi} \partial^3_{XYY} \Psi_- + 5 \partial^2_{XX} \Psi_+ \right\}, \tag{12.77}$$

$$\nabla^2 \theta_3 - \partial^2_{xx} \mathcal{P}_3 = \cos \pi z \left\{ \frac{\cos 2\pi z}{12\pi^2} |C|^2 \Psi_+ + \frac{1}{3\pi^2} \partial_T \Psi_+ \right.$$

$$\left. + \frac{2}{3\pi^2} \partial^2_{XX} \Psi_+ \right\} + \sin 2\pi z \{\ldots\}, \tag{12.78}$$

where $k_c = \pi/\sqrt{2}$ and

$$\nabla^2 = \partial^2_{xx} + \partial^2_{zz}, \quad \Psi_\pm = C \exp(ik_c x) \pm \overline{C} \exp(-ik_c x).$$

Now reduce the preceding equations to a single equation for $\mathcal{P}_3$ by taking ∇^2 of (12.77), eliminating $\nabla^2 \theta_3$ using (12.78) and taking the limit $\epsilon \to 0$ to eliminate higher-order terms. The result is

$$\nabla^6 \mathcal{P}_3 - Ra_0 \partial^2_{xx} \mathcal{P}_3 = \Phi,$$

$$\Phi = \sin 2\pi z \{\ldots\} + \cos \pi z \cos 2\pi z \left\{ \frac{Ra_0}{12\pi^2} |C|^2 \Psi_+ \right\}$$

$$+ \cos \pi z \left\{ -\frac{Ra_0}{2} \Psi_+ + \frac{Ra_0}{3\pi^2} \partial_T \Psi_+ + \frac{3}{2} \partial^4_{YYYY} \Psi_+ \right.$$

$$\left. + 3\sqrt{2} \pi i \partial^3_{XYY} \Psi_- + \left(\frac{2Ra_0}{3\pi^2} - \frac{15\pi^2}{2} \right) \partial^2_{XX} \Psi_+ \right\}.$$

Now the solvability condition is

$$\int_{-1/2}^{1/2} dz \cos \pi z \int_{-\pi/k_c}^{\pi/k_c} dx e^{-ik_c x} \Phi = 0,$$

which yields

$$\frac{\partial C}{\partial T} - \frac{4}{3}\left(\frac{\partial}{\partial X} - \frac{i}{\sqrt{2\pi}}\frac{\partial^2}{\partial Y^2}\right)^2 C = \frac{3\pi^2}{2}C - \frac{1}{8}|C|^2 C,$$

where Ra_0 has been replaced by $Ra_c \equiv 27\pi^4/4$. Setting $C = W/\epsilon$ and $\epsilon^2 = (Ra - Ra_c)/Ra_c$, we obtain (10.55).

Exercise 10.4. (Newell and Whitehead, 1969). First simplify equation (10.55) using the transformations

$$W \to 2\sqrt{3}\pi \epsilon W, \quad T \to \frac{2}{3\pi^2}T, \quad X \to \frac{2\sqrt{2}}{3\pi}X,$$

$$K \to \frac{3\pi}{2\sqrt{2}}K, \quad Y \to \frac{2^{3/4}}{\sqrt{3\pi}}Y.$$

The result is

$$\frac{\partial W}{\partial T} - \left(\frac{\partial}{\partial X} - \frac{i}{\sqrt{2\pi}}\frac{\partial^2}{\partial Y^2}\right)^2 W = W(1 - |W|^2),$$

a steady finite-amplitude solution of which is

$$W_s = (1 - K^2)^{1/2} e^{iKX}.$$

Now substitute $W = W_s + u(X, Y, T)$ into the preceding envelope equation and linearize the resulting equation for u to obtain

$$\frac{\partial u}{\partial T} - \left(\frac{\partial}{\partial X} - \frac{i}{\sqrt{2\pi}}\frac{\partial^2}{\partial Y^2}\right)^2 u = (1 - 2p^2)u - p^2 \bar{u} e^{2iKX},$$

where $p^2 \equiv 1 - K^2$ and the overbar denotes complex conjugation. As in § 10.3.2, we seek a solution as a sum of products of functions that vary periodically in X and Y and a function of X that has the same periodicity as the stationary solution. Thus we write

$$u = e^{iKX} e^{\lambda T}\left[A e^{i(MX+NY)} + B e^{-i(MX+NY)}\right],$$

where M and N are wavenumbers and λ is the growth rate. Now substitute the preceding form into the linear equation for u, divide out the common factor $e^{iKX} e^{\lambda T}$ and set the coefficients of $e^{i(MX+NY)}$ and $e^{-i(MX+NY)}$ separately to zero. This gives two simultaneous linear equations for A and B that have a solution only if the determinant of the coefficient matrix is zero. That condition yields a quadratic equation for λ. Choosing the solution with a positive growth rate corresponding to instability, we find

$$\lambda = -p^2 - M^2 + K^2 - Q^2 + (p^4 + 4M^2 Q^2)^{1/2}$$

where $Q = K + N^2/\sqrt{2\pi}$. Now $\lambda(M, N; K)$ has relative maxima at

(I) $M = 0$, $N^2 = -\sqrt{2\pi}K$; $\lambda = K^2$;

(II) $M^2 = \dfrac{(3K^2 - 1)(K^2 + 1)}{4K^2}$, $N = 0$; $\lambda = \dfrac{(3K^2 - 1)^2}{4K^2}$.

These two relative maxima correspond to two different types of instability, which Newell and Whitehead (1969) call Type I and Type II. Because $N^2 = -\sqrt{2\pi}K$ for the Type I instability, it is only possible when $K < 0$. This corresponds to a wave vector $\mathbf{k}_c + \mathbf{K}$ inside the critical circle, i.e., to rolls with a wavelength longer than the critical wavelength (Figure 12.6). The instability involves the growth of rolls that are oblique to the critical roll and whose wavevector lies on the critical circle. Type II instability occurs for $K^2 > 1/3$ and corresponds to the Eckhaus (sideband) instability (Eckhaus, 1965).

Exercise 10.5. From (3.20), the maximum growth rate of the isoviscous ($\gamma = 1$) RTI is $s \sim g\Delta\rho h_0/\eta$, where the constant of proportionality has been ignored. In the context of a TBL of thickness δ, $h_0 \sim \delta$ and $\Delta\rho \sim \alpha\rho_0\Delta T$, whence $s \sim g\alpha\delta\Delta T/\nu$, where ν is the kinematic viscosity. The relative rate of thickening of the conductive BL is $\dot{\delta}/\delta = 1/(2t)$. The growth rate s becomes comparable to $\dot{\delta}/\delta$ when $g\alpha(\kappa t)^{1/2}\Delta T/\nu \sim t^{-1}$, yielding a critical time

$$t_c \sim \frac{1}{\kappa}\left(\frac{\nu\kappa}{g\alpha\Delta T}\right)^{2/3}$$

which agrees with (10.79) apart from a numerical factor. A similar argument was applied to a fluid with strongly temperature-dependent viscosity by Ke and Solomatov (2004).

Exercise 10.6. (a) The predicted time of instability is $t_c = 700$ s, which is between the times of the first two images of Figure 10.5. (b) Use the solution for the conductive temperature profile in an impulsively heated half-space with initial temperature T_0, viz.,

$$T = T_0 + \Delta T \operatorname{erfc}\left(\frac{z}{2\sqrt{\kappa t}}\right).$$

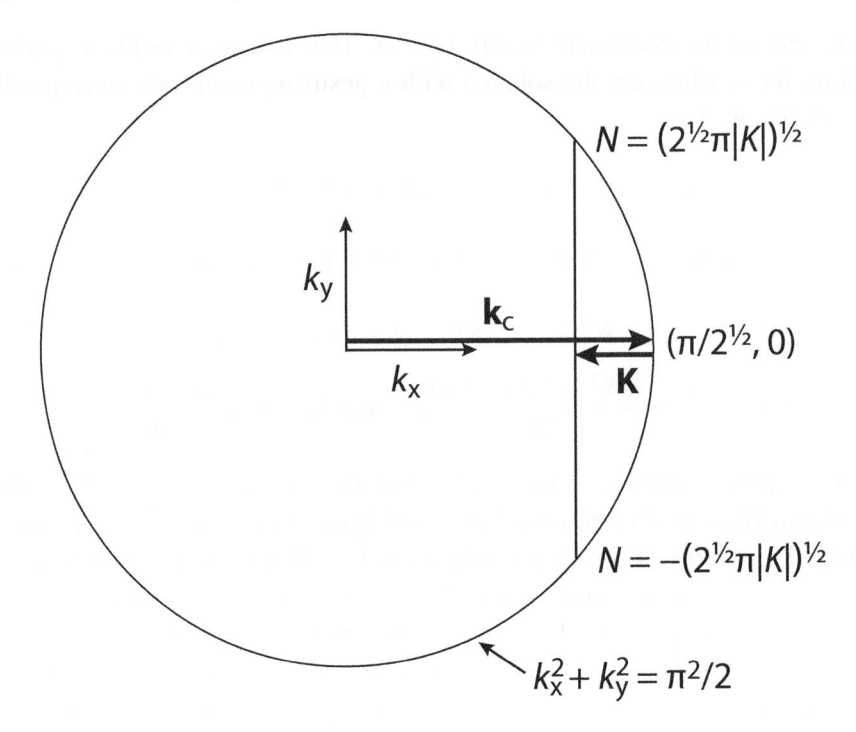

Figure 12.6 Diagram illustrating the Type I instability of Newell and Whitehead (1969). The critical roll with wavevector $(\pi/\sqrt{2}, 0)$ lies at the rightmost extremity of the critical circle $k_x^2 + k_y^2 = \pi^2/2$. The critical wavevector $\mathbf{k}_c$ and the wavevector $\mathbf{K}$ of the finite-amplitude steady solution are shown. Instability corresponds to oblique rolls that lie on the critical circle with wavenumbers $N = \pm(2^{1/2}\pi|K|)^{1/2}$ in the direction (y) perpendicular to the critical roll. The scale for all wavenumbers is the inverse of the layer depth, i.e., the transformation (12) has not been applied. Figure redrawn from Figure 3 of Newell and Whitehead (1969) by permission of Cambridge University Press.

where erfc is the complementary error function. At the left side of the image, the $24.35°C \equiv T_1$ isotherm is at a height $z \approx 0.017$ m. The value of the complementary error function at this isotherm is

$$\text{erfc}\left(\frac{z}{2\sqrt{\kappa t}}\right) = \frac{T_1 - T_0}{\Delta T} = 0.1016.$$

Thus the time is

$$t = \frac{1}{4\kappa}\left[\frac{z}{\text{erfc}^{-1}(0.1016)}\right]^2,$$

where erfc^{-1} is the inverse of erfc. Now $\text{erfc}^{-1}(0.1016) = 1.158$, whence the preceding formula gives $t = 464$ s. The most probable cause of the discrepancy is that the heated lower surface of the experimental tank does not reach its final temperature T_{hot} immediately, but only after 1–2 minutes.

Exercise 11.1. We can work with the equations for layer 2 with no loss of generality. The dimensionless momentum equation for this layer is (11.32b) with $n = 2$, or

$$\nabla^2 \mathbf{u} = \nabla p - \theta \mathbf{k}$$

where the subscript 2 has been suppressed to simplify the notation. For a 2-D flow in the x-z plane, the components of the preceding equation are

$$\nabla^2 u = p_x, \tag{12.79a}$$

$$\nabla^2 w = p_z - \theta \tag{12.79b}$$

where subscripts denote partial derivatives. Now take ∂_z of (12.79a) and ∂_x of (12.79b) and subtract the two resulting equations. The result is

$$\nabla^2 (w_x - u_z) = -\theta_x. \tag{12.80}$$

Take ∂_x of the preceding equation and use the continuity equation $u_x = -w_z$ to obtain

$$\nabla^4 w = -\theta_{xx}.$$

Because $\theta_{xx} = 0$ on the boundary when $\theta = 0$ there, $\nabla^4 w = 0$ on the boundary.

Exercise 11.2 (Richter and Johnson, 1974) In the notation of our general treatment of compositionally layered convection, the case in question corresponds to $\gamma = 1$ and $a = 1/2$. The equations satisfied by $W_n(z)$ ($n = 1$ or 2) are (11.44) with $\gamma = 1$, or

$$(sRa + k^2 - D^2)(D^2 - k^2)^2 W_n = k^2 Ra W_n. \tag{12.81}$$

The general solution that is symmetric about $z = 0$ and that satisfies the boundary conditions $W = D^2 W = D^4 W$ on both boundaries $z = \pm 1/2$ is

$$W_1 = \sum_{m=1}^{3} A_m \sinh q_m \left(\frac{1}{2} + z\right) \tag{12.82a}$$

$$W_2 = \sum_{m=1}^{3} B_m \sinh q_m \left(\frac{1}{2} - z\right), \tag{12.82b}$$

where the coefficients q_m satisfy

$$(sRa + k^2 - q_m^2)(q_m^2 - k^2)^2 = k^2 Ra. \tag{12.83}$$

290   Solutions to Exercises

The matching conditions (11.37), (11.39) and (11.41) are satisfied by choosing $B_m = A_m$. The matching condition (11.38) on the horizontal velocity requires

$$\sum_{m=1}^{3} q_m \cosh \frac{q_m}{2} A_m = 0. \tag{12.84}$$

The matching condition (11.42) on the heat flux, after subtraction of k^4 times (12.84), requires

$$\sum_{m=1}^{3} q_m^3 (q_m^2 - 2k^2) \cosh \frac{q_m}{2} A_m = 0. \tag{12.85}$$

The matching condition (11.40) on the normal stress, after addition of $6k^2 s$ times (12.84), requires

$$\sum_{m=1}^{3} \left(2sq_m^3 \cosh \frac{q_m}{2} - Bk^2 \sinh \frac{q_m}{2} \right) A_m = 0. \tag{12.86}$$

Now (12.84), (12.85) and (12.86) are a system of three homogeneous linear equations for A_1–A_3 and have a nontrivial solution only if the determinant of the coefficient matrix vanishes. Dividing each element in a column of that matrix by the same quantity does not change the condition for the vanishing of the determinant, and so we divide each element in column m by $q_m \cosh(q_m/2)$. The elements of the matrix $\mathbf{M}$ whose determinant must vanish are then

$$M_{1m} = 1, \quad M_{2m} = q_m^2(q_m^2 - 2k^2), \quad M_{3m} = 2sq_m^2 - \frac{Bk^2}{q_m} \tanh \frac{q_m}{2},$$

for $m = 1, 2, 3$. The preceding expressions agree with equation (18) of Richter and Johnson (1974) once account is taken of the different nondimensionalization used. To verify the solution cited, first substitute the values of k, s and Ra into (12.83), and solve the resulting equation to obtain $q_1 = -4.96935 - 3.15228i$, $q_2 = -4.30404 + 1.6706i$ and $q_3 = -0.90333 - 4.09726i$. Then substitute these values and the values of k, s and B into the expression for $\det(\mathbf{M})$.

Exercise 11.3. (McKenzie et al., 1974). Conservation of vertical mass flux requires

$$Wd \sim w\delta. \tag{12.87}$$

The surface heat flux integrated across the cell must equal the rate of internal heat generation, or

$$k \frac{\Delta T}{\delta} d \sim \rho H d^2. \tag{12.88}$$

Balancing vertical advection and diffusion in the horizontal BL, we obtain

$$W\frac{\Delta T}{\delta} \sim \kappa \frac{\Delta T}{\delta^2}. \tag{12.89}$$

Now the vorticity equation in the vertical plume is

$$\frac{\partial^2 \omega}{\partial x^2} \sim \frac{g\alpha}{\nu}\frac{\partial T}{\partial x},$$

which implies

$$\omega \sim g\alpha \Delta T \delta / \nu. \tag{12.90}$$

Because the vorticity $\omega \approx 0$ in the uniformly upwelling core, the vorticity in the plume is related to w by

$$\omega \sim w/\delta \tag{12.91}$$

(this is different from the case of RBC, where $\omega \sim w/d$ because the core rotates as a rigid body). Solving (12.87), (12.88), (12.89), (12.90) and (12.91), we find

$$\delta \sim d Ra_H^{-1/5}, \quad w \sim \frac{\kappa}{d} Ra_H^{2/5}, \quad W \sim \frac{\kappa}{d} Ra_H^{1/5}, \quad \Delta T \sim \frac{d^2 H}{c_p \kappa} Ra_H^{-1/5},$$

where the Rayleigh number is

$$Ra_H = \frac{\alpha g H d^5}{c_p \kappa^2 \nu}.$$

References

Aharonov, E., Whitehead, J. A., Kelemen, P. B., and Spiegelman, M. 1995. Channeling instability of upwelling melt in the mantle. *J. Geophys. Res.*, **100**, 20, 433–420, 450.

Albers, M., and Christensen, U. R. 2001. Channeling of plume flow beneath mid-ocean ridges. *Earth Planet. Sci. Lett.*, **187**, 207–220.

Androvandi, S. 2009. *Convection multi-échelles à haut nombre de Rayleigh dans un fluide dont la viscosité dépend fortement de la température: Application au manteau terrestre*. Ph.D. thesis, Institut de Physique du Globe de Paris, Paris, France.

Ansari, A., and Morris, S. 1985. The effects of a strongly temperature-dependent viscosity on Stokes's drag law: experiments and theory. *J. Fluid Mech.*, **159**, 459–476.

Asaadi, N., Ribe, N. M., and Sobouti, F. 2011. Inferring nonlinear mantle rheology from the shape of the Hawaiian swell. *Nature*, **473**, 501–504.

Backus, G. E. 1958. A class of self-sustaining dissipative spherical dynamos. *Ann. Phys.*, **4**, 372–447.

Backus, G. E. 1967. Converting vector and tensor equations to scalar equations in spherical co-ordinates. *Geophys. J. R. Astr. Soc.*, **13**, 71–101.

Bai, Q., Mackwell, S. J., and Kohlstedt, D. L. 1991. High-temperature creep of olivine single crystals, 1. Mechanical results for buffered samples. *J. Geophys. Res.*, **96**, 2441–2463.

Barcilon, V., and Lovera, O. M. 1989. Solitary waves in magma dynamics. *J. Fluid Mech.*, **204**, 121–133.

Barcilon, V., and Richter, F. M. 1986. Nonlinear waves in compacting media. *J. Fluid Mech.*, **164**, 429–448.

Barenblatt, G. I. 1996. *Scaling, Self-Similarity, and Intermediate Asymptotics*. Cambridge: Cambridge University Press.

Bassi, G., and Bonnin, J. 1988. Rheological modelling and deformation instability of lithosphere under extension. *Geophys. J.*, **93**, 485–504.

Batchelor, G. K. 1967. *An Introduction to Fluid Dynamics*. Cambridge: Cambridge University Press.

Batchelor, G. K. 1970. Slender-body theory for particles of arbitrary cross-section in Stokes flow. *J. Fluid Mech.*, **44**, 419–440.

Bayly, B. 1982. Geometry of subducted plates and island arcs viewed as a buckling problem. *Geology*, **10**, 629–632.

Bellahsen, N., Faccenna, C., and Funiciello, F. 2005. Dynamics of subduction and plate motion in laboratory experiments: insights into the "plate tectonics" behavior of the Earth. *J. Geophys. Res.*, **110**, B01401.

Bercovici, D. 1993. A simple model of plate generation from mantle flow. *Geophys. J. Int.*, **114**, 635–650.

Bercovici, D. 1994. A theoretical model of cooling viscous gravity currents with temperature-dependent viscosity. *Geophys. Res. Lett.*, **21**, 1177–1180.

Bercovici, D. 1998. Generation of plate tectonics from lithosphere-mantle flow and void-volatile self-lubrication. *Earth Planet. Sci. Lett.*, **154**, 139–151.

Bercovici, D., and Kelly, A. 1997. The non-linear initiation of diapirs and plume heads. *Phys. Earth Planet. Int.*, **101**, 119–130.

Bercovici, D., and Lin, J. 1996. A gravity current model of cooling mantle plume heads with temperature-dependent buoyancy and viscosity. *J. Geophys. Res.*, **101**, 3291–3309.

Bercovici, D., and Long, M. D. 2014. Slab rollback instability and supercontinent dispersal. *Geophys. Res. Lett.*, **41**, 6659–6666.

Bercovici, D., and Ricard, Y. 2003. Energetics of a two-phase model of lithospheric damage, shear localization and plate-boundary formation. *Geophys. J. Int.*, **152**, 581–596.

Bercovici, D., Ricard, Y., and Schubert, G. 2001a. A two-phase model for compaction and damage 1. General theory. *J. Geophys. Res.*, **106**, 8887–8906.

Bercovici, D., Ricard, Y., and Schubert, G. 2001b. A two-phase model for compaction and damage 3. Applications to shear localization and plate boundary formation. *J. Geophys. Res.*, **106**, 8925–8940.

Bercovici, D., and Rudge, J. F. 2016. A mechanism for mode selection in melt band instabilities. *Earth Planet. Sci. Lett.*, **433**, 139–145.

Bercovici, D., Schubert, G., and Tackley, P. J. 1993. On the penetration of the 660 km phase change by mantle downflows. *Geophys. Res. Lett.*, **20**, 2599–2602.

Bevis, M. 1986. The curvature of Wadati-Benioff zones and the torsional rigidity of subducting plates. *Nature*, **323**, 52–53.

Biot, M. A. 1954. Theory of stress-strain relations in anisotropic viscoelasticity and relaxation phenomena. *J. Appl. Phys.*, **25**, 1385–1391.

Blake, J. R. 1971. A note on the image system for a Stokeslet in a no-slip boundary. *Proc. Camb. Phil. Soc.*, **70**, 303–310.

Bloor, M. I. G., and Wilson, M. J. 2006. An approximate analytic solution method for the biharmonic problem. *Proc. R. Soc. Lond. A*, **462**, 1107–1121.

Bolton, E. W., and Busse, F. H. 1985. Stability of convection rolls in a layer with stress-free boundaries. *J. Fluid Mech.*, **150**, 487–498.

Buckingham, E. 1914. On physically similar systems; illustrations of the use of dimensional equations. *Phys. Rev.*, **4**, 345–376.

Buckmaster, J. D., Nachman, A., and Ting, L. 1975. The buckling and stretching of a viscida. *J. Fluid Mech.*, **69**, 1–20.

Budiansky, B., and Sanders, J. L. 1967. On the 'best' first-order linear shell theory. Pages 129–140 of: *W. Prager Anniversary Volume*. New York: Macmillan.

Buffett, B. A. 2006. Plate force due to bending at subduction zones. *J. Geophys. Res.*, **111**, B09405.

Buffett, B. A., and Becker, T. W. 2012. Bending stress and dissipation in subducted lithosphere. *J. Geophys. Res.*, **117**, B05413.

Buffett, B. A., Gable, C. W., and O'Connell, R. J. 1994. Linear stability of a layered fluid with mobile surface plates. *J. Geophys. Res.*, **99**, 19,885–19,900.

Busse, F. H. 1967a. On the stability of two-dimensional convection in a layer heated from below. *J. Math. Phys.*, **46**, 140–150.

Busse, F. H. 1967b. The stability of finite amplitude cellular convection and its relation to an extremum principle. *J. Fluid Mech.*, **30**, 625–649.

Busse, F. H. 1981. On the aspect ratio of two-layer mantle convection. *Phys. Earth Planet. Int.*, **24**, 320–324.

Busse, F. H. 1984. Instabilities of convection rolls with stress-free boundaries near threshold. *J. Fluid Mech.*, **146**, 115–125.

Busse, F. H., and Frick, H. 1985. Square-pattern convection in fluids with strongly temperature-dependent viscosity. *J. Fluid Mech.*, **150**, 451–465.

Busse, F. H., Richards, M. A., and Lenardic, A. 2006. A simple model of high Prandtl and high Rayleigh number convection bounded by thin low-viscosity layers. *Geophys. J. Int.*, **164**, 160–167.

Busse, F. H., and Schubert, G. 1971. Convection in a fluid with two phases. *J. Fluid Mech.*, **46**, 801–812.

Butler, S. L. 2009. The effects of buoyancy on shear-induced melt bands in a compacting porous medium. *Phys. Earth Planet. Int.*, **173**, 51–59.

Butler, S. L. 2010. Porosity localizing instability in a compacting porous layer in a pure shear flow and the evolution of porosity band wavelength. *Phys. Earth Planet. Int.*, **182**, 30–41.

Butler, S. L. 2012. Numerical models of shear-induced melt band formation with anisotropic matrix viscosity. *Phys. Earth Planet. Int.*, **200–201**, 28–36.

Butterworth, N. P., Quevedo, L., Morra, G., and Müller, R. D. 2012. Influence of overriding plate geometry and rheology on subduction. *Geochem. Geophys. Geosyst.*, **13**, Q06W15.

Caldwell, J. G., Haxby, W. F., Karig, D. E., and Turcotte, D. L. 1976. On the applicability of a universal elastic trench profile. *Earth Planet. Sci. Lett.*, **31**, 239–246.

Canright, D., and Morris, S. 1993. Buoyant instability of a viscous film over a passive fluid. *J. Fluid Mech.*, **255**, 349–372.

Chandrasekhar, S. 1981. *Hydrodynamic and Hydromagnetic Stability*. New: Dover.

Čížková, H., and Bina, C. R. 2013. Effects of mantle and subduction-interface rheologies on slab stagnation and trench rollback. *Earth Planet. Sci. Lett.*, **379**, 95–103.

Cloetingh, S., and Burov, E. 2011. Lithospheric folding and sedimentary basin evolution: a review and analysis of formation mechanisms. *Basin Res.*, **23**, 257–290.

Connolly, J. A. D., and Podladchikov, Y. Y. 1998. Compaction-driven fluid flow in viscoelastic rock. *Geodin. Acta*, **11**, 55–84.

Connolly, J. A. D., and Podladchikov, Y. Y. 2017. An analytical solution for solitary porosity waves: dynamic permeability and fluidization of nonlinear viscous and viscoplastic rock. Pages 285–306 of: Gleeson, T., and Ingebritsen, S. E. (eds.), *Crustal Permeability*. Chichester: John Wiley & Sons, Ltd.

Conrad, C. P., and Molnar, P. 1997. The growth of Rayleigh-Taylor-type instabilities in the lithosphere for various rheological and density structures. *Geophys. J. Int.*, **129**, 95–112.

Conrad, C. P., and Molnar, P. 1999. Convective instability of a boundary layer with temperature- and strain-rate-dependent viscosity in terms of 'available buoyancy'. *Geophys. J. Int.*, **139**, 51–68.

Cox, R. G. 1970. The motion of long slender bodies in a viscous fluid Part 1. General theory. *J. Fluid Mech.*, **44**, 791–810.

Crosby, A., and Lister, J. R. 2014. Creeping axisymmetric plumes with strongly temperature-dependent viscosity. *J. Fluid Mech.*, **745**, R2.

Dannberg, J., Eilon, Z., Faul, U., Gassmöller, R., Moulik, P., and Myhill, R. 2017. The importance of grain size to mantle dynamics and seismological observations. *Geochem. Geophys. Geosyst.*, **18**, 3034–3061.

Davaille, A. 1999. Two-layer thermal convection in miscible viscous fluids. *J. Fluid Mech.*, **379**, 223–253.

Davaille, A., and Jaupart, C. 1994. Onset of thermal convection in fluids with temperature-dependent viscosity: application to the oceanic mantle. *J. Geophys. Res.*, **99**, 19,853–19,866.

Davaille, A., and Limare, A. 2015. Laboratory studies of mantle convection. Pages 73–144 of: Bercovici, D. (ed.), *Treatise on Geophysics*, 2nd edn., vol. 7. Amsterdam: Elsevier.

Davaille, A., and Vatteville, J. 2005. On the transient nature of mantle plumes. *Geophys. Res. Lett.*, **32**, L14309.

Drew, D. A., and Segel, L. A. 1971. Averaged equations for two-phase flows. *Stud. Appl. Maths.*, **50**, 205–257.

Duretz, T., Schmalholz, S. M., and Gerya, T. V. 2012. Dynamics of slab detachment. *Geochem. Geophys. Geosyst.*, **13**, Q03020.

Dvorkin, J., Nur, A., Mavko, G., and Ben-Avraham, Z. 1993. Narrow subducting slabs and the origin of backarc basins. *Tectonophys.*, **227**, 63–79.

Dziewonski, A. M., and Anderson, D. L. 1981. Preliminary reference Earth model. *Phys. Earth Planet. Int.*, **25**, 297–356.

Eckhaus, W. 1965. *Studies in Non-Linear Stability Theory*. Springer Tracts in Natural Philosophy, vol. 6. Berlin: Springer-Verlag.

England, P., and McKenzie, D. 1983. Correction to: a thin viscous sheet model for continental deformation. *Geophys. J. R. Astr. Soc.*, **73**, 523–532.

Farrell, W. E. 1972. Deformation of the Earth by surface loads. *Rev. Geophys. Space Phys.*, **10**, 761–797.

Feighner, M., and Richards, M. A. 1995. The fluid dynamics of plume-ridge and plume-plate interactions: an experimental investigation. *Earth Planet. Sci. Lett.*, **129**, 171–182.

Fenner, R. T. 1975. On local solutions to non-Newtonian slow viscous flows. *Int. J. Non-Linear Mech.*, **10**, 207–214.

Fleitout, L., and Froidevaux, C. 1983. Tectonic stresses in the lithosphere. *Tectonics*, **2**, 315–324.

Fletcher, C. A. J. 1984. *Computational Galerkin Methods*. New York: Springer.

Fletcher, R. C. 1974. Wavelength selection in the folding of a single layer with power-law rheology. *Am. J. Sci.*, **274**, 1029–1043.

Fletcher, R. C. 1977. Folding of a single viscous layer: exact infinitesimal-amplitude solution. *Tectonophys.*, **39**, 593–606.

Fletcher, R. C., and Hallet, B. 1983. Unstable extension of the lithosphere: a mechanical model for Basin-and-Range structure. *J. Geophys. Res.*, **88**, 7457–7466.

Forte, A. M., and Mitrovica, J. X. 1996. A new inference of mantle viscosity based on a joint inversion of post-glacial rebound data and long-wavelength geoid anomalies. *Geophys. Res. Lett.*, **23**, 1147–1150.

Forte, A. M., Dziewonski, A. M., and Woodward, R. L. 1993. Aspherical structure of the mantle, tectonic plate motions, nonhydrostatic geoid, and topography of the core-mantle boundary. Pages 135–166 of: Le Mouël, J.-L., Smylie, D. E., and Herring, T. (eds), *Dynamics of the Earth's Deep Interior and Earth Rotation*. Geophys. Monogr., vol. 72. Washington: Am. Geophys. Union.

Forte, A. M., and Peltier, W. R. 1987. Plate tectonics and aspherical Earth structure: the importance of poloidal-toroidal coupling. *J. Geophys. Res.*, **92**, 3645–3679.

Forte, A. M., and Peltier, W. R. 1991. Viscous flow models of global geophysical observables 1. Forward problems. *J. Geophys. Res.*, **96**, 20,131–20,159.

Forte, A. M., Peltier, W. R., and Dziewonski, A. M. 1991. Inferences of mantle viscosity from tectonic plate velocities. *Geophys. Res. Lett.*, **18**, 1747–1750.

Fowler, A. C. 1985a. Fast thermoviscous convection. *Stud. Appl. Maths.*, **72**, 189–219.

Fowler, A. C. 1985b. A mathematical model of magma transport in the asthenosphere. *Geophys. Astrophys. Fluid Dyn.*, **33**, 63–96.

Fowler, A. C. 1990. A compaction model for melt transport in the Earth's asthenosphere, part I, The basic model. Pages 3–14 of: Ryan, M. P. (ed.), *Magma Transport and Storage*. New York: John Wiley & Sons.

Fowler, A. C. 2011. *Mathematical Geoscience*. Berlin: Springer-Verlag.

Fowler, A. C., Howell, P. D., and Khaleque, T. S. 2016. Convection of a fluid with strongly temperature and pressure dependent viscosity. *Geophys. Astrophys. Fluid Dyn.*, **110**, 130–165.

Frank, F. C. 1968. Curvature of island arcs. *Nature*, **220**, 363.

Gantmacher, F. R. 1960. *Matrix Theory*. Vol. 1. Providence: AMS Chelsea Publishing.

Gavrilov, S. V., and Boiko, A. N. 2012. Waves in the diapir tail as a mechanism of hotspot pulsation. *Izvestiya Phys. Solid Earth*, **48**, 550–553.

Gebhardt, D. J., and Butler, S. L. 2016. Linear analysis of melt band formation in a mid-ocean ridge corner flow. *Geophys. Res. Lett.*, **43**, 3700–3707.

Gerardi, G., and Ribe, N. 2018. Boundary-element modeling of two-plate interaction at subduction zones: scaling laws and application to the Aleutian subduction zone. *J. Geophys. Res. Solid Earth*, **123**, 5227–5248.

Goldenveizer, A. L. 1963. Derivation of an approximate theory of shells by means of asymptotic integration of the equations of the theory of elasticity. *Prikl. Mat. Mech.*, **27**, 593–608.

Gomilko, A. M., Malyuga, V. S., and Meleshko, V. V. 2003. On steady Stokes flow in a trihedral rectangular corner. *J. Fluid Mech.*, **476**, 159–177.

Gratton, J., and Minotti, F. 1990. Self-similar viscous gravity currents: phase-plane formalism. *J. Fluid Mech.*, **210**, 155–182.

Gratton, J., Minotti, F., and Mahajan, S. M. 1999. Theory of creeping gravity currents of a non-Newtonian liquid. *Phys. Rev. E*, **60**, 6960.

Green, A. E., and Zerna, W. 1992. *Theoretical Elasticity*. 2nd edn. Mineola: Dover.

Griffiths, R. W. 1986. Thermals in extremely viscous fluids, including the effects of temperature-dependent viscosity. *J. Fluid Mech.*, **166**, 115–138.

Griffiths, R. W., and Campbell, I. H. 1991. Interaction of mantle plume heads with the Earth's surface and onset of small-scale convection. *J. Geophys. Res.*, **96**, 18,295–18,310.

Grimshaw, R. H. J., Helfrich, K. R., and Whitehead, J. A. 1992. Conduit solitary waves in a visco-elastic medium. *Geophys. Astrophys. Fluid Dyn.*, **65**, 127–147.

Hadamard, J. 1911. Mouvement permanent lent d'une sphère liquide et visqueuse dans un liquide visqueux. *C. R. Acad. Sci. Paris*, **152**, 1735–1738.

Hager, B. H. 1984. Subducted slabs and the geoid: constraints on mantle rheology and flow. *J. Geophys. Res.*, **89**, 6003–6015.

Hager, B. H., Clayton, R. W., Richards, M. A., Comer, R. P., and Dziewonski, A. M. 1985. Lower mantle heterogeneity, dynamic topography and the geoid. *Nature*, **313**, 541–545.

Hager, B. H., and O'Connell, R. J. 1979. Kinematic models of large-scale flow in the Earth's mantle. *J. Geophys. Res.*, **84**, 1031–1048.

Hager, B. H., and O'Connell, R. J. 1981. A simple global model of plate tectonics and mantle convection. *J. Geophys. Res.*, **86**, 4843–4867.

Hager, B. H., and Richards, M. A. 1989. Long-wavelength variations in Earth's geoid: physical models and dynamical implications. *Phil. Trans. R. Soc. Lond. A*, **328**, 309–327.

Happel, J., and Brenner, H. 1991. *Low Reynolds Number Hydrodynamics*. 2nd edn. Dordrecht: Kluwer Academic.

Harig, C., Molnar, P., and Houseman, G. A. 2008. Rayleigh–Taylor instability under a shear stress free top boundary condition and its relevance to removal of mantle lithosphere from beneath the Sierra Nevada. *Tectonics*, **27**, TC6019.

Harris, S. 1996. Conservation laws for a nonlinear wave equation. *Nonlinearity*, **9**, 187–208.

Haskell, N. A. 1935. The motion of a viscous fluid under a surface load. *Physics*, **6**, 265–269.

Hauri, E. H., Whitehead, J. A. Jr, and Hart, S. R. 1994. Fluid dynamic and geochemical aspects of entrainment in mantle plumes. *J. Geophys. Res.*, **99**, 24,275–24,300.

Helmholtz, H. von. 1868. Zur Theorie der stationären Ströme in reibenden Flüssigkeiten. *Verh. des naturh.-med. Vereins zu Heidelberg*, **5**, 1–7.

Hesse, M. A., Schiemenz, A. R., Liang, Y., and Parmentier, E. M. 2011. Compaction-dissolution waves in an upwelling mantle column. *Geophys. J. Int.*, **187**, 1057–1075.

Hewitt, I. J., and Fowler, A. C. 2009. Melt channelization in ascending mantle. *J. Geophys. Res.*, **114**, B06210.

Hier-Majumder, S., Ricard, Y., and Bercovici, D. 2006. Role of grain boundaries in magma migration and storage. *Earth Planet. Sci. Lett.*, **248**, 735–749.

Hinch, E. J. 1991. *Perturbation Methods*. Cambridge: Cambridge University Press.

Höink, T., and Lenardic, A. 2010. Long wavelength convection, Poiseuille–Couette flow in the low-viscosity asthenosphere and the strength of plate margins. *Geophys. J. Int.*, **180**, 23–33.

Holtzman, B. K., Groebner, N. J., Zimmerman, M. E., Ginsberg, S. B., and Kohlstedt, D. L. 2003. Stress-driven melt segregation in partially molten rocks. *Geochem. Geophys. Geosyst.*, **4**, 8607.

Houseman, G. A., and Molnar, P. 1997. Gravitational (Rayleigh–Taylor) instability of a layer with non-linear viscosity and convective thinning of continental lithosphere. *Geophys. J. Int.*, **128**, 125–150.

Howard, L. N. 1964. Convection at high Rayleigh number. Pages 1109–1115 of: Görtler, H. (ed.), *Proc. 11th Int. Congr. Appl. Mech.* Berlin: Springer.

Hsui, A. T., and Tang, X.-M. 1988. A note on the weight and the gravitational torque of a subducting slab. *J. Geodyn.*, **10**, 1–8.

Hsui, A. T., Tang, X.-M., and Toksöz, M. N. 1990. On the dip angle of subducting plates. *Tectonophys.*, **179**, 163–175.

Huppert, H. E. 1982. The propagation of two-dimensional and axisymmetric viscous gravity currents over a rigid horizontal surface. *J. Fluid Mech.*, **121**, 43–58.

Husson, L., and Ricard, Y. 2004. Stress balance above subduction: application to the Andes. *Earth Planet. Sci. Lett.*, **222**, 1037–1050.

Ismail-Zadeh, A., and Tackley, P. 2010. *Computational Methods for Geodynamics*. Cambridge: Cambridge University Press.

Ito, G. 2001. Reykjanes 'V'-shaped ridges originating from a pulsing and dehydrating mantle plume. *Nature*, **411**, 681–684.

Jackson, J. D. 1975. *Classical Electrodynamics*. 2nd edn. New York: John Wiley & Sons.

Jarvis, G. T., and McKenzie, D. P. 1980. Convection in a compressible fluid with infinite Prandtl number. *J. Fluid Mech.*, **96**, 515–583.

Jaupart, C., Molnar, P., and Cottrell, E. 2007. Instability of a chemically dense layer heated from below and overlain by a deep less viscous fluid. *J. Fluid Mech.*, **572**, 433–469.

Jeffreys, H. 1930. The instability of a compressible fluid heated below. *Proc. Camb. Phil. Soc.*, **26**, 170–172.

Jeong, J.-T., and Moffatt, H. K. 1992. Free-surface cusps associated with flow at low Reynolds number. *J. Fluid Mech.*, **241**, 1–22.

Jimenez, J., and Zufiria, J. A. 1987. A boundary-layer analysis of Rayleigh–Bénard convection at large Rayleigh number. *J. Fluid Mech.*, **178**, 53–71.

Johnson, A. M., and Fletcher, R. C. 1994. *Folding of Viscous Layers*. New York: Columbia University Press.

Johnson, R. E. 1980. Slender-body theory for slow viscous flow. *J. Fluid Mech.*, **75**, 705–714.

Kameyama, M. 2016. Linear analysis on the onset of thermal convection of highly compressible fluids with variable physical properties: implications for the mantle convection of super-Earths. *Geophys. J. Int.*, **204**, 1164–1178.

Kameyama, M., Miyagoshi, T., and Ogawa, M. 2015. Linear analysis on the onset of thermal convection of highly compressible fluids: implications for the mantle convection of super-Earths. *Geophys. J. Int.*, **200**, 1064–1075.

Katopodes, F. V., Davis, A. M. J., and Stone, H. A. 2000. Piston flow in a two-dimensional channel. *Phys. Fluids*, **12**, 1240–1243.

Katz, R. F., Spiegelman, M., and Holtzman, B. 2006. The dynamics of melt and shear localization in partially molten aggregates. *Nature*, **442**, 676–679.

Kaufmann, G., and Lambeck, K. 2000. Mantle dynamics, postglacial rebound and the radial viscosity profile. *Phys. Earth Planet. Int.*, **121**, 303–327.

Kaus, B. J. P., and Becker, T. W. 2007. Effects of elasticity on the Rayleigh–Taylor instability: implications for large-scale geodynamics. *Geophys. J. Int.*, **168**, 843–862.

Ke, Y., and Solomatov, V. S. 2004. Plume formation in strongly temperature-dependent viscosity fluids over a very hot surface. *Phys. Fluids*, **16**, 1059–1063.

Keller, J. B., and Rubinow, S. I. 1976. An improved slender-body theory for Stokes flow. *J. Fluid Mech.*, **99**, 411–431.

Keller, T., and Katz, R. F. 2016. The role of volatiles in reactive melt transport in the asthenosphere. *J. Petrol.*, **57**, 1073–1108.

Kemp, D. V., and Stevenson, D. J. 1996. A tensile, flexural model for the initiation of subduction. *Geophys. J. Int.*, **125**, 73–94.

Kerr, R. C., and Lister, J. R. 1987. The spread of subducted lithospheric material along the mid-mantle boundary. *Earth Planet. Sci. Lett.*, **85**, 241–247.

Kevorkian, J., and Cole, J. D. 1996. *Multiple Scale and Singular Perturbation Methods*. New York: Springer.

Kim, M. C., and Choi, C. K. 2006. The onset of buoyancy-driven convection in fluid layers with temperature-dependent viscosity. *Phys. Earth Planet. Int.*, **155**, 42–47.

Kim, S., and Karrila, S. J. 1991. *Microhydrodynamics: Principles and Selected Applications*. Boston: Butterworth-Heinemann.

King, S. D., and Masters, G. 1992. An inversion for radial viscosity structure using seismic tomography. *Geophys. Res. Lett.*, **19**, 1551–1554.

Koch, D. M., and Koch, D. L. 1995. Numerical and theoretical solutions for a drop spreading below a free fluid surface. *J. Fluid. Mech.*, **287**, 251–278.

Koch, D. M., and Ribe, N. M. 1989. The effect of lateral viscosity variations on surface observables. *Geophys. Res. Lett.*, **16**, 535–538.

Korenaga, J., and Jordan, T. H. 2003a. Physics of multiscale convection in Earth's mantle: onset of sublithospheric convection. *J. Geophys. Res.*, **108**, 2333.

Korenaga, J., and Jordan, T. H. 2003b. Linear stability analysis of Richter rolls. *Geophys. Res. Lett.*, **30**, 2157.
Kraus, H. 1967. *Thin Elastic Shells*. New York: John Wiley & Sons.
Lachenbruch, A. H., and Nathenson, M. 1974. *Rise of a variable viscosity fluid in a steadily spreading wedge-shaped conduit with accreting walls*. Open File Rep. 74-251. U. S. Geol. Surv., Menlo Park, CA.
Ladyzhenskaya, O. A. 1963. *The Mathematical Theory of Viscous Incompressible Flow*. New York: Gordon and Breach.
Lamb, H. 1945. *Hydrodynamics*. New York: Dover.
Landau, L. D. 1944. On the problem of turbulence. *C. R. Acad. Sci. U. R. S. S.*, **44**, 311–314.
Landau, L. D., and Lifshitz, E. 1986. *Theory of Elasticity*. 3rd edn. Oxford: Butterworth-Heinemann.
Landuyt, W., and Ierley, G. 2012. Linear stability analysis of the onset of sublithospheric convection. *Geophys. J. Int.*, **189**, 19–28.
Langlois, W. E., and Deville, M. O. 2014. *Slow Viscous Flow*. New York: Springer.
Laravie, J. A. 1975. Geometry and lateral strain of subducted plates in island arcs. *Geology*, **3**, 484–486.
Larson, R. G. 1999. *The Structure and Rheology of Complex Fluids*. New York: Oxford University Press.
Leahy, G. M., and Bercovici, D. 2010. Reactive infiltration of hydrous melt above the mantle transition zone. *J. Geophys. Res.*, **115**, B08406.
Le Bars, M., and Davaille, A. 2002. Stability of thermal convection in two superimposed miscible viscous fluids. *J. Fluid Mech.*, **471**, 339–363.
Le Bars, M., and Davaille, A. 2004. Large interface deformation in two-layer thermal convection of miscible viscous fluids. *J. Fluid Mech.*, **499**, 75–110.
Lee, C., and King, S. 2011. Dynamic buckling of subducting slabs reconciles geological and geophysical observations. *Earth Planet. Sci. Lett.*, **312**, 360–370.
Lee, E. H. 1955. Stress analysis in visco-elastic bodies. *Quart. Appl. Math.*, **13**, 183–190.
Lee, S. H., and Leal, L. G. 1980. Motion of a sphere in the presence of a plane interface. Part 2. An exact solution in bipolar co-ordinates. *J. Fluid Mech.*, **98**, 193–224.
Lemery, C., Ricard, Y., and Sommeria, J. 2000. A model for the emergence of thermal plumes in Rayleigh–Bénard convection at infinite Prandtl number. *J. Fluid Mech.*, **414**, 225–250.
Lenardic, A., Richards, M. A., and Busse, F. H. 2006. Depth-dependent rheology and the horizontal length scale of mantle convection. *J. Geophys. Res.*, **111**, B07404.
Le Pichon, X., Francheteau, J., and Bonnin, J. 1973. *Plate Tectonics*. Amsterdam: Elsevier.
Lev, E., and Hager, B. H. 2008. Rayleigh–Taylor instabilities with anisotropic lithospheric viscosity. *Geophys. J. Int.*, **173**, 806–814.
Li, Z., and Ribe, N. M. 2012. Dynamics of free subduction from 3-D boundary element modeling. *J. Geophys. Res.*, **117**, B06408.
Ligi, M., Cuffaro, M., Chierici, F., and Calafato, A. 2008. Three-dimensional passive mantle flow beneath mid-ocean ridges: an analytical approach. *Geophys. J. Int.*, **175**, 783–805.
Lister, J. R., and Kerr, R. C. 1989a. The effect of geometry on the gravitational instability of a region of viscous fluid. *J. Fluid Mech.*, **202**, 577–594.
Lister, J. R., and Kerr, R. C. 1989b. The propagation of two-dimensional and axisymmetric viscous gravity currents at a fluid interface. *J. Fluid Mech.*, **203**, 215–249.
Liu, X., and Zhong, S. 2013. Analyses of marginal stability, heat transfer and boundary layer properties for thermal convection in a compressible fluid with infinite Prandtl number. *Geophys. J. Int.*, **194**, 125–144.

Longman, I. M. 1962. A Green's function for determining the deformation of the earth under surface mass loads. *J. Geophys. Res.*, **67**, 845–850.

Loper, D. E., and Stacey, F. D. 1983. The dynamical and thermal structure of deep mantle plumes. *Phys. Earth Planet. Int.*, **33**, 305–317.

Lorentz, H. A. 1907. Ein allgemeiner Satz, die Bewegung einer reibenden Flüssigkeit betreffend, nebst einigen Anwendungen desselben. *Abhand. Theor. Phys.*, **1**, 23–42.

Love, A. E. H. 1967. *Some Problems of Geodynamics*. New York: Dover.

Mahadevan, L., Bendick, R., and Liang, H. 2010. Why subduction zones are curved. *Tectonics*, **29**, TC6002.

Maiden, M. D., and Hoefer, M. A. 2016. Modulations of viscous fluid conduit periodic waves. *Proc. R. Soc. Lond. A*, **472**, 20160533.

Malkus, W. V. R., and Veronis, G. 1958. Finite amplitude cellular convection. *J. Fluid Mech.*, **4**, 225–260.

Manga, M. 1996. Mixing of heterogeneities in the mantle: effect of viscosity differences. *Geophys. Res. Lett.*, **23**, 403–406.

Manga, M. 1997. Interactions between mantle diapirs. *Geophys. Res. Lett.*, **24**, 1871–1874.

Manga, M., and Stone, H. A. 1993. Buoyancy-driven interaction between two deformable viscous drops. *J. Fluid Mech.*, **256**, 647–683.

Manga, M., Stone, H. A., and O'Connell, R. J. 1993. The interaction of plume heads with compositional discontinuities in the earth's mantle. *J. Geophys. Res.*, **98**, 19,979–19,990.

Manga, M., Weeraratne, D., and Morris, S. J. S. 2001. Boundary-layer thickness and instabilities in Bénard convection of a liquid with a temperature-dependent viscosity. *Phys. Fluids*, **13**, 802–805.

Mangler, W. 1948. Zusammenhang zwischen ebenen und rotationssymmetrischen Grenzschichten in kompressiblen Flüssigkeiten. *Z. Angew. Math. Mech.*, **28**, 97–103.

Marotta, A. M., and Mongelli, F. 1998. Flexure of subducted slabs. *Geophys. J. Int.*, **132**, 701–711.

Marsh, B. D. 1978. On the cooling of ascending andesitic magma. *Phil. Trans. R. Soc. Lond.*, **288**, 611–625.

Marsh, B. D., and Carmichael, I. S. E. 1974. Benioff zone magmatism. *J. Geophys. Res.*, **79**, 1196–1206.

McAdoo, D. C., Caldwell, J. G., and Turcotte, D. L. 1978. On the elastic-perfectly plastic bending of the lithosphere under generalized loading with application to the Kuril Trench. *Geophys. J. R. Astr. Soc.*, **54**, 11–26.

McKenzie, D. 1988. The symmetry of convective transitions in space and time. *J. Fluid Mech.*, **191**, 287–339.

McKenzie, D. P. 1967. Some remarks on heat flow and gravity anomalies. *J. Geophys. Res.*, **72**, 6261–6273.

McKenzie, D. P. 1969. Speculations on the consequences and causes of plate motions. *Geophys. J. R. Astr. Soc.*, **18**, 1–32.

McKenzie, D. P. 1977. Surface deformation, gravity anomalies and convection. *Geophys. J. R. Astr. Soc.*, **48**, 211–238.

McKenzie, D. P. 1984. The generation and compaction of partially molten rock. *J. Petrology*, **25**, 713–765.

McKenzie, D. P., and Bowin, C. 1976. The relationship between bathymetry and gravity in the Atlantic ocean. *J. Geophys. Res.*, **81**, 1903–1915.

McKenzie, D. P., Roberts, J. M., and Weiss, N. O. 1974. Convection in the earth's mantle: towards a numerical simulation. *J. Fluid Mech.*, **62**, 465–538.

Medvedev, S. E., and Podladchikov, Y. Y. 1999. New extended thin-sheet approximation for geodynamic applications – I. Model formulation. *Geophys. J. Int.*, **136**, 567–585.

Meleshko, V. V. 1996. Steady Stokes flow in a rectangular cavity. *Proc. R. Soc. Lond. A*, **452**, 1999–2022.

Michaut, C., and Bercovici, D. 2009. A model for the spreading and compaction of two-phase viscous gravity currents. *J. Fluid Mech.*, **630**, 299–329.

Mitrovica, J. X., and Forte, A. M. 2004. A new inference of mantle viscosity based upon joint inversion of convection and glacial isostatic adjustment data. *Earth Planet. Sci. Lett.*, **225**, 177–189.

Mitrovica, J. X., Hay, C. C., Morrow, E., Kopp, R. E., Dumberry, M., and Stanley, S. 2015. Reconciling past changes in Earth's rotation with 20th century global sea-level rise: resolving Munk's enigma. *Sci. Adv.*, **1**, e1500679.

Mitrovica, J. X., and Peltier, W. R. 1995. Constraints on mantle viscosity based upon the inversion of post-glacial uplift data from the Hudson Bay region. *Geophys. J. Int.*, **122**, 353–377.

Mittelstaedt, E., and Ito, G. 2005. Plume-ridge interaction, lithospheric stresses, and the origin of near-ridge volcanic lineaments. *Geochem. Geophys. Geosyst.*, **6**, Q06002.

Moffatt, H. K. 1964. Viscous and resistive eddies near a sharp corner. *J. Fluid Mech.*, **18**, 1–18.

Molnar, P., and Houseman, G. A. 2004. The effects of buoyant crust on the gravitational instability of thickened mantle lithosphere at zones of intracontinental convergence. *Geophys. J. Int.*, **158**, 1134–1150.

Molnar, P., and Houseman, G. A. 2013. Rayleigh–Taylor instability, lithospheric dynamics, surface topography at convergent mountain belts, and gravity anomalies. *J. Geophys. Res. Solid Earth*, **118**, 2544–2557.

Molnar, P., and Houseman, G. A. 2015. Effects of a low-viscosity lower crust on topography and gravity at convergent mountain belts during gravitational instability of mantle lithosphere. *J. Geophys. Res.*, **120**, 537–551.

Molnar, P., Houseman, G. A., and Conrad, C. P. 1998. Rayleigh–Taylor instability and convective thinning of mechanically thickened lithosphere: effects of non-linear viscosity decreasing exponentially with depth and of horizontal shortening of the layer. *Geophys. J. Int.*, **133**, 568–584.

Mondal, P., and Korenaga, J. 2018. A propagator matrix method for the Rayleigh–Taylor instability of multiple layers: a case study on crustal delamination in the early Earth. *Geophys. J. Int.*, **212**, 1890–1901.

Morra, G., Chatelain, P., Tackley, P., and Koumoutsakos, P. 2007. *Large scale three-dimensional boundary element simulation of subduction*. Pages 1122–1129 of: Computational Science - ICCS 2007. New York: Springer.

Morra, G., Chatelain, P., Tackley, P., and Koumoutsakos, P. 2009. Earth curvature effects on subduction morphology: modeling subduction in a spherical setting. *Acta Geotech.*, **4**, 95–105.

Morra, G., Quevedo, L., and Müller, R. D. 2012. Spherical dynamic models of top-down tectonics. *Geochem Geophys Geosyst.*, **13**, Q03005.

Morra, G., and Regenauer-Lieb, K. 2006. A coupled solid-fluid method for modelling subduction. *Philos. Mag.*, **86**, 3307–3323.

Morra, G., Regenauer-Lieb, K., and Giardini, D. 2006. Curvature of island arcs. *Geology*, **34**, 877–880.

Morris, S. 1980. *An asymptotic method for determining the transport of heat and matter by creeping flows with strongly variable viscosity; fluid dynamic problems motivated by island arc volcanism*. Ph.D. thesis, The Johns Hopkins University, Baltimore, MD.

Morris, S. 1982. The effects of a strongly temperature-dependent viscosity on slow flow past a hot sphere. *J. Fluid Mech.*, **124**, 1–26.

Morris, S., and Canright, D. 1984. A boundary-layer analysis of Bénard convection in a fluid of strongly temperature-dependent viscosity. *Phys. Earth Planet. Int.*, **36**, 355–373.

Morris, S. J. S. 2008. Viscosity stratification and the horizontal scale of end-driven cellular flow. *Phys. Fluids*, **20**, 063103.

Muskhelishvili, N. I. 1953. *Some Basic Problems in the Mathematical Theory of Elasticity*. Groningen: P. Noordhoff.

Nayfeh, A. 1973. *Perturbation Methods*. New York: John Wiley & Sons.

Neil, E. A., and Houseman, G. A. 1999. Rayleigh–Taylor instability of the upper mantle and its role in intraplate orogeny. *Geophys. J. Int.*, **138**, 89–107.

Newell, A. C., Passot, T., and Lega, J. 1993. Order parameter equations for patterns. *Ann. Rev. Fluid Mech.*, **25**, 399–453.

Newell, A. C., Passot, T., and Souli, M. 1990. The phase diffusion and mean drift equations for convection at finite Rayleigh numbers in large containers. *J. Fluid Mech.*, **220**, 187–252.

Newell, A. C., and Whitehead, J. A. Jr. 1969. Finite bandwidth, finite amplitude convection. *J. Fluid Mech.*, **38**, 279–303.

Niordson, F. I. 1985. *Shell Theory*. Amsterdam: North-Holland.

Nobili, C., and Otto, F. 2017. Limitations of the background field method applied to Rayleigh–Bénard convection. *J. Math. Phys.*, **58**, 093102.

Novozhilov, V. V. 1959. *The Theory of Thin Shells*. Groningen: Noordhoff.

Nyblade, A. A., and Sleep, N. H. 2003. Long lasting epeirogenic uplift from mantle plumes and the origin of the Southern African Plateau. *Geochem. Geophys. Geosyst.*, **4**, 1105.

O'Connell, R. J., Gable, C. W., and Hager, B. H. 1991. Toroidal-poloidal partitioning of lithospheric plate motion. Pages 535–551 of: Sabadini, R. et al. (ed), *Glacial Isostasy, Sea Level and Mantle Rheology*. Dordrecht: Kluwer Academic.

Ohta, K., Yagi, T., Hirose, K., and Ohishi, Y. 2017. Thermal conductivity of ferropericlase in the Earth's lower mantle. *Earth Planet. Sci. Lett.*, **465**, 29–37.

Ohta, K., Yagi, T., Taketoshi, N., Hirose, K., Komabayashi, T., Baba, T., Ohishi, Y., and Hernlund, J. 2012. Lattice thermal conductivity of $MgSiO_3$ perovskite and post-perovskite at the core–mantle boundary. *Earth Planet. Sci. Lett.*, **349–350**, 109–115.

Ohtani, E., Mizobata, H., Kudoh, Y., Nagase, T., Arashi, H., Yurimoto, H., and Miyagi, I. 1997. A new hydrous silicate, a water reservoir, in the upper part of the lower mantle. *Geophys. Res. Lett.*, **24**, 1047–1050.

Olson, P. 1990. Hot spots, swells and mantle plumes. Pages 33–51 of: Ryan, M. P. (ed.), *Magma Transport and Storage*. New York: John Wiley & Sons.

Olson, P., and Christensen, U. 1986. Solitary wave propagation in a fluid conduit within a viscous matrix. *J. Geophys. Res.*, **91**, 6367–6374.

Olson, P., and Corcos, G. M. 1980. A boundary-layer model for mantle convection with surface plates. *Geophys. J. R. Astr. Soc.*, **62**, 195–219.

Olson, P., and Singer, H. 1985. Creeping plumes. *J. Fluid Mech.*, **158**, 511–531.

Olson, P., Schubert, G., and Anderson, C. 1993. Structure of axisymmetric mantle plumes. *J. Geophys. Res.*, **98**, 6829–6844.

Ortoleva, P., Chadam, J., Merino, E., and Sen, A. 1987. Geochemical self-organization II: the reactive-infiltration instability. *Am. J. Sci.*, **287**, 1008–1040.

Otto, F., and Seis, C. 2011. Rayleigh–Bénard convection: improved bounds on the Nusselt number. *J. Math. Phys.*, **52**, 083702.

Palm, E. 1960. On the tendency towards hexagonal cells in steady convection. *J. Fluid Mech.*, **8**, 183–192.

Palm, E., Ellingsen, T., and Gjevik, B. 1967. On the occurrence of cellular motion in Bénard convection. *J. Fluid Mech.*, **30**, 651–661.

Panasyuk, S. V., and Hager, B. H. 2000. Inversion for mantle viscosity profiles constrained by dynamic topography and the geoid, and their estimated errors. *Geophys. J. Int.*, **143**, 821–836.

Parsons, B., and Daly, S. 1983. The relationship between surface topography, gravity anomalies, and the temperature structure of convection. *J. Geophys. Res.*, **88**, 1129–1144.

Parsons, B., and McKenzie, D. 1978. Mantle convection and the thermal structure of the plates. *J. Geophys. Res.*, **83**, 4485–4496.

Parsons, B., and Molnar, P. 1976. The origin of outer topographic rises asssociated with trenches. *Geophys. J. R. Astr. Soc.*, **45**, 707–712.

Pekeris, C. L. 1935. Thermal convection in the interior of the Earth. *Mon. Not. R. Astr. Soc. Geophys. Suppl.*, **3**, 343–367.

Peltier, W. R. 1974. The impulse response of a Maxwell Earth. *Rev. Geophys. Space Phys.*, **12**, 649–669.

Peltier, W. R. 1989. Mantle viscosity. Pages 389–478 of: Peltier, W. R. (ed.), *Mantle Convection: Plate Tectonics and Global Dynamics*. The Fluid Mechanics of Astrophysics and Geophysics, vol. 4. Montreux: Gordon and Breach.

Peltier, W. R. 1996. Mantle viscosity and ice-age ice sheet topography. *Science*, **273**, 1359–1364.

Peltier, W. R. 2004. Global glacial isostasy and the surface of the ice-age Earth: the ICE-5G (VM2) model and GRACE. *Ann. Rev. Earth Planet. Sci.*, **32**, 111–149.

Peltier, W. R., Drummond, R. A., and Tushingham, A. M. 1986. Post-glacial rebound and transient lower mantle rheology. *Geophys. J. R. Astr. Soc.*, **87**, 79–116.

Peltier, W. R., Jarvis, G. T., Forte, A. M., and Solheim, L. P. 1989. The radial structure of the mantle general circulation. Pages 765–816 of: Peltier, W. R. (ed.), *Mantle Convection: Plate Tectonics and Global Dynamics*. The Fluid Mechanics of Astrophysics and Geophysics, vol. 4. Montreux: Gordon and Breach.

Peng, G. G., and Lister, J. R. 2014. The initial transient and approach to self-similarity of a very viscous buoyant thermal. *J. Fluid Mech.*, **744**, 352–375.

Pozrikidis, C. 1990. The deformation of a liquid drop moving normal to a plane wall. *J. Fluid Mech.*, **215**, 331–363.

Pozrikidis, C. 1992. *Boundary Integral and Singularity Methods for Linearized Viscous Flow*. Cambridge: Cambridge University Press.

Ramalho, R., Helffrich, G., Cosca, M., Vance, D., Hoffmann, D., and Schmidt, D. N. 2010. Episodic swell growth inferred from variable uplift of the Cape Verde hotspot islands. *Nature Geosci.*, **3**, 774–777.

Rasenat, S., Busse, F. H., and Rehberg, I. 1989. A theoretical and experimental study of double-layer convection. *J. Fluid Mech.*, **199**, 519–540.

Rayleigh, J. W. S. 1945. *The Theory of Sound*. 2nd edn. 2 vols. New York: Dover.

Renardy, Y., and Joseph, D. D. 1985. Oscillatory instability in a Bénard problem of two fluids. *Phys. Fluids*, **28**, 788–793.

Renardy, Y., and Renardy, M. 1985. Perturbation analysis of steady and oscillatory onset in a Bénard problem with two similar liquids. *Phys. Fluids*, **28**, 2699–2708.

Ribe, N. M. 1989. Mantle flow induced by back-arc spreading. *Geophys. J. R. Astr. Soc.*, **98**, 85–91.

Ribe, N. M. 1992. The dynamics of thin shells with variable viscosity and the origin of toroidal flow in the mantle. *Geophys. J. Int.*, **110**, 537–552.

Ribe, N. M. 1998. Spouting and planform selection in the Rayleigh–Taylor instability of miscible viscous fluids. *J. Fluid Mech.*, **234**, 315–336.

Ribe, N. M. 2001. Bending and stretching of thin viscous sheets. *J. Fluid Mech.*, **433**, 135–160.

Ribe, N. M. 2002. A general theory for the dynamics of thin viscous sheets. *J. Fluid Mech.*, **457**, 255–283.

Ribe, N. M. 2003. Periodic folding of viscous sheets. *Phys. Rev. E*, **68**, 036305.

Ribe, N. M. 2010. Bending mechanics and mode selection in free subduction: a thin-sheet analysis. *Geophys. J. Int.*, **180**, 559–576.

Ribe, N. M. 2015. Analytical approaches to mantle dynamics. Pages 145–196 of: Schubert, G., and Bercovici, D. (eds.), *Treatise on Geophysics*, 2nd edn., vol. 7. Oxford: Elsevier.

Ribe, N. M., and Christensen, U. 1999. The dynamical origin of Hawaiian volcanism. *Earth Planet. Sci. Lett.*, **171**, 517–531.

Ribe, N. M., Christensen, U. R., and Theissing, J. 1995. The dynamics of plume-ridge interaction, 1: Ridge-centered plumes. *Earth Planet. Sci. Lett.*, **134**, 155–168.

Ribe, N. M., and Davaille, A. 2013. Dynamical similarity and density (non-) proportionality in experimental tectonics. *Tectonophys.*, **608**, 1371–1379.

Ribe, N. M., and Delattre, W. L. 1998. The dynamics of plume-ridge interaction-III. The effects of ridge migration. *Geophys. J. Int.*, **133**, 511–518.

Ribe, N. M., Stutzmann, E., Ren, Y., and van der Hilst, R. 2007. Buckling instabilities of subducted lithosphere beneath the transition zone. *Earth Planet. Sci. Lett.*, **254**, 173–179.

Ricard, Y. 2015. Physics of mantle convection. Pages 23–71 of: Schubert, G., and Bercovici, D. (eds.), *Treatise on Geophysics*, 2nd. edn., vol. 7. Oxford: Elsevier.

Ricard, Y., Bercovici, D., and Schubert, G. 2001. A two-phase model for compaction and damage 2. Applications to compaction, deformation, and the role of interfacial surface tension. *J. Geophys. Res.*, **106**, 8907–8924.

Ricard, Y., Fleitout, L., and Froidevaux, C. 1984. Geoid heights and lithospheric stresses for a dynamic earth. *Ann. Geophys.*, **2**, 267–286.

Ricard, Y., and Froidevaux, C. 1986. Stretching instabilities and lithospheric boudinage. *J. Geophys. Res.*, **91**, 8314–8324.

Ricard, Y., and Husson, L. 2005. Propagation of tectonic waves. *Geophys. Res. Lett.*, **32**, L17308.

Ricard, Y., Vigny, C., and Froidevaux, C. 1989. Mantle heterogeneities, geoid, and plate motion: a Monte Carlo inversion. *J. Geophys. Res.*, **94**, 13,739–13,754.

Richards, M. A., and Hager, B. H. 1984. Geoid anomalies in a dynamic Earth. *J. Geophys. Res.*, **89**, 5987–6002.

Richards, M. A., Hager, B. H., and Sleep, N. H. 1988. Dynamically supported geoid highs over hotspots: observation and theory. *J. Geophys. Res.*, **93**, 7690–7708.

Richardson, C. N., Lister, J. R., and McKenzie, D. 1996. Melt conduits in a viscous porous matrix. *J. Geophys. Res.*, **101**, 20,423–20,432.

Richter, F. M. 1973a. Convection and the large-scale circulation of the mantle. *J. Geophys. Res.*, **78**, 8735–8745.

Richter, F. M. 1973b. Dynamical models for sea floor spreading. *Rev. Geophys. Space Phys.*, **11**, 223–287.

Richter, F. M., and Johnson, C. E. 1974. Stability of a chemically layered mantle. *J. Geophys. Res.*, **79**, 1635–1639.

Richter, F. M., and McKenzie, D. P. 1984. Dynamical models for melt segregation from a deformable matrix. *J. Geol.*, **92**, 729–740.

Richter, F. M., and Parsons, B. 1975. On the interaction of two scales of convection in the mantle. *J. Geophys. Res.*, **80**, 2529–2541.

Roberts, G. O. 1979. Fast viscous Bénard convection. *Geophys. Astrophys. Fluid Dyn.*, **12**, 235–272.

Roberts, P., Schubert, G., Zhang, K., Liao, X., and Busse, F. H. 2007. Instabilities in a fluid layer with phase changes. *Phys. Earth Planet. Int.*, **165**, 147–157.

Royden, L. H., and Husson, L. 2006. Trench motion, slab geometry and viscous stresses in subduction systems. *Geophys. J. Int.*, **167**, 881–905.

Rudge, J. F. 2014. Analytical solutions of compacting flow past a sphere. *J. Fluid Mech.*, **746**, 466–497.

Rudge, J. F., and Bercovici, D. 2015. Melt-band instabilities with two-phase damage. *Geophys. J. Int.*, **201**, 640–651.

Rudge, J. F., Bercovici, D., and Spiegelman, M. 2011. Disequilibrium melting of a two phase multicomponent mantle. *Geophys. J. Int.*, **184**, 699–718.

Rudolph, M. L., Lekić, V., and Lithgow-Bertelloni, C. 2015. Viscosity jump in Earth's mid-mantle. *Science*, **350**, 1349–1352.

Runcorn, S. K. 1967. Flow in the mantle inferred from the low degree harmonics of the geopotential. *Geophys. J. R. Astr. Soc.*, **14**, 375–384.

Rybczynski, W. 1911. Über die fortschreitende Bewegung einer flüssigen Kugel in einem zähen Medium. *Bull. Int. Acad. Sci. Cracov.*, **1911A**, 40–46.

Sanchez-Palencia, E. 1990. Passages à la limite de l'élasticité tri-dimensionnelle à la théorie asymptotique des coques minces. *C. R. Acad. Sci. Paris I*, **309**, 909–916.

Sayag, R., and Worster, M. G. 2013. Axisymmetric gravity currents of power-law fluids over a rigid horizontal surface. *J. Fluid Mech.*, **716**, R5.

Schettino, A., and Tassi, L. 2012. Trench curvature and deformation of the subducting lithosphere. *Geophys. J. Int.*, **188**, 18–34.

Schmalholz, S. M. 2011. A simple analytical solution for slab detachment. *Earth Planet. Sci. Lett.*, **304**, 45–54.

Schmalholz, S. M., and Mancktelow, N. S. 2016. Folding and necking across the scales: a review of theoretical and experimental results and their applications. *Solid Earth*, **7**, 1417–1465.

Schmeling, H. 2000. Partial melting and melt segregation in a convecting mantle. Pages 141–178 of: Bagdassarov, N., Laporte, D., and Thompson, A. (eds.), *Physics and Chemistry of Partially Molten Rocks*. Norwell: Kluwer Academic.

Scholz, C. H., and Page, R. 1970. Buckling in island arcs. *EOS (Am. Geophys. Union Trans.)*, **51**, 429.

Schrank, C. E., Karrech, A., Boutelier, D. A., and Regenauer-Lieb, K. 2017. A comparative study of Maxwell viscoelasticity at large strains and rotations. *Geophys. J. Int.*, **211**, 252–262.

Schubert, G., Olson, P., Anderson, C., and Goldman, P. 1989. Solitary waves in mantle plumes. *J. Geophys. Res.*, **94**, 9523–9532.

Schubert, G., and Turcotte, D. L. 1971. Phase changes and mantle convection. *J. Geophys. Res.*, **76**, 1424–1432.

Schubert, G., Turcotte, D. L., and Olson, P. 2001. *Mantle Convection in the Earth and Planets*. Cambridge: Cambridge University Press.

Schubert, G., Yuen, D. A., and Turcotte, D. L. 1975. Role of phase transitions in a dynamic mantle. *Geophys. J. R. Astr. Soc.*, **42**, 705–735.

Scott, D. R., and Stevenson, D. J. 1984. Magma solitons. *Geophys. Res. Lett.*, **11**, 1161–1164.

Scott, D. R., and Stevenson, D. J. 1986. Magma ascent by porous flow. *J. Geophys. Res.*, **91**, 9283–9296.

Scott, D. R., Stevenson, D. J., and Whitehead, J. A. Jr. 1986. Observations of solitary waves in a deformable pipe. *Nature*, **319**, 759–761.

Segel, L. 1969. Distant side-walls cause slow amplitude modulation of cellular convection. *J. Fluid Mech.*, **38**, 203–224.

Shankar, P. N. 1993. The eddy structure in Stokes flow in a cavity. *J. Fluid Mech.*, **250**, 371–383.

Shankar, P. N. 2005. Eigenfunction expansions on arbitrary domains. *Proc. R. Soc. Lond. A*, **461**, 2121–2133.

Shiels, C., and Butler, S. L. 2015. Couette and Poiseuille flows in a low viscosity asthenosphere: effects of internal heating rate, Rayleigh number, and plate representation. *Phys. Earth Planet. Int.*, **246**, 31–40.

Shishkina, O., Emran, M. S., Grossman, S., and Lohse, D. 2017. Scaling relations in large-Prandtl-number natural thermal convection. *Phys. Rev. Fluids*, **2**, 103502.

Simpson, G., and Weinstein, M. I. 2008. Asymptotic stability of ascending solitary magma waves. *SIAM J. Math. Anal.*, **40**, 1337–1391.

Sleep, N. H. 1974. Segregation of a magma from a mostly crystalline mush. *Bull. Geol. Soc. Am.*, **85**, 1225–1232.

Sleep, N. H. 1987. Lithospheric heating by mantle plumes. *Geophys. J. R. Astr. Soc.*, **91**, 1–11.

Sleep, N. H. 1996. Lateral flow of hot plume material ponded at sublithospheric depths. *J. Geophys. Res. Solid Earth*, **101**, 28,065–28,083.

Smith, R. B. 1975. Unified theory of the onset of folding, boudinage and mullion structure. *Geol. Soc. Am. Bull.*, **86**, 1601–1609.

Smith, R. B. 1977. Formation of folds, boudinage, and mullions in non-Newtonian materials. *Geol. Soc. Am. Bull.*, **88**, 312–320.

Solomatov, V. S. 1995. Scaling of temperature- and stress-dependent viscosity convection. *Phys. Fluids*, **7**, 266–274.

Sotin, C., and Parmentier, E. M. 1989. On the stability of a fluid layer containing a univariant phase transition: application to planetary interiors. *Phys. Earth Planet. Int.*, **55**, 10–25.

Sparrow, E. M., Husar, R. B., and Goldstein, R. J. 1970. Observations and other characteristics of thermals. *J. Fluid Mech.*, **41**, 793–800.

Spiegelman, M. 1993. Flow in deformable porous media. Part 1 Simple analysis. *J. Fluid Mech.*, **247**, 17–38.

Spiegelman, M. 2003. Linear analysis of melt band formation by simple shear. *Geochem. Geophys. Geosyst.*, **4**, 8615.

Spiegelman, M., Kelemen, P. B., and Aharonov, E. 2001. Causes and consequences of flow organization during melt transport: the reaction infiltration instability in compactible media. *J. Geophys. Res.*, **106**, 2061–2077.

Spiegelman, M., and McKenzie, D. 1987. Simple 2-D models for melt extraction at mid-ocean ridges and island arcs. *Earth Planet. Sci. Lett.*, **83**, 137–152.

Šrámek, O., Ricard, Y., and Bercovici, D. 2007. Simultaneous melting and compaction in deformable two-phase media. *Geophys. J. Int.*, **168**, 964–982.

Stengel, K. C., Oliver, D. S., and Booker, J. R. 1982. Onset of convection in a variable-viscosity fluid. *J. Fluid Mech.*, **120**, 411–431.

Stevenson, D. J. 1989. Spontaneous small-scale melt segregation in partial melts undergoing deformation. *Geophys. Res. Lett.*, **16**, 1067–1070.

Stevenson, D. J., and Turner, J. S. 1977. Angle of subduction. *Nature*, **270**, 334–336.

Stimson, M., and Jeffrey, G. B. 1926. The motion of two spheres in a viscous fluid. *Proc. R. Soc. Lond. A*, **111**, 110–116.

Stokes, G. G. 1845. On the theories of the internal friction of fluids and of the equilibrium and motion of elastic solids. *Trans. Camb. Phil. Soc.*, **8**, 287–347.

Stokes, G. G. 1851. On the effect of the internal friction of fluids on the motion of pendulums. *Trans. Camb. Phil. Soc.*, **9**, 1–99.

Straus, J. M. 1972. Finite amplitude doubly diffusive convection. *J. Fluid Mech.*, **56**, 353–374.

Takei, Y., and Hier-Majumder, S. 2009. A generalized formulation of interfacial tension driven fluid migration with dissolution/precipitation. *Earth Planet. Sci. Lett.*, **288**, 138–148.

Takei, Y., and Holtzman, B. K. 2009. Viscous constitutive relations of solid-liquid composites in terms of grain-boundary contiguity: 3. Causes and consequences of viscous anisotropy. *J. Geophys. Res.*, **114**, B06207.

Takei, Y., and Katz, R. F. 2013. Consequences of viscous anisotropy in a deforming, two-phase aggregate. Part 1. Governing equations and linearized analysis. *J. Fluid Mech.*, **734**, 424–455.

Takei, Y., and Katz, R. F. 2015. Consequences of viscous anisotropy in a deforming, two-phase aggregate. Why is porosity-band angle lowered by viscous anisotropy? *J. Fluid Mech.*, **784**, 199–224.

Tanimoto, T. 1997. Bending of a spherical lithosphere – axisymmetric case. *Geophys. J. Int.*, **129**, 305–310.

Tanimoto, T. 1998. State of stress within a bending spherical shell and its implications for subducting lithosphere. *Geophys. J. Int.*, **134**, 199–206.

Taylor-West, J., and Katz, R. F. 2015. Melt-preferred orientation, anisotropic permeability and melt-band formation in a deforming, partially molten aggregate. *Geophys. J. Int.*, **203**, 1253–1262.

Tovish, A., Schubert, G., and Luyendyk, B. P. 1978. Mantle flow pressure and the angle of subduction: non-Newtonian corner flows. *J. Geophys. Res.*, **83**, 5892–5898.

Turcotte, D. L. 1967. A boundary-layer theory for cellular convection. *Int. J. Heat Mass Transfer*, **10**, 1065–1074.

Turcotte, D. L. 1974. Membrane tectonics. *Geophys. J. R. Astr. Soc.*, **36**, 33–42.

Turcotte, D. L., McAdoo, D. C., and Caldwell, J. G. 1978. An elastic-perfectly plastic analysis of the bending of the lithosphere at a trench. *Tectonophys.*, **47**, 193–205.

Turcotte, D. L., and Oxburgh, E. R. 1967. Finite amplitude convection cells and continental drift. *J. Fluid Mech.*, **28**, 29–42.

Turcotte, D. L., and Schubert, G. 2014. *Geodynamics*. 3rd edn. Cambridge: Cambridge University Press.

Turcotte, D. L., Willemann, R. J., Haxby, W. W., and Norberry, J. 1981. Role of membrane stresses in the support of planetary topography. *J. Geophys. Res.*, **86**, 3951–3959.

Umemura, A., and Busse, F. H. 1989. Axisymmetric convection at large Rayleigh number and infinite Prandtl number. *J. Fluid Mech.*, **208**, 459–478.

Van Ark, E., and Lin, J. 2004. Time variation in igneous volume flux of the Hawaii-Emperor hot spot seamount chain. *J. Geophys. Res.*, **109**, B11401.

Van Dyke, M. 1975. *Perturbation Methods in Fluid Mechanics*. Stanford: Parabolic Press.

Vasilyev, O. V., Ten, A. A., and Yuen, D. A. 2001. Temperature-dependent viscous gravity currents with shear heating. *Phys. Fluids*, **13**, 3664–3674.

Vermeersen, L. L. A., and Sabadini, R. 1997. A new class of stratified viscoelastic models by analytical techniques. *Geophys. J. Int.*, **129**, 531–570.

Vidal, V., and Bonneville, A. 2004. Variations of the Hawaiian hot spot activity revealed by variations in the magma production rate. *J. Geophys. Res.*, **109**, B03104.

von Mises, R. 1927. Bemerkungen zur Hydrodynamik. *Z. Angew. Math. Mech.*, **7**, 425–431.

Vynnycky, M., and Masuda, Y. 2013. Rayleigh–Bénard convection at high Rayleigh number and infinite Prandtl number: asymptotics and numerics. *Phys. Fluids*, **25**, 113602.

Wakiya, S. 1975. Application of bipolar coordinates to the two-dimensional creeping motion of a liquid. II. Some problems for two circular cylinders in viscous fluid. *J. Phys. Soc. Japan*, **39**, 1603–1607.

Watts, A. B. 1978. An analysis of isostasy in the world's oceans 1. Hawaiian-Emperor Seamount Chain. *J. Geophys. Res.*, **83**, 5989–6004.

Watts, A. B. 2001. *Isostasy and Flexure of the Lithosphere*. Cambridge: Cambridge University Press.

Watts, A. B., and Talwani, M. 1974. Gravity anomalies seaward of deep-sea trenches and their tectonic implications. *Geophys. J. R. Astr. Soc.*, **36**, 57–90.

Wdowinski, S., O'Connell, R. J., and England, P. 1989. A continuum model of continental deformation above subduction zones: application to the Andes and the Aegean. *J. Geophys. Res.*, **94**, 10,331–10,346.

Weinstein, S. A., and Olson, P. L. 1992. Thermal convection with non-Newtonian plates. *Geophys. J. Int.*, **111**, 515–530.

Wessel, P. 1996. Analytical solutions for 3-D flexural deformation of semi-infinite elastic plates. *Geophys. J. Int.*, **124**, 907–918.

White, D. B. 1981. *Experiments with convection in a variable viscosity fluid*. Ph.D. thesis, University of Cambridge.

White, D. B. 1988. The planforms and onset of convection with a temperature-dependent viscosity. *J. Fluid Mech.*, **191**, 247–286.

Whitehead, J. A. Jr., Dick, H. B. J., and Schouten, H. 1984. A mechanism for magmatic accretion under spreading centers. *Nature*, **312**, 146–148.

Whitehead, J. A. Jr., and Helfrich, K. R. 1986. The Korteweg-de Vries equation from laboratory conduit and magma migration equations. *Geophys. Res. Lett.*, **13**, 545–546.

Whitehead, J. A. Jr., and Helfrich, K. R. 1988. Wave transport of deep mantle material. *Nature*, **335**, 59–61.

Whitehead, J. A. Jr., and Luther, D. S. 1975. Dynamics of laboratory diapir and plume models. *J. Geophys. Res.*, **80**, 705–717.

Whitham, G. B. 1974. *Linear and Non-Linear Waves*. Sydney: Wiley-Interscience.

Whittaker, R. J., and Lister, J. R. 2006a. Steady axisymmetric creeping plumes above a planar boundary. Part 1. A point source. *J. Fluid Mech.*, **567**, 361–378.

Whittaker, R. J., and Lister, J. R. 2006b. Steady axisymmetric creeping plumes above a planar boundary. Part 2. A distributed source. *J. Fluid Mech.*, **567**, 379–397.

Whittaker, R. J., and Lister, J. R. 2008a. The self-similar rise of a buoyant thermal in very viscous flow. *J. Fluid Mech.*, **606**, 295–324.

Whittaker, R. J., and Lister, J. R. 2008b. Slender-body theory for steady sheared plumes in very viscous fluid. *J. Fluid Mech.*, **612**, 21–44.

Wiggins, C., and Spiegelman, M. 1993. Magma migration and magmatic solitary waves in 3-D. *Geophys. Res. Lett.*, **22**, 1289–1292.

Willemann, R. J., and Davies, G. F. 1982. Bending stresses in subducted lithosphere. *Geophys. J. R. Astr. Soc.*, **71**, 215–224.

Worster, M. G. 1986. The axisymmetric laminar plume: asymptotic solution for large Prandtl number. *Stud. Appl. Maths.*, **75**, 139–152.

Xu, B., and Ribe, N. M. 2016. A hybrid boundary-integral/thin-sheet equation for subduction modelling. *Geophys. J. R. Astr. Soc.*, **206**, 1552–1562.

Yale, M. M., and Phipps Morgan, J. 1998. Asthenosphere flow model of hotspot-ridge interactions: a comparison of Iceland and Kerguelen. *Earth Planet. Sci. Lett.*, **161**, 45–56.

Yamazaki, D., and Karato, S.-I. 2001. Some mineral physics constraints on the rheology and geothermal structure of Earth's lower mantle. *Am. Mineralogist*, **86**, 385–391.

Yarushina, V. M., and Podladchikov, Y. Y. 2015. (De)compaction of porous viscoelastoplastic media: model formulation. *J. Geophys. Res. Solid Earth*, **120**, 4146–4170.

Yarushina, V. M., Podladchikov, Y. Y., and Connolly, J. A. D. 2015. (De)compaction of porous viscoelastoplastic media: solitary porosity waves. *J. Geophys. Res. Solid Earth*, **120**, 4843–4862.

Yuen, D. A., and Schubert, G. 1976. Mantle plumes: a boundary-layer approach for Newtonian and non-Newtonian temperature-dependent rheologies. *J. Geophys. Res.*, **81**, 2499–2510.

Zhong, S. 1996. Analytic solutions for Stokes' flow with lateral variations in viscosity. *Geophys. J. Int.*, **124**, 18–28.

Index

adiabatic
   compression, 101, 103, 222
   reference state, 221
   temperature gradient, 224–225, 235
Airy isostatic compensation, 90
anelastic liquid equations, 219–224
asthenosphere, 43, 215
available buoyancy, 196, 229

bending moment, 151
   effective, 157, 159–161
biharmonic equation, 39, 48, 51, 53, 57, 85, 88, 95, 140
   general solution, 51
biharmonic function, 56
biorthogonality, 53
boudinage, *see* necking
boundary layer
   compaction, 186
   thermal
      canonical geometries, 97f, 96–97
      equations, 97–98
      hot sphere moving in a fluid with temperature-dependent viscosity, 3, 117–118
      hot sphere moving in an isoviscous fluid, 2f, 3, 18–20, 118
      isoviscous stagnation flow, 102f, 101–104
      plume adjoining an isothermal wall, 111–112
      plume from a point source of buoyancy, 105–111
      plume with temperature-dependent viscosity, 112–117
      Rayleigh–Bénard convection, 210–212, 214–215
      solution using variable transformations, 98–100
      variable-viscosity stagnation flow, 3–4, 119
      vorticity, 96
boundary-element method (BEM), 6, 65–67, 67f, 169–171
boundary-integral representation, 63–65, 172
   for an immersed fluid drop, 64–65

Boussinesq
   approximation, 197–198, 219
   equations, 198, 223–224
Buckingham's Π-Theorem, *see* dimensional analysis
buckling, *see* folding
bulk modulus, *see* elasticity
buoyancy number, 233
Busse balloon, 205, 206f

compaction, *see* two-phase flow
complex-variable method for 2-D Stokes flow, 56–57, 85
conductive heat transfer, 23f, 22–24
convection
   buoyant thermal, 25–26, 26f
   compressible, 219–226, 225f
   forced, 1–4, 11–14, 18–20, 100–101, 117
   in a compositionally layered mantle, 57, 230–235
      oscillatory, 230, 234–235, 234f
   internally heated, 240
   Rayleigh–Bénard, 51, 54, 198f, 197–218
   small-scale, 228–230
   with a low-viscosity channel, 138
   with a phase transition, 235–239
   with temperature-dependent viscosity, 20–21, 229f, 226–230
corner flow, 45–51
correspondence principle
   between Maxwell viscoelasticity and linear elasticity, 87–88, 91
   between slow viscous flow and linear incompressible elasticity (Stokes–Rayleigh analogy), 86–88, 147–148
creeping flow, *see* slow viscous flow

Darcy's law, 181
diapir, xiii, 1–4, 2f, 11–14, 18–20, 100, 117–118
differential geometry of thin sheets, 149f, 151f, 148–151
   Christoffel symbols, 149f, 150, 173, 174
   contravariant base vectors, 149f, 148–150

covariant base vectors, 148–150
covariant differentiation, 150–151
curvature tensor, 149
Gaussian curvature, 149–150
mean curvature, 149–150
metric tensor, 149–150
dimensional analysis
  Buckingham's Π-theorem, 9–13, 20
    creeping plume, 17–18
    dimensionless group, 10–11
    free subduction, 12–13
    plume-ridge interaction, 132
  nondimensionalization, 13–17, 20–21
    compressible convection, 220–224
    conduit solitary waves, 143
    convection in a compositionally layered mantle, 230
    convection with a phase transition, 237–238
    creeping plume, 15–17
    heat transport in stagnation flow, 102
    isoviscous plume from a point source, 107–108
    low-viscosity channel in an end-driven layer, 140–141
    lubrication theory, 123
    plume with temperature-dependent viscosity, 113–114
    Rayleigh–Bénard convection, 199
dissipation number, 222–224
  definition, 222
double-layer potential, 64
driven cavity flow, 52f, 52–54
dynamical similarity, 10–11

elasticity
  bulk modulus, 86–87
  constitutive law for a linear elastic solid, 86–87
  loading of an elastic lithosphere, 88–91, 95
  Poisson's ratio, 87
  shear modulus, 86–87
  surface loading of a stratified elastic sphere, 95
  theory of thin elastic shells, 159–161
  thin elastic plate, 161–163
    flexural rigidity, 162
  thin spherical shell, 163–165
  Young's modulus, 87
exponential of a matrix, see propagator matrix
extended thin-sheet approximation, 169

finite-time singularity, 136
flexural rigidity, see elasticity
flow driven by internal loads, see slow viscous flow
folding, 67f, 217, 217f, 279–283, 282f
  periodic, 167–169, 168f
Fourier series, 52–53, 204
Fourier transform, 48, 58–60, 135
Fredholm integral equation of the second kind, 65, 173
free-air gravity anomaly, 90–91
  admittance for elastic plate flexure model, 89f, 89, 91

Galerkin method, 204–205, 209
Gegenbauer polynomial, 56
geoid, see gravitational potential
geometrical similarity, 10–11, 170
glacial isostatic adjustment, 86, 91, 197
Goursat representation, 56–57, 85
Grüneisen parameter, 222
gravitational potential
  anomaly due to a sinker, 74f, 73–74
  equation satisfied by, 33, 75
  integral representation, 77
  kernel for an isoviscous mantle, 79f, 78–80
Green function, 57–63, 71, 73–80, 91, 109

Hawaiian plume, 116–117, 145
Hawaiian swell, 7, 131
Howard's scaling, see scaling analysis

Iceland, 145
intermediate asymptotics, 22–24
  with respect to parameters, 28–31

kinematic condition on a material interface, 16, 125, 193, 195, 232, 236
Korteweg–de Vries equation, 188

Laplace transform, 87–88
Legendre polynomial, 56, 93
linear stability analysis
  convection in a compositionally layered mantle, 230, 233–235, 234f, 239–240
  convection with a phase transition, 235, 237–239, 239f
  compressible convection, 224–226, 225f
  convection with temperature-dependent viscosity, 227, 227f
  of finite-amplitude convection rolls, 205, 210, 218
  Rayleigh–Bénard convection, 199–200
  Rayleigh–Taylor instability, 135–136, 194–195
  small-scale convection, 229
lithosphere, xiii, 1–8, 5f, 65–67, 67f, 88–90, 89f, 127–132, 136–138, 137f, 147, 161–173, 170f, 195–197
long-wave analysis of thin-layer flows
  interaction of convection with a passive lithosphere, 137f, 136–138
  low-viscosity channel in an end-driven layer, 137f, 138–142, 141f
  thermal boundary-layer instability, 134f, 133–136
Lorentz reciprocal theorem, see slow viscous flow
low Reynolds number flow, see slow viscous flow
lubrication theory, 6, 28, 112, 122f, 121–132
  low-viscosity channel in an end-driven layer, 138–140
  plume–plate interaction, 7f, 128–130, 130f
  plume–ridge interaction, 7f, 132–133, 133f
  viscous gravity current, 124–125

Mach number, 221–223
   definition, 221
Mangler's transformation, 99–100
Maxwell viscoelastic solid, *see* viscoelasticity
method of matched asymptotic expansions
   geodynamical applications, 105
   intermediate matching principle, 104–105, 119
   isoviscous stagnation flow, 101–104
   plume from a point source of buoyancy, 107–111
   Prandtl's matching principle, 104
   second-order model equation, 119
   variable-viscosity stagnation flow, 119
minimal polynomial, 82
multipole expansion, 109

necking, 163, 282–283
Newell–Whitehead–Segel equation, 208
non-Newtonian fluid, 4, 45–48, 84, 128–131, 163
Nusselt number, 12, 105, 118, 119, 212–215, 228
   local, 12, 18, 100, 101

order-parameter equation, 200–201
   amplitude equation for convection rolls, 201–204
   envelope equation for modulated convection rolls, 207f, 206–208
   finite-amplitude convection rolls, 204–206
   phase diffusion equation, 208–210
Oseen tensor, 60

Papkovich–Fadle eigenfunction, xiv, 53
Péclet number, 3, 12, 14, 19–20, 100–102, 118, 212
   definition, 3
pinch-and-swell instability, *see* necking
plume, *see also* boundary layer, 7f, 15f, 106f
plume–lithosphere interaction, 6–8, 7f
   plume–plate interaction, 130f, 128–131
   plume–ridge interaction, 131–133, 133f
Poiseuille (pipe) flow, 43, 84, 142–143
   plane, 43, 124
Poisson's equation, 33, 34, 58, 75
Poisson's ratio, *see* elasticity
poloidal scalar, 39–42, 75, 78, 84, 85, 194, 198f, 202, 218
   boundary conditions on, 41
   equation satisfied by, 40, 42
poloidal-toroidal decomposition, 39–43
Prandtl number, 12, 111, 210, 221–223
   definition, 12
propagator matrix, 80–82

Rayleigh number, 26, 54, 105, 113, 199–202, 205–207, 212–213, 215–218, 222, 227f, 227–232, 229f, 234–235, 238–239, 239f
   convection with internal heating, definition, 223
   Rayleigh–Bénard convection, definition, 199
Rayleigh–Taylor instability, 29f, 28–31, 192–197
reactive infiltration instability, *see* two-phase flow
Reynolds number, 12, 14, 19, 33, 96, 100, 123
   definition, 12
Richter rolls, 285
ridge, mid-ocean
   corner flow model, 45–48
   Fourier transform approach, 48

scaling analysis, xii, xiii, 27, 28, 114, 119, 165, 168, 195, 229
   compressible convection, 220–224
   convection at high Rayleigh number, 210–212
   convection with temperature-dependent viscosity, 228
   corner flow in Rayleigh–Bénard convection, 215
   free subduction, 169–171, 170f
   hot sphere moving in a fluid with temperature-dependent viscosity, 117–118
   hot sphere moving in an isoviscous fluid, 19–21
   Howard's scaling for convection at high Rayleigh number, 216–218, 229
   isoviscous plume from a point source, 107
   low-viscosity channel in an end-driven layer, 141–142
   plume–plate interaction, 129
   plume–ridge interaction, 132
   Rayleigh–Taylor instability, 30–31
   shallow viscous sheet, 155–156
   viscous gravity current, 125–126
scaling law, *see* scaling analysis
self-similarity, *see* similarity solution
separation of variables, 27–28, 46, 48–49, 51–56, 75, 89–90, 92–93, 99, 119, 126, 131, 194, 199, 204, 215, 233
shear modulus, *see* elasticity
sheet, *see* thin viscous sheet
shell, *see* thin elastic shell
similarity solution, 22, 72, 112
   buoyant thermal, 25–26, 26f
   conductive heat transfer, 23f, 22–24
   corner flow in Rayleigh–Bénard convection, 215
   corner flow models for subduction zones and ridges, 45–48
   first kind, 27
   impulsive cooling of a deforming half-space, 27–28
   plume with temperature-dependent viscosity, 114–116
   plume–plate interaction, 130–131
   second kind, 27
   thermal boundary layer, 99, 118
   viscous eddies, 48–51
   viscous gravity current, 126–127
sine transform, 146
single-layer potential, 64
singular solutions of the Stokes equations
   force dipole, 60
   in the presence of a boundary, 62f, 61–62
   line force, 60
   point force (Stokeslet), 58–60
   point source, 61

rotlet, 60
source doublet, 61
stresslet, 60
singularity method, 66–69, 69f
slender-body theory, *see* slow viscous flow
slow viscous flow
  corner flow, 45–48
  governing equations, 33–34
  instantaneity, 34
  Stokes flow (= linear slow viscous flow)
    bispherical coordinates, 56f, 55–56
    boundary-integral representation, 63–65
    Cartesian coordinates, 51–54
    flow driven by internal loads, 72–84
    linearity, 34
    Lorentz reciprocal theorem, 38, 63
    method of eigenfunction expansions, 53–54
    minimum dissipation theorem, 37–38, 141–142
    polar coordinates, 45–51
    reversibility, 35f, 36f, 34–36
    singular solutions, 57–63
    slender-body theory, 70f, 68–72, 107, 110
    spherical coordinates, 44, 54–55
    Stokes's paradox, 60, 62, 68–69, 105–106, 112–113
    superposition method, 52–54
    uniqueness, 38
    viscous eddies, 48–51
solitary waves
  conduit, 142–145, 145f
  Korteweg–de Vries equation, 188
  two-phase, 186–189
solvability condition, 203–204, 208, 210, 218
spherical harmonic
  solid, 54–55
  surface, 55, 73, 75, 77, 80, 174
stagnation
  flow, 2f, 4, 102f, 101–104, 119
  streamline, 6–7, 7f
Stokes flow, *see* slow viscous flow
Stokes's paradox, *see* slow viscous flow
Stokes–Hadamard–Rybczynski solution for a sphere, 19, 44, 100
Stokes–Rayleigh analogy, *see* correspondence principle
streamfunction
  for plane flow, 39, 46, 48, 49, 51, 53, 54, 57, 98, 112, 132, 139, 140, 213–215
    equation satisfied by, 39
  Stokes, 39, 44, 55, 107, 108, 259
    equation satisfied by, 40
stress function, 95
stress resultant, 156
  effective, 157, 159–161
subduction, xiii, 1, 4–6, 5f, 49, 50, 146, 165, 195
  boundary-element modeling, 65–67
  corner flow model, 4–5, 5f, 45f, 45–48
  dimensional analysis, 12–13
  folding instability, 168f, 169

  thin-sheet scaling analysis, 169–171, 170f
  surface tension, 6, 178–180, 185

thin elastic shell
  constitutive relations, 161
  equilibrium equations, 160
  spherical shell, 163–165
  theory in lines-of-curvature coordinates, 159–161
thin viscous sheet
  constitutive relations, 157–159
  differential geometry, *see* differential geometry of thin sheets
  equilibrium equations, 156–157
  flexural stiffness, 170
  immersed, 169–173
    boundary-integral/thin-sheet model, 171–173, 171f
    scaling analysis for free subduction, 169–171, 170f
  inextensional deformation, 154
  membrane deformation, 154
  membrane flow in a flat sheet, 163
  shallow sheet
    inextensional scaling, 155–156
    membrane scaling, 155–156
    solution for a normally loaded sheet, 152–155
  theory in general coordinates, xiv, 147–159
  two-dimensional sheet, 165–169
toroidal scalar, 39, 40, 42, 84
  equation satisfied by, 40, 42
two-phase flow, 175–191
  average interface curvature, 176–177
  bulk viscosity, 184
  compaction
    forced, 191
    gravitational, 185–186
    length, definition, 186
  conservation laws
    energy, 181–183
    mass, 177–178
    momentum, 178–181
    summary, 183
  damage, 183
  Darcy velocity, 176
  Darcy's law, 181
  interaction force, 178–181
  interstitial velocity, 176
  permeability, 181
    Kozeny–Carman relation, 187
  reactive infiltration instability, 191
  solitary waves, 186–189
  surface tension, 178–180, 185
    Laplace static equilibrium condition, 180
    Marangoni force, 179
  uniform fluidization, 186
  viscous dissipation, 182

unidirectional flow, 43

viscoelasticity, 86
  constitutive relation for a linear Maxwell solid, 87
  relation to linear elasticity (correspondence principle), 87–88
  surface loading of a stratified viscoelastic sphere, 91–95
viscosity
  in the constitutive relation for a compressible fluid, 220
  in the constitutive relation for an incompressible fluid, 33

viscous dissipation
  in compressible convection, 220, 224
  in deforming thin sheets, 154, 155f, 168f
  in two-phase flow, 182
viscous eddies, 48–51
von Mises's transformation, 98–99
vorticity, 19, 57, 96, 100, 111, 210, 215
  equation, 134, 211
  poloidal, 40
  toroidal, 40–42

Young's modulus, *see* elasticity